Intelligent Instruction by Computer
Theory and Practice

Edited by
Marshall J. Farr and Joseph Psotka
U.S. Army Research Institute for Behavioral and Social Sciences
Alexandria, Virginia

Taylor & Francis
Washington • Philadelphia • London

USA	Publishing Office:	Taylor & Francis New York, Inc. 1101 Vermont Ave., Suite 200, Washington, DC 20005
	Sales Office:	Taylor & Francis Inc. 1900 Frost Road, Bristol PA 19007-1598
UK		Taylor & Francis Ltd. 4 John St., London WC1N 2ET

INTELLIGENT INSTRUCTION BY COMPUTER: Theory and Practice

Copyright © 1992 Taylor & Francis New York Inc. and Mentor Systems, Inc. All rights reserved. Printed in the United States of America. Except as permitted under the United States Copyright Act of 1976, no part of this publication may be reproduced or distributed in any form or by any means, or stored in a database or retrieval system, without the prior written permission of the publisher.

Chapters 1, 2, 4, 7, 10, and 11 originally appeared in *Machine-Mediated Learning,* Vol. 3, published by Taylor & Francis.

1 2 3 4 5 6 7 8 9   E B E B   9 8 7 6 5 4 3 2

This book was set in Times Roman by Hemisphere Publishing Corporation. The editors were Susan M. Connell and S. Michele Nix, the production supervisor was Peggy M. Rote, and the typesetter was Sandra F. Watts. Cover design by Kathleen Ernst. Printing and binding by Edwards Brothers.

A CIP catalog record for this book is available from the British Library.

Library of Congress Cataloging-in-Publication Data

Intelligent instruction by computer: theory and practice / edited by
  Marshall J. Farr and Joseph Psotka.
     p.   cm.
  Includes bibliographical references and index.
  1. Technology—Computer-assisted instruction. 2. Intelligent
  tutoring systems. I. Farr, Marshall J. II. Psotka, Joseph.
T65.5.C65I58   1992
  607.8—dc20   91-31890
ISBN 0-8448-1707-4 (case)   CIP
ISBN 0-8448-1687-6 (paper)

Intelligent Instruction by Computer

Contents

Preface ... xi

**A Thematic Introduction to Intelligent Instruction:
Progress Toward Practice** ... 1
Marshall J. Farr and Joseph Psotka

Introducing Intelligent Tutoring Systems ... 2
The Influence of Intelligent Tutoring Systems ... 4
Where the ITSs in This Book Are Being Used ... 5
Organizational Scheme of This Book ... 7
Evaluations of Intelligent Tutoring Systems ... 11
Ordering of Chapters ... 11
References ... 12

Part I
Theoretical Approaches

Chapter 1 Apprenticeship Training in the Workplace: Computer-Coached Practice Environment as a New Form of Apprenticeship ... 15
Susanne P. Lajoie and Alan Lesgold

Computer-Coached Practice Environments ... 18
A Closer Look at SHERLOCK ... 21
Evidence that a Computer-Coached Practice Environment Works ... 28
Future Enhancements ... 31
Acknowledgments ... 34
Notes ... 35
References ... 35

Chapter 2 Automated Tutoring in Interactive Environments: A Task-Centered Approach ... 37
Ursula Wolz, Kathleen R. McKeown, and Gail E. Kaiser

Introduction ... 37
Tutoring in Interactive Environments ... 39
Tutoring Strategies ... 40
Knowledge that Affects the Choice of Strategy ... 46

	Choosing a Tutoring Strategy	48
	The Use of Tutoring Strategies in GENIE	51
	Related Work	57
	Summary and Conclusions	60
	Acknowledgments	61
	Notes	61
	References	61
Chapter 3	**Explanation, Analogy, and Transfer in an Intelligent Tutoring System Paradigm** *Peter Pirolli*	**65**
	Performance and Skill Acquisition in Programming: The GRAPES Ideal Model	67
	The Transfer of Procedural Knowledge: A Production System Analysis	69
	Analogy from Examples	73
	The X Model	75
	The Role of Self-Explanations in Learning to Solve Problems	79
	Summary	81
	Acknowledgments	82
	References	83

Part II
Tools and Environments

Chapter 4	**Control for Intelligent Tutoring Systems: A Comparison of Blackboard Architectures and Discourse Management Networks** *William R. Murray*	**87**
	Introduction	87
	Discourse Management Networks	90
	The Blackboard Architecture	96
	Scenario: Controlling a Programming Language Tutor	100
	Summary	103
	Acknowledgments	104
	References	104
Chapter 5	**Two Approaches to Simulation Composition for Training** *Douglas M. Towne and Allen Munro*	**105**
	The Student Interface	106
	Deep Simulations	107
	IMTS Deep Model Approach	108
	Development and Applications of Derivative Simulations	114
	Surface Simulations	116

	Expanded Training Range	124
	Acknowledgments	125
	References	125

Chapter 6 Integrating Intelligent Tutoring, Computer-Based Training, and Interactive Video in a Prototype Maintenance Trainer   127
Mark L. Miller and Scott R. Lucado

Introduction	127
Technical Approach	129
Design Methodology	132
Implementation	137
Results	143
Acknowledgments	149
References	149

Part III
Implementations and Evaluations

Chapter 7 Semantically Constrained Exploration and Heuristic Guidance 153
Henry Hamburger and Akhtar Lodgher

Introduction	153
Semantically Constrained Exploration	154
Heuristic Guidance	163
The Syllabus	168
Future Directions, Testing, Acknowledgments	175
References	176

Chapter 8 Transfer, Adaptation, and Use of Intelligent Tutoring Technology: The Case of Grace   179
Wayne D. Gray and Michael E. Atwood

Introduction	179
Overview	180
Features of an ACT* Tutor	185
TAU of the Ideal Student Model	187
The TAU of Knowledge Tracing	191
The TAU of Curriculum Specification	192
The TAU of Interface Design	193
Conclusions and Summary	198
Postscript	200
Acknowledgments	202
References	202

Chapter 9	**Design, Development, and Implementation of an Intelligent Tutoring System for Training Radar Mechanics to Troubleshoot**	**205**
	Laura C. Kurland, Robert Alan Granville, and Dawn M. MacLaughlin	
	Introduction	205
	Front-End Research: Cognitive Task Analysis	206
	Addressing Instructional Needs with MACH-III	208
	What the User Sees	208
	Technical Overview of MACH-III	214
	Integrating MACH-III into the Existing Curriculum	224
	Initial Field Test Results	227
	Conclusion	235
	Acknowledgments	236
	References	237
Chapter 10	**Intelligent Conduct of Fire Trainer: Intelligent Technology Applied to Simulator-Based Training**	**239**
	Denis Newman, Mario Grignetti, Mark Gross, and L. Dan Massey	
	The PATRIOT System	240
	Current Instruction with the PATRIOT Conduct of Fire Trainer (PCOFT)	242
	Design of an Intelligent Training System	243
	Conclusions for the Design and Use of ITS Technology	248
	Acknowledgments	249
	References	249
Chapter 11	**An Intelligent System for Training Space Shuttle Flight Controllers in Satellite Deployment Procedures**	**251**
	R. Bowen Loftin, Lui Wang, Paul Baffes, and Grace Hua	
	Introduction	251
	Background	252
	System Overview	253
	Detailed Description of System Components	254
	Special Features of the System	257
	Development and Delivery	258
	Conclusions	259
	Acknowledgments	259
	References	260

Part IV
Implications for Future Research and Development

Chapter 12   **An Emerging Technology Gears Up**   **265**
Elliot Soloway

     As ITS Matures, Technology Matures   265
     What to Do Next   267
     References   268

Index   269
About the Editors and Contributors   275

Preface

This volume will be valuable to everyone involved with education and training. It records the dramatic new prospects for computers in school instruction, the workplace, and high technology research facilities. The text offers teachers and trainers a vision of how their professions will be fundamentally altered by these new systems, as well as how their roles will be changed. Researchers in many fields will find the implications of the work reported here fascinating. Psychologists will see ripe new fields of inquiry; computer scientists will discover in practical detail how the fruits of their work are being exploited in useful ways that affect the broadest segments of society. The challenges and opportunities exposed by these developments in intelligent instruction by computer are manifold and meaty. We hope this volume encourages a new generation of researchers and practitioners to join in the exploration of this new world of information.

This volume really began almost ten years ago when Marshall Farr directed a program at the Office of Naval Research that supported advanced research in cognitive science, and Joseph Psotka supported innovative applications in training technology for the U.S. Army Research Institute (ARI). Although none of the contributors in this volume was directly funded by these programs (with the notable exception of Elliot Soloway, who contributed Chapter 12) there is no question that much of the work described here has a direct link to that early research. Yet, what a difference! The research has matured and taken on a life of its own. And still, the progress continues at an astonishing pace. Who could have believed only such a short time ago that authoring systems of the power and convenience of HyperCard and ToolBook would now be so widely available? Who could have believed that semantic networks could have such power and usefulness for ordering instruction? Who could have believed that within a few years the most wildly optimistic dreams about a cheap, voluminous, powerful notebook computer with sharp HDTV video, communicating with the entire world, could have turned into reality? Yet it all has. This text is a record of this tumultuous change.

We would like to thank all the contributors to this volume and those whose pioneering research laid the foundation for this work. The preparation of an edited volume such as this is a labor marked by long hours of boring preparation, much frustrated and floundering coordination over telephone and via electronic mail, and a few brief hours of sustaining satisfaction. We hope that the readers share only the pleasures.

We thank many other people who helped make this work possible. During the gestation period of this book, Joseph Psotka was supported by a year-long fellowship from the Secretary of the Army, which he spent with great intellectual nourishment at Ben Shneiderman's laboratory and at the University of Maryland Institute for Advanced

Computer Studies. This imbued him with greater vigor for the long evenings and many weekends it took to pull all this information together. In his absence from ARI, Melissa Holland, Zita Simutis, and Bob Seidel continued the work and tested the strength of their magnanimity and support. A forthcoming book on hypertext, artificial intelligence, and instruction will owe them much for its existence.

Marshall Farr also did his share of the work as a "moonlighter," since his full working days were occupied in writing a book on the transfer of skills and knowledge for ARI as a senior resident research associate, mentored by Dr. O.-K. Park and Dr. Seidel under the auspices of the National Research Council.

Joseph Psotka, Ph.D.
U.S. Army Research Institute for the Behavioral and Social Sciences

Marshall J. Farr, Ph.D.
Senior Associate, National Research Council
U.S. Army Research Institute for the Behavioral and Social Sciences

A Thematic Introduction to Intelligent Instruction: Progress Toward Practice

MARSHALL J. FARR

Farr-Sight Co.
Arlington, VA
Senior Research Associate, National Research Council
Washington, DC

JOSEPH PSOTKA

U.S. Army Research Institute for the Behavioral and Social Sciences
Alexandria, VA

Reviewing the progress that the field of computer-based intelligent instruction has made over the past decade is extraordinarily gratifying. Putting together a book like this always affords a nostalgic opportunity for reflection and projection. What has happened in the last decade and where are we all going? Ten years ago the two of us happened to be in related positions. (Farr was then managing the cognitive science and innovative instructional programs for the Office of Naval Research (ONR), and Psotka was responsible for a program of computer-based instructional technology research, with Harry O'Neil, at the Army Research Institute (ARI).) Together, we began a three-year research program, combining basic research with more exploratory development of productive tools.

Essentially, the program that we fashioned aimed to stimulate the development of environments and tools for the advancement of intelligent instruction by computer. Half a dozen groups of researchers at various universities and research laboratories were funded with a view to finding what works and what could be moved from research into more practical development. At the time, we foresaw at least one, or perhaps even three, large development efforts growing out of this research. The goals were to provide testbeds and examples of how this advanced technology of artificial intelligence (AI), expert systems, instructional design, and the advancing power of workstations designed around Lisp could be fashioned into optimal instructional systems. We expected results in about five years. Not surprisingly, as research managers, we were overly optimistic in our expectations.

Now, 10 years later, our goals have finally been realized. The field around us is exploding with many fine implementations, while at the same time opening challenging new areas of research. Solid evaluations are beginning to herald a new decade of instructional promise for this technology. The research project was, and remains, an exciting enterprise, and it is thrilling to see how many of those visionary achievements of a decade ago have influenced everyday products or proven applications. What is most encouraging about these developments, especially for those of us who are psychologists

---

This order of names is alphabetical. Both individuals contributed equally as co-authors.

and cognitive scientists, are the implications of intelligent systems for the augmentation of human learning, reasoning, and thought.

History informs us that human artifacts continue to reshape the course of human thought—as in the past, when fire brought us new weapons and food, or wheels enlarged our horizons. However, it is entirely novel that the artifacts we created are so intimately connected to the structure and functioning of human thought that they may be considered to be embodiments of mind. This is an idea implicit in by-now familiar terms such as "knowledge engineering." But nowhere do these knowledge structures or expert systems that emerge from knowledge engineering so directly implement human reasoning as in education and training environments.

Introducing Intelligent Tutoring Systems

This volume addresses the systems that instruct and train—via a computer—intelligently. The systems are the latest embodiments of the wedding of artificial intelligence (AI) and computer-assisted instruction (CAI). Some people call this partnership intelligent CAI (ICAI). Of late the term "intelligent tutoring system" (ITS) has often been employed instead of ICAI, in part to maintain a kind of compatibility with "intelligent systems," a term increasingly being used by computer scientists and engineers to mean artificially intelligent systems. Calling these systems ITSs also makes it clearer that what is typically meant is a one-on-one tutorial interaction that is sophisticated and complex in ways that traditional or "dumb" CAI is not.

An ITS goes well beyond conventional CAI by drawing on a body of domain expertise embedded in the system as an "expert system" of rules. This expertise is presented to the learner under the control of some appropriate pedagogical strategies tailored to his or her changing states of knowledge and understanding. Ideally, the student's progress toward greater expertise is continuously monitored and diagnosed by an evolving student model. Although the ITS is placed into an existing setting for instruction—school, university, home, workplace, etc., as is much of existing CAI—it may fundamentally alter this environment in ways we are only beginning to understand.

At the very least, ITSs are difficult to design and implement; they tax the intellectual and technical resources and patience of a host of different disciplinary specialists: cognitive scientists, instructional designers, computer scientists, experimental psychologists, programmers, and subject-matter experts. It is little wonder that most ITSs to date have centered on only a few computer programming languages (e.g., LOGO, LISP, and COBOL) and bodies of knowledge that comprise relatively circumscribed and manageable domains.

Fortunately, this restricted focus is now changing, and many of the ITSs discussed in this book are extending the range of ITS domains to subject maters that have so far been difficult, if not intractable. Domain tractability means that the subject matter or domain of concern can be reasonably represented by sets of explicit rules that are easily created and ordered. For example, computer programming and medical diagnosis, by virtue of being somewhat limited in their scope, yet being highly interesting, have been popular targets of ITSs.

Fuzziness of the Knowledge Domain

The knowledge domain can be expressed as a loose or wavy continuum that extends from ambiguous, poorly analyzed content areas, such as the arts and professions, to more explicit domains such as technical and engineering knowledge, all the way to carefully circumscribed and formal systems that are the goals of science, and on to the actual domains of logic and mathematics. Aspects of ITSs cover this entire spectrum: ITSs are created through art, engineering, science, and mathematics.

As editors of this volume, we faced the formidable problem of how much to assume that the typical reader would know about AI, in general, and ITSs, in particular. We chose to assume enough knowledge by the reader to justify our approach in writing what follows. This, we believe, allows us to explain meaningfully how the chapters in this book deal with fundamental and significant issues and how they relate thematically to each other.

For us to make the necessary connections for you, the reader, we will on occasion refer in this introductory chapter to the theme or thrust of the specific chapters, although we fully realize that you probably have not yet read them. We will attempt to provide enough relevant and contextual information at these times so that an intelligent and persevering reader can grasp our basic points well enough to follow our arguments. We believe that a serious reading of the remainder of this introduction will significantly facilitate a more complete and in-depth understanding of the basic AI, cognitive, and instructional issues that must be confronted to truly appreciate the current scope and potential of ITSs.

Each of the chapters in this book can be placed somewhere on the dimension of the kind of domain knowledge used to construct the ITS: how formally based on principled structures it is, in contrast to more informal structures derived from specific forms of task analysis. At one extreme are the highly abstract formalisms of subtraction used by Hamburger and Lodgher (Chapter 7) or the explicit production rules for COBOL used by Gray and Atwood (Chapter 8).

At the other extreme fall the heuristic classifications derived for consulting systems by Wolz et al. (Chapter 2). In between lies the more or less formal systems design of the explicit simulation of a functional radar in Kurland et al. (Chapter 9), the expert hierarchies mapped out after years of intensive analysis by Lajoie and Lesgold (Chapter 1), a computer program by Newman and his colleagues (Chapter 10), or of virtually any mechanical or electrical device, as in the work of Towne and Munro (Chapter 5). This middle range also encompasses the controlled decomposition of text, rules, and simulations in the BlackBoard architecture of Murray (Chapter 4) and the tightly coupled simulation of real NASA systems performed by Loftin and his associates (Chapter 11).

Knowledge Representation

What emerges as the most compelling and powerful component of these analyses of many domains is the usefulness of a suite of analytical tools for knowledge analysis and representation. Whether it is the two-person dialogues monitored by Kurland et al., or the LISP coding and problem solving by Pirolli, or the rifle maintenance by Miller, knowledge is being broken into smaller digestible ("bite sized pieces," as Lajoie and Lesgold call them).

The pieces constitute goals, actions, and errors (or bugs), and they are fundamentally important for the development of ITSs. What makes them particularly important or useful is that they can be analyzed and described in plain natural language by the experts themselves, or by knowledge engineers who know little about computer programming. But they also have the force of being easily convertible into forms that the computer can manipulate.

These goals, actions, and errors can be created using traditional rational task analysis (cf. [1]), or more sophisticated analyses using such tools as GOMS [2], ACT* [3], or X or SOAR, discussed by Pirolli in this collection. They can then be relatively easily converted into the formalisms that run on computers: production systems (Pirolli, this volume), rules, frames, blackboards (Murray, this volume), or predicate logic. In fact, these computer implementations carry a task analysis further than traditional analysis, because they check its consistency in an ultimate sense: they have to work and run cleanly.

These computer programs of knowledge representations provide ultimate feedback on the quality of the analysis and the coherence and tractability of the domain. The development of these powerful knowledge representation and analysis tools has radically improved the speed and effectiveness of transitioning intelligent instruction by computer into the real world (cf. Soloway, this volume).

Instructional Design

Instructional design was never very systematic even when "systems" was its middle name. Yet the future surely portends a design process, by and with the computer, that is fundamental and integral to intelligent instruction by computer. Design is an essential component of all human problem solving [4], learning, and instruction. With only a little luck and a good amount of effort, computers used as part of any human activity will be major parts of engineering and design stations that will help shape the production of all human artifacts, especially knowledge—a point well made by Soloway in this volume.

It will not be too long before we will see computers, not just as stand-alone workstations, but as integral parts of large and fast technological webs. These webs will be focal areas for the design of books, film, communication, and information of all kinds, especially instruction. When linked together over high-speed, high-bandwidth, fiber-optic networks, computers will support interactive, widely distributed design in a system John Seely Brown, of Xerox's "thinktank" at the Palo Alto Research Center, has called "collaboratories." In that future era, ITSs will not have to be taken to the workplace, but instead will be built *into* the workplace, distributed over great distances—comprising a collab computer.

The Influence of Intelligent Tutoring Systems

Fortunately, ITSs are beginning to come of age, in the sense that they are undergoing their first real scrutiny in the user world and are beginning to influence the broad spectrum of learning, teaching, and training. The result is a healthy confrontation with reality that, in characteristically competitive ways, is reshaping our consensus on designs, features, and goals. Finding out how these systems are best used provides more than measures of their effectiveness; it provides real feedback to reshape their theoretical foundations.

"Situated" Symbol Systems

Intelligent tutoring systems are becoming "situated" symbol systems; that is, they represent mental models, and they work in the real world. They are the complex symbolic expressions of our best thinking, based on careful analyses of people engaged in thinking, learning, discovering, and problem solving. If we use these symbol systems to improve thinking, we have created a recursive structure that is open to continued improvement. It is in the evaluation of an interacting human and ITS (as a situated symbol system) that the reflection of our state of knowledge about knowing and learning is most easily perceived.

The chapters in this volume provide our first analyzed insight into people interacting with ITSs in ways that their designers, understandably, failed utterly to foresee. It is in these novel and innovative uses that ITSs show their potential for growth. The variety of these possible human–machine interactions are exciting demonstrations of the vast, untapped potential of these nurturing learning contexts to reshape the culture of all our learning environments (school, home, and workplace) across the multidimensional spectrum of human situations and experiences.

Where the ITSs in This Book Are Being Used

Some of the work described in these chapters was conducted with students in elementary schools (Chapter 7) and college (Chapter 2). Other chapters address implementations in the workplace: Air Force technicians working in a depot (Chapter 1); Army officers learning Air Defense tactics (Chapter 10); army technicians who, as uninformed high school graduates, are learning to repair the complex electronics of a radar system (Chapter 9); employees in a large, modern telephone company (NYNEX) learning to program in COBOL (Chapter 8); and NASA flight controllers learning difficult deployment procedures (Chapter 11). Even Murray's work and the system designed by Towne and Munro are largely driven by experience with systems used by adults on the job.

The lessons learned from these implementations in so many different settings provide insights that are uniquely valuable for improving the next generation of ITSs. The chapter by Murray shows that comparisons among alternative strategies and structures for ITSs can now be made on the basis of theoretical perspectives, as well as from real experience.

On the more practical end of the continuum, Miller's innovative development bridges the wide gap between more traditional instructional design and authoring tools for computer-based instruction (CBI) and advanced expert-systems shells. This unique work provides a path to the future, a future where authoring environments and hypertext tools turn computers into computer-aided design (CAD) workstations for instruction that provide a seamless integration of the best of CBI and the best of ICAI. The next generation of ITS will undoubtedly learn much from the analyses and ecological investigations of the systems now being described for the first time in the chapters in this text.

The Effectiveness of Intelligent Tutoring Systems

Much has been made of the need to evaluate ITS to determine once and for all whether this technology is cost-effective and deserving of widespread implementation. This question has remained largely unanswerable over the last decade with respect to CBI, and there seems little reason to believe that the same question addressed to ITSs will fare any better. Whenever evaluations are performed, there are always many ambiguities and unanswered questions. Each time it is a specific implementation that is being evaluated, not the whole class. So the same questions keep emerging: if it fails or succeeds, did it do so on the strength of its own specific design, on the domain chosen, on the amount of money and effort used to create it, and so forth? Or were there, in fact, some general-category characteristics that can be used to go beyond this one effort? Given the many difficulties of achieving great generality, it is at least enlightening to see some specific evaluations and carefully hedged conclusions.

As we have already noted, many of the ITS and design environments and tools described in this volume are undergoing evaluation in several settings. Perhaps the most recent evaluation is that of the HAWK MACH-III radar maintenance tutor described by Kurland et al. An official Army evaluation on site at the Army's Air Defense School is collecting information on how well this advanced technology is performing in schoolhouse classrooms. We already have some preliminary evidence that it is effective (Chapter 9). In the evaluation by Kurland et al., a group of 11 Army soldiers and Marines used MACH III in their regular course on the receiver and transmitter of the HAWK radar. Compared to another group without the ITS, their performance was one standard deviation higher ($t = 1.88$; $df = 20$; $p = .04$).

It is important to note that a careful study of educational interventions found that the largest single effect occurs through one-on-one tutoring: a two-standard-deviation effect. The next largest effect is that of mastery learning, a one-standard deviation effect [5]. Thus one standard deviation is as high an effect as any that has been evaluated. A similar relationship holds in the ongoing evaluation by the Army. However, the Army's evaluation is much more concerned with the cost-effectiveness of the system, and its "usability": how well it fulfills the critical training needs for specific skills. Since MACH III was mainly designed to test the feasibility of building such a complicated AI-based system, its acceptance by a demanding and critical user community was a rewarding conclusion to the effort.

Preliminary results from the ongoing Army evaluation are in keeping with these exploratory results, again finding superior performance on a number of measures by groups trained with a hypertext interface to a qualitative simulation. However, real-world evaluations are always clouded by uncontrollable factors. This superior performance of the computer-based MACH III over paper-based troubleshooting tasks is somewhat problematical because of factors such as the small sample size, ceiling effects in the paper-and-pencil tests, lack of double-blind scoring in the practical troubleshooting exercises, and difficulties in controlling past experience or training time. However, one aspect of the comparison is clearcut and important.

Table 1 presents, for the ongoing Army evaluation data, the mean number of troubleshooting tasks completed by the two groups of 13 trainee soldiers and marines on the radar transmitter over the same amount of training time. During the troubleshooting

Table 1

Mean Number of Troubleshooting Tasks Completed
(Per Trainee)

Troubleshooting Environment	Group	
	Control (Paper)	Experimental (MACH III)
Paper	22	—
MACH III (Computer)	—	38
Radar	11	12
Total number of tasks	33	50

phase of training, the groups rotated between performing troubleshooting on the radar and troubleshooting with their respective supplemental methods of instruction. In other words, the control group of students rotated between the radar and the desk (where they were troubleshooting on paper); and the experimental group rotated between the radar and the MACH III computer. It would have been better to use MACH III to replace some hands-on troubleshooting on the real radar, but the Army is not about to use an untried training system in such an important way. Perhaps that will be the next step.

Table 1 notes that, with the support of the computer, a much larger number of troubleshooting opportunities for learning can be created and delivered. In addition, it appears that training on the real radar is speeded up because trainees are better able to troubleshoot. Finally, the two groups of trainees were asked whether paper (for the group training with paper-based troubleshooting tasks) or the MACH III computer (for the group learning troubleshooting using the MACH III) could be used to train troubleshooting skills instead of on the radar. The MACH III trainees declared that a lot of practical exercise time on the radar could be saved by practicing troubleshooting using the MACH III instead. For anyone familiar with the hands-on emphasis of this training culture, this is truly an extraordinary statement!

Of course, all is not rosy and simple in such a complicated enterprise. Once instructors began using the system on a regular basis, they soon found areas they wanted to change and improve. Although no training time was lost, the computers crashed occasionally and inexplicably. The stunning visuals and computer graphics in the simulation were used to produce charts for overhead projection, but a high resolution, large-screen projector for the dynamic graphics was even more desirable, and out of reach.

The interface was easy to use for students and instructors, but adding new problems was too difficult. After using the system for a time, the instructors began to have a preferred sequence of troubleshooting problems and wanted to be able to fix that sequence automatically in the computer. But even that small change demanded programming skills beyond their capability. And, of course, the instructors wanted the whole radar in the simulation, in even greater detail, with representations that looked like the real radar rather than just symbolic pipes and devices.

Many of the shortcomings of the ITS were easily foreseeable, and yet they are bound to be revisited by other systems. Some of the problems were just a matter of inadequate resources to do everything. They are a natural consequence of systems coming out of the laboratory to confront the real world. Although any one of the deficiencies may have been foreseen, all of them could not have been avoided because we simply lack adequate experience with how ITSs will be used.

Systems that work beautifully in the laboratory, where everyone understands their limitations and uses workarounds or self-edited interactions, will not be tolerated by normal users. And so we must build up our knowledge of how these systems will be used as best we can. This knowledge must be built up through situated evaluation. With that knowledge in hand, ITSs will be used profitably to free the instructor to do what he or she does best, and to empower the trainee with knowledge structures that facilitate faster and better performance and understanding.

Organizational Scheme of This Book

Several different organizations could be imposed on the chapters to address the issues surrounding the development and implementation of an ITS. The simplest organization might be in terms of its characteristic components: the expert model, student model, interface, and so on. However, the important research issues that differentiate these chapters do not really depend on differences in these significant components, since the goal for most of these systems is to build a useful, working tutor. Given this concern, the research issues' differences have derived from the decisions these investigators have made to manage the complexity of the knowledge they communicate.

Managing the Complexity of Knowledge

The issues around the complexity of knowledge representation can be divided into two groups—one dealing with the knowledge domain of the instruction itself, and the other addressing the knowledge-representation techniques employed to enter the knowledge into the computer. The interactions of these two sets of issues constitute the source of many of the difficult decisions that must be made in constructing an ITS. With human tutors, explanations are sometimes supplemented by demonstrations where appropriate and possible. With ITSs, as this volume reveals, techniques that blend text and simulations are used judiciously to meet the educational objectives. The following sections provide an overview of some of these techniques to introduce and relate their important aspects. Our intention is to make them more understandable when they are encountered in the chapters that follow.

Representation Techniques

Invariably, tutors and the knowledge representation techniques they use try to make the material being taught less complex than it really is so that novices can begin the hard climb to expertise. In both the Chapters 1 and 9, the electronic equipment the authors discuss is very complex. Yet their goal is to have trainees with nontechnical backgrounds understand the system well enough to troubleshoot it within a reasonable time frame. Their techniques—the use of hypertext browsers of hierarchies of components and procedures, combined with textual explanations, simulations that illuminate hidden parts and invisible processes, and "rulesystems" that prune the space of possible actions and states in any given context—are all at the cutting edge of the technology. And all of these techniques can be found in some form in the other tutors.

Hierarchical Trees

All of these systems use some form of hierarchical classification of knowledge structures. The hierarchies in some are based on text (as in discourse networks), or implicit goals and plans (Chapter 2), or troubleshooting steps (Chapter 1). This technique of a systematic componential analysis into hierarchically related byte-sized chunks appears to hold great promise, not only for simplifying the presentation of knowledge, but also for computing on this knowledge directly. One great advantage of hierarchical trees is that there exist many well-understood algorithms for searching and retrieving pieces of these trees.

Computations on these trees benefit from many years of computational graph theory. However, we are still at an early stage of understanding how to use these computations and knowledge representations. In effect, we are creating a new language that talks about nodes and links, and the syntax of this language is still evolving. Because no common formulation of the nodes and links connecting the abstractions has yet been created, all of the knowledge bases have a unique structure that is not strictly comparable.

Hypertext

An exciting new development that has encouraged the growth of hierarchical text and simulation systems is the widespread growth of hypertext techniques and systems. Hypertext is generally described as nonlinear text, in which an assortment of multi-windowing, hot-button, and search techniques is combined. The purpose is to allow people to explore or browse through complex text structures, in any order they please. Generally, text, pictorial, and simulation presentations are treated equally throughout hypertext materials. Kurland and associates (Chapter 9) wrap their expert system for troubleshooting with a pleasing interface based on hypertext.

Learning from Self-Reflection

One of the potential strengths of a hypertext system that maps out the structure of the materials to be learned is that students can see directly, and review to gain insight, their own paths through the "space of knowledge" after successful learning. Structuring this space of knowledge is critical for its effective use in learning from self-reflection. Lajoie and Lesgold (Chapter 1), for example, use their understanding of expert problem solving to prune the effective problem space (EPS) and increase the value of the solution trace.

Semantic Constraints Created by the Environment

This simplification of the problem space is the core of the work by Hamburger and Lodgher (Chapter 7). Within their systems, the essential choices of the features of "Coinland" (a simulation environment using images of coins) were all constrained by this goal. The selection of the guiding metaphor, of using the coins to pay and make change, stems from a desire for a simple mapping between (a) the rules and constraints of subtraction principles, and (b) the constraints that operate in the environment (semantic constraints). The design of the environment prunes the general problem space of declarative knowledge; and rules and functions individualize the local student model.

Practice Versus Discovery

A distinction that is related to the choice of text versus simulation for the presentation of knowledge contrasts the ability of each system to provide practice on a knowledge or skill, as distinct from their ability to impart new knowledge or to encourage conceptual creation and change. This distinction between providing practice or encouraging new learning through discovery is an important pedagogical concern. Wolz et al. (Chapter 2) provide a clean classification that emphasizes the respective roles of procedural environments for practicing and extending a skill into new domains, as compared to declarative explanations for introducing, reminding, clarifying, and elucidating the functions and plans that the user needs.

This is a thread that runs through many of the chapters in this collection, perhaps most clearly in Hamburger and Lodgher's notion of semantically constrained environments, where students' abstract knowledge and skills of subtraction can be implemented in a way that is both naturally supported and restricted by the environment. The largely procedural constraints are controlled by the declarative structure of the Coinland world. In Chapter 3, Pirolli presents, in a very natural way, an analysis for using the declarative structure of goals to determine the kinds of procedures that are important.

Simulations, Hierarchies, and Blackboard Architectures

The issue of whether to teach through verbal (declarative) explanation, or via simulations, is strongly a function of whether the instructional objectives include skills for which simulation is the best prescription. Ideally, wherever possible, we would want explanation *and* simulation to be both complementary and supplementary, to afford maximal learning and understanding. To achieve optimal instruction, research and development needs to be directed at defining whether and when the declarative versus the procedural approach is more appropriate.

From a pedagogical viewpoint, we do not always need to make the learning hierarchies explicit—especially when simulation is the primary medium of instruction. "Rulesystems" then become convenient. Two chapters in this book address this focus. In the first one, Loftin et al. (Chapter 11) chose to dissect their knowledge domains into efficient production rules that are modular and independent. But even then, most of the work in designing these systems comes from the detailed decomposition of knowledge necessary to create computational forms.

Murray (Chapter 4) provides a convincing argument for the best control mechanism to use to choose most flexibly among various tutorial strategies. (See the following discussion of "blackboard architectures.") Murray also discusses the case for planning instruction by going from an initial overview of the material and lesson plan to "successive layers or hierarchies of detail that build on and refine earlier material."

Hierarchies in simulation is also a focus of the work by Newman and his colleagues (Chapter 10) and by Kurland et al. (Chapter 9): they fashion, separately, hierarchies for the components of a device, contrasted with the knowledge and procedures of using the device. Similarly, to map out the problem space, Lajoie and Lesgold performed a remarkably detailed analysis of the device structure and the procedures and skills employed to actually manipulate the device.

We referred earlier to Murray's "blackboard architectures" in connection with mechanisms to control the sequencing and layering of instruction. This type of architecture, argues Murray, promotes the use of flexible, hierarchically organized procedures. These procedures serve to amplify the power of discourse management networks in critical ways. Murray's work paves the way for researchers to analyze the construction of these AI knowledge bases. This would permit heuristics to be replaced with more systematic principles and consistent grammars for structuring smart tutoring systems.

Graphic Authoring

Intelligent graphic authoring environments show great promise for improving the efficiency of simulations. The Intelligent Maintenance Training System (IMTS), developed and studied by Towne and Munro, provides a direct manipulation environment for composing and interacting with complex graphical simulations. Behaviors are defined at the object level by authors using a "behavior editor" rather than a programming language.

A generic expert on device troubleshooting analyzes the behavior of the simulated system—that "behavior" is derived from the individually defined behaviors of the objects that compose a simulation. This expert analysis serves as the basis for intelligent instruction, expert protocols, and context-sensitive advice to students performing troubleshooting exercises using the IMTS.

Evaluations of Intelligent Tutoring Systems

Evaluations of ITS have been promised for years, but none has been forthcoming until the work reported in this book. In these papers, real evaluation data are presented. Not only do we find that ITS can dramatically improve test results with very complicated tasks (Chapter 1), but also that trainers, teachers, and students really like the systems (Chapters 7, 10, 11). However, more than these straightforward results are involved. Analyses of ITS in the workplace and school are redefining our understanding of how teachers and students can cooperatively use these tools to enhance learning. This in turn is improving our capability to design systems that can, in fact, promote and support cooperative learning and generally improve the culture of learning within these environments. The need for this improvement is implicit in the findings by Newman et al., as well as by Lajoie and Lesgold: the growth of knowledge within on-the-job training environments is often unmotivated and exceedingly slow.

These findings reflect pejoratively on widely accepted ideas about the advantages of learning by doing and apprenticeship training. The widespread nature of unmotivated performance of students in public school is something documented widely and warrants little comment here. If ITSs can even modestly alter this attitude toward learning, our investment in intelligent tutoring may be returned many times over.

Ordering of Chapters

This volume begins with the chapter by Lajoie and Lesgold to set the stage properly for the reader. We believe that it functions as a basic exemplar: most of the issues are clearly presented, and several different representation approaches are intertwined and explained. The remaining chapters have been placed in three groups: (a) those that are demonstrations of a particular theoretical approach, (b) those that deal mainly with the development and improvement of tools and environments, and (c) those that demonstrate the power of well-engineered implementations.

Wolz et al. and Pirolli (Chapters 2 and 3) address larger theoretical issues that help to explain and enlighten the examples provided by the later chapters. They cover in greater detail the issues of human-computer interaction and knowledge representation. Chapters 4, 5, and 6, by Murray, Towne and Munro, and Miller extend powerful, existing tools. Murray deals with blackboard architectures for lesson planning. Towne and Munro analyze the power of their graphic simulation editor. And Miller integrates an expert-system shell with a powerful CAI-authoring environment.

Chapters 7-11—by Hamburger and Lodgher; Gray and Atwood; Kurland et al.; Newman et al.; and Loftin et al.—reveal how these components can produce working intelligent systems for computer-guided instruction. Together, they offer an excellent overview of the state of intelligent tutoring systems as they begin to emerge into the real world. In the final chapter, Soloway provides a provocative review, in his inimitable way, of the implications of all this work for future research and development. It is a skillful summary that suggests an exciting and promising future.

References

1. R. M. Gagne, *The conditions of learning* (4th ed.), New York: Holt, Rinehart & Winston, 1985.
2. S. K. Card, T. P. Moran, and A. Newell, *The psychology of human-computer interaction*, Hillsdale, NJ: Erlbaum, 1983.
3. J. R. Anderson, *The architecture of cognition*, Cambridge, MA: Harvard University Press, 1983.
4. D. N. Perkins, *Knowledge as design*, Hillsdale, NJ: Erlbaum, 1986.
5. B. S. Bloom, The 2 sigma problem: The search for methods of group instruction as effective as one-to-one tutoring, *Educational Researcher,* 13:4–16, 1984.

I
Theoretical Approaches

Chapter 1

Apprenticeship Training in the Workplace: Computer-Coached Practice Environment as a New Form of Apprenticeship

SUSANNE P. LAJOIE
ALAN LESGOLD

Learning Research and Development Center
University of Pittsburgh
Pittsburgh, Pennsylvania 15260

> **Abstract** *Air Force technicians who practice with SHERLOCK, a computer-based coached practice environment, show marked improvement in difficult troubleshooting skills. SHERLOCK's strategy is to provide holistic practice in a realistic context, supported by tailored coaching on request. This article compares the SHERLOCK approach to other recent cognitive apprenticeship proposals.*[1]

It is widely recognized that many people in our society fail to acquire adequate skills from their schooling. A less-recognized problem is the problem of "inert knowledge."[1] Time after time it has been demonstrated that students who are successful in school—who have demonstrated on tests that they know schooled facts and concepts—nevertheless fail to use these facts and concepts on appropriate occasions [1, 2].

Technical training, even though it is intended to enable skillful practice in a particular job, is not exempt from this problem. We developed SHERLOCK, a computer-based coached practice environment, as a new approach to training specific cognitive skills. SHERLOCK is an intelligent coached practice environment. It has much in common with intelligent tutoring systems, but is not as much driven by a dynamic student model. Also, its intelligence is used primarily to respond to student requests rather than to intervene actively. Extensive field testing shows that SHERLOCK works. Air Force trainees who used SHERLOCK learned the difficult skills of troubleshooting the electronic equipment they use to make diagnoses in faulty devices—to diagnose the diagnoser [3, 4]. In fact, after less experienced technicians practiced on SHERLOCK for about twenty to twenty-five hours, they compared in their ability to troubleshoot test station failures with colleagues who have four more years of on-the-job experience [4]. It seems worthwhile to reflect on SHERLOCK as a "success model" within which we can identify effective strategies for tackling difficult instructional problems. For example:

- How can we motivate the learning typically conducted in schools?
- How can we ensure that schooled knowledge will be usable rather than "inert?"
- How can we enable skillful performance in complex technical jobs?

Present address for Susanne P. Lajoie is Applied Cognitive Science Laboratory, McGill University, 3700 McTravish Street, Montreal, Quebec H3A 1Y2, Canada.

In this article, we consider what SHERLOCK can tell us about one particular proposal for improving the instruction of cognitive skills, "cognitive apprenticeship" [5, 6, 7]. Collins and Brown and their colleagues have proposed a "new apprenticeship" as a model for the instruction of cognitive skills. In their view, such a model enables students—the apprentices—to use their school knowledge in the world because it teaches knowledge *in the context of its use.* They identify *modeling, coaching,* and *fading* as the essential activities of masters in relation to apprentices. In apprenticeship, the learners perform with support (coaching) that is gradually withdrawn (fading). Cognitive processes can be modeled for students just as physical skills like tailoring and weaving are modeled for apprentices. One complexity of cognitive apprenticeship, however, is that the components of skill are often exercised only mentally and thus are less observable than physical skills. In a traditional apprenticeship setting, many of the skills required for performing a task are physical, such as sewing a hem. In writing, reading, or diagnosing faulty equipment, the skills and procedures may take place "in the head;" they are not as easy to observe [7, 8]. Ways of externalizing cognitive processes, therefore, are essential. Finally, apprenticeship learning is "situated;" it is embedded in a context of activity rather than being taught as abstract knowledge to be applied later.

Collins et al. identified three "success models" of teaching cognitive processes: one for teaching the processes of comprehending texts, one for writing, and one for solving mathematics problems [6, 9, 10, 11]. These success models emphasize the role of the teacher as an expert practitioner who can model skilled performance for the students. Brown et al. developed the theme of cognitive apprenticeship with a somewhat different emphasis [5]. They regard knowledge as fundamentally situated in the context of a specific activity within a specific culture. They suggest a theory of "situated learning" in which knowledge is socially constructed.

If "cognitive apprenticeship" is to be a useful model, we must identify which of its features are essential to successful learning. For example, among the features of cognitive apprenticeship as it is described and exemplified are the following:

- Learning is situated in a social context similar to those in which the skills will be used;
- Both novice and master are active participants in the learning environment;
- Cognitive processes are externalized and displayed for inspection.

Further, the success models have an element of role reversal, in which novices both evaluate others' performances and are themselves evaluated by others. This role reversal might promote self-esteem and the ability to tolerate feedback and to learn from it. Palincsar and Brown used this role reversal concept in their producer/critic paradigm, called reciprocal teaching, in which students take turns either producing an answer or critiquing a peer's solution [9]. Scardamalia and Bereiter fostered collaborative efforts in which students critiqued both the teacher's and other student's work [10]. Schoenfeld had students propose math problems for the teacher to solve impromptu [11]. The teacher's sometimes floundering efforts modeled realistic problem solving, with its false paths and revisions. When students take on the role of the teacher, evaluating other students, they can come to understand the levels of competence that can, with practice, be reached. SHERLOCK has modest capability to support comparisons of performances, and more such capability is being added in the new version now being prepared.

The presence of students at various stages of performance can support the sense of learning as an incremental process and the externalization of expert and novice thought processes. In the "success models," the environment of the classroom supported these processes. An ideal apprenticeship learning situation might have only limited pressure for product output, allowing trainees to work at their own pace, to provide help to each other when necessary, and to learn skills in the social and practical context in which they will be used. Ideally, a group of apprentices, each at different stages of expertise, would be present. Such a group enables the apprentice to observe the problem-solving process of more capable peers and to compare his or her own strategy and tactics with those of others at different stages of competence. The apprentice is encouraged to see learning as an incrementally staged process and has benchmarks for his or her own progress [6]. This situation is generally not fully practical; real-world apprenticeship is constrained by the need for efficient production.

Granting, even, that cognitive apprenticeship can be successful, there are still elements of the approach that need to be tested more completely and stated more explicitly before the approach will generalize across subject matters, teachers, and students. SHERLOCK can be viewed as a formalization of apprenticeship training for complex skills. As a computer tool, it forced us to specify the principles it embodies: those that will have to be taught to teachers if they are to become cognitive "masters." And such computer tools as SHERLOCK may supplement a classroom teacher, who cannot coach every student at once. In its current form the cognitive apprenticeship approach may demand too much of the teacher's monitoring skills and intuitions.

Traditional apprenticeship training was relatively unconcerned with accelerating learning, since the apprenticeship performed useful work. Apprenticeship training encompasses years. Learning in such situations is often slow and unfocused. Masters may vary in their experience as teachers, and novices may vary in their ability to learn in what is essentially a discovery learning environment. Apprenticeship learning is largely student driven. This is good when it suits the student's learning style and when a discovery approach to knowledge acquisition is feasible in the work environment.

But learning is facilitated by a discovery approach in some situations and not in others. In a complex domain, "learning by discovery" may take years, even though skill could be fostered earlier by focused instruction [12, 13, 14]. It is likely, too, that masters do not always recognize the current performance level of their apprentices. For these reasons, and for economic reasons, apprentices may be performing fragmentary tasks that are below their capability even after they are capable of performing more complex and complete tasks, alone or with assistance. Novices must observe the master attentively, comparing their skills to the master or other apprentices in the setting who are at varying levels of competence. The master must recognize when the novice needs assistance (coaching) and decide when assistance is no longer needed (fading). This puts a heavy burden on both learner and teacher—more than some can handle well.

Although novices are afforded the opportunity to observe skilled performances before they are expected to enact them, the traditional master's main agenda is to practice his trade or profession—to produce a product rather than to teach a novice new skills [15]. Much of the instruction then is modulated by the economic or product-oriented goals of the master's shop rather than by any search for the work situations that could provide the most instructional value, in terms of guidelines on when to coach and when not to coach, and undercontrolled, with educational goals subordinate to production goals.

The cognitive apprenticeship approach to instruction shares several instructional components with traditional apprenticeship approaches: observation, coaching, fading of assistance, shared problem solving between teacher and student, and situated learning in the context of subsequent knowledge use. Where cognitive apprenticeship approaches go beyond the traditional approach is in developing ways to facilitate the externalization of expert problem solving activity and the student's skills of learning. But does cognitive apprenticeship depend on the special qualities of exceptional master teachers?

Like expert cognition, expert teaching is often inaccessible to study. SHERLOCK can be viewed as an externalization of expertise: it is specified by rules and is replayable. It can be studied. There can then be debate about whether and to what extent it embodies the approaches labeled cognitive apprenticeship. Furthermore, to the extent that it is effective, SHERLOCK can provide information about the efficacy of those approaches. Specifically, we believe that SHERLOCK provides an opportunity to answer questions such as the following:

When should coaching be provided and when should it be faded? The expert teacher probably depends on intuition. SHERLOCK uses explicit models of the student's competence to drive the coaching and fading of feedback.

How can coaching be individualized for a student? The expert teacher may deal with many students at once, or at best rely on a general sense of a student's ability. SHERLOCK gives different levels of help according to the student's current state of achievement.

Experience with SHERLOCK alone cannot directly address questions about the social environment of learning. For example, what principles should guide students' evaluation of their peers? How should a cooperative environment be maintained (if it should)? In field testing, however, SHERLOCK was embedded in a social situation as well as a context of application. In discussing possible future developments of SHERLOCK, we suggest means by which SHERLOCK can be used to foster such aspects of socially situated learning as peer tutoring and role reversal.

Computer-Coached Practice Environments

SHERLOCK is a specific example of a form of job-situated training that we call a *computer-coached practice environment*. It was developed for the Air Force as a research vehicle, to allow us to address the issues raised above. In the remainder of this article, we discuss why such an approach was needed, how the approach was developed, and how well it worked.

Background

SHERLOCK was developed for a particular job in the Air Force, that of F-15 manual avionics technician. People in this job repair electronic navigation equipment from F-15 aircraft, using a *test station* to isolate the sources of failures. With this test station, the diagnosis task is relatively easy, and trainees have no difficulty learning the basic job from extant "schoolhouse" instruction. However, the test station itself can fail. Indeed, it fails relatively often. It is composed of massive amounts (40 to 100 ft.^{3}) of discrete electronic components soldered to printed circuit cards. Furthermore, the special tools for diagnosing its failures are minimal (there are confidence tests that can be run on the test station, but they tend not to isolate faults very completely or efficiently). So while the everyday job of these technicians is to use good tools to make simple diagnoses, their occasional job is to use minimal tools to make complex diagnoses. It was this second, more difficult, chore that we set out to teach with SHERLOCK. SHERLOCK is designed for airmen who have just completed technical school and have started to apprentice in their assigned F-15 manual avionics shops. It teaches trainees skill in troubleshooting the manual avionics test station by coaching them through realistic fault isolation problems.

The technical training for the manual avionics job is modeled after schooling. Much of the training is didactic, and the rest involves following specific directions in using the test station, much as a student in a high school science lab might follow directions. There is little direct instruction in problem solving, even though fixing the test station (at least) is a problem solving task. Basically, technical schools like those for this job teach troubleshooting by focusing on formal principles such as the electrical laws that underlie the electronics instrumentation domain. There is evidence, however, that such a theoretical approach fails to produce good troubleshooters. Morris and Rouse reviewed more than thirty training studies and concluded that the closer the training was to actual technical performance in terms of task-related specificity, completeness, and realism, the greater was the training effectiveness [16]. As Gott and Davis suggest, teaching functional concepts of electronics systems, grounded in experience, might facilitate on-the-job learning in avionics better than teaching theoretical principles that are not situated in experience [17].

The electronics troubleshooting training that technicians receive in the Air Force does not stop at technical school. The purpose of preliminary schooling is to make the technician able to do useful work on the job. The harder parts of the job, such as fixing the test station, are meant to be acquired through on-the-job experience. It is on the job that novices have some opportunity to observe experts diagnose and repair faults in the navigation equipment or the test stations. Thus, in a sense, the F-15 manual avionics shop could be a natural apprenticeship setting, where novices observe as more capable peers solve troubleshooting problems, developing mental models of the troubleshooting process and later exercising those models by attempting harder diagnoses under supervision. Regrettably, though, the organizational rules make this a solitary job. One summons help when necessary, but one usually works alone.

Because there is much work to be done, there is little time for watching experts work. In particular, when the test station breaks, an expert is usually assigned to repair it. Often, the novice and journeyman technicians have work at other test stations at this time, so they cannot observe the expert performance. There are also few opportunities to practice the complex skills of test station diagnosis. Troubleshooting the test station cannot be practiced efficiently in the shop, because (1) failures of the test station are relatively rare; (2) failures of the test station are diverse and do not occur in an instructionally optimal sequence; (3) inexpert troubleshooting of the test station is often unsafe; (4) inexpert troubleshooting of the test station is time consuming; and, more generally, (5) the goals of training conflict with the goal of repairing test station faults quickly. Even if the conditions were otherwise ideal for apprenticeship, the on-the-job learner would still face the difficulty that many of the skills and procedures of the jobs are invisible because they take place "in the head." SHERLOCK was developed to make the cognitive processes of test station troubleshooting more observable so they could be learned and practiced.

What is a Computer-Coached Practice Environment? A practice environment provides opportunities for practicing cognitive activity in a realistic setting, with computer-supplied support when needed. The practice environment in SHERLOCK is a realistic computer simulation of a job environment; trainees acquire and practice skills in a context similar to those in which they will be used. So, for example, SHERLOCK's simulated work environment matches that of the real job. Each control and display panel of the test station, with knobs and dials, is accurately simulated. The schematic diagrams, instructions for testing units from aircraft, and other documentation, while slightly simplified, retain many aspects of those used in the real job: the symbols in diagrams are the same, the layout of testing algorithms is the same, the indexing schemes are the same, etc. We do not believe that it is critical to have the simulation entirely based in computer graphics; in fact, the next version of SHERLOCK will instead use interactive video for part of the simulated work environment. What is critical, we believe, is the combination of intelligent coaching with realistic simulation of the actual work environment.

How is SHERLOCK Different from Other Cognitive Apprenticeship Settings? SHERLOCK's fundamental purpose is to provide an efficient and realistic, but artificial, means of practicing, with support and feedback, a skill that cannot be efficiently practiced in its real application context. There are several factors that make practice in a computer-coached environment more effective than either ordinary on-the-job experience or the classroom approach to cognitive apprenticeship. A human coach could also provide efficient practice, but adequate coaches and work environments are often not available.

What differentiates a computer-coached practice environment from shop experience is efficiency and focus. In the shop, a technician sees only the equipment failures that happen to occur while he is there, and often he cannot participate in the most useful troubleshooting cases. A computer-coached practice environment can quickly and safety give trainees experience with sequences of fault diagnosis tasks selected for instructional purposes. In the shop, trainees receive advice of varying quality when they need help solving a problem. A computer-coached practice environment can provide instruction when the trainee needs it and can tailor the help provided so that outside help is not an unnecessary crutch. In the shop, complex troubleshooting problems may take as long as two shifts (sixteen hours); airmen on the second shift may have to begin troubleshooting where the previous shift stopped. Problems that take twelve hours on the job can be completed in less than a hour on SHERLOCK. In the shop, time is spent waiting for parts, dismantling components, and carrying out long sequences of performances that are already well learned. Using SHERLOCK, all the activity is focused on the problem solving that is to be learned. SHERLOCK can ensure "coverage" of the set of instructive experiences as well as provide appropriate levels of support for those trainees who need assistance. The practice environment preserves the full context of what technicians experience in the shop, presenting a problem-oriented interface along with support when needed.

SHERLOCK shares with the cognitive apprenticeship models that we reviewed the principles of observation, practice, coaching, fading, and externalization of problem-solving processes. Similarly, SHERLOCK shares with other approaches a basis in extensive knowledge of the subject matter and of common individual differences in subject-matter knowledge. One advantage, however, of a computer-coached practice environment over other forms of cognitive apprenticeship is that a computer coach provides support or coaching based on a detailed picture of the trainee's performance, which many human teachers lack. SHERLOCK models performance and competence dynamically and uses the models to determine the level of assistance that the trainee needs. SHERLOCK also features a strong link to empirical research, extensive involvement of domain experts in developing instruction, a pedagogical approach of holistic instruction of the problem-solving activity, and the use of the computer as coach.

Empirical Research Behind SHERLOCK

SHERLOCK's instructional goals reflect the cognitive components of troubleshooting proficiency identified in empirical research. Extensive cognitive task analyses of the F-15 manual avionics test station job compared more skilled with less skilled airmen and identified several skill components that differentiated them [3, 18]. Proficient troubleshooters were skilled at generating and testing fault isolation plans, and at using appropriate methods and strategies to confirm their hypotheses. They did not, however, differ from less skilled troubleshooters on isolated skill components such as logic gate comprehension. The skill differences emerged only in the context of a troubleshooting problem.

A Closer Look at SHERLOCK

What Does SHERLOCK Look Like?

The real work environment consists of (1) the test station, which is divided into twelve drawers, (2) whichever unit from the airplane is being tested, and (c) a cabling package called the *test package* which connects the unit to the station. Each drawer of the station is itself a large entity, containing a number of printed circuit cards and having a complex front panel of knobs, switches, and indicator displays. Figure 1 shows a wide-angle view of the SHERLOCK work environment. It depicts the test station, including its twelve drawers; the specific navigation unit from the plane that is being tested; and the test package which interconnects the station with that unit. Each of the drawer displays can be expanded (by pointing to the drawer with a mouse and pressing the mouse button) to show its front display and control panel.[2] Tests can be performed by retrieving schematic circuit diagrams from a data base (by pointing to entries in realistic tables of contents and indices) and then indicating on the diagrams where the test probes of measurement instruments should be placed. The trainee's basic job, in real life and while using SHERLOCK, is to isolate where the faults are in the airplane units. However, SHERLOCK is designed to simulate a high rate of test station and test package failures, since finding faults in these devices is the specific problem airmen in the job seem to have.

SHERLOCK provides a simulation of a real-world apprenticeship setting, where all the proper tools and equipment are available for efficient troubleshooting. Although the test station as simulated by SHERLOCK is reduced in complexity, all of the critical components contained in the four different types of Air Force manual avionics test stations are represented. To some extent, the test station that SHERLOCK simulates is generic, an abstraction of those used by the Air Force for maintenance of various electronic systems for several different types of military aircraft. Since the critical test station components, technical manuals, schematics, and checkout technical orders are accurately portrayed by SHERLOCK, it has the ability to replicate fault isolation problems that actually occur with real test stations.

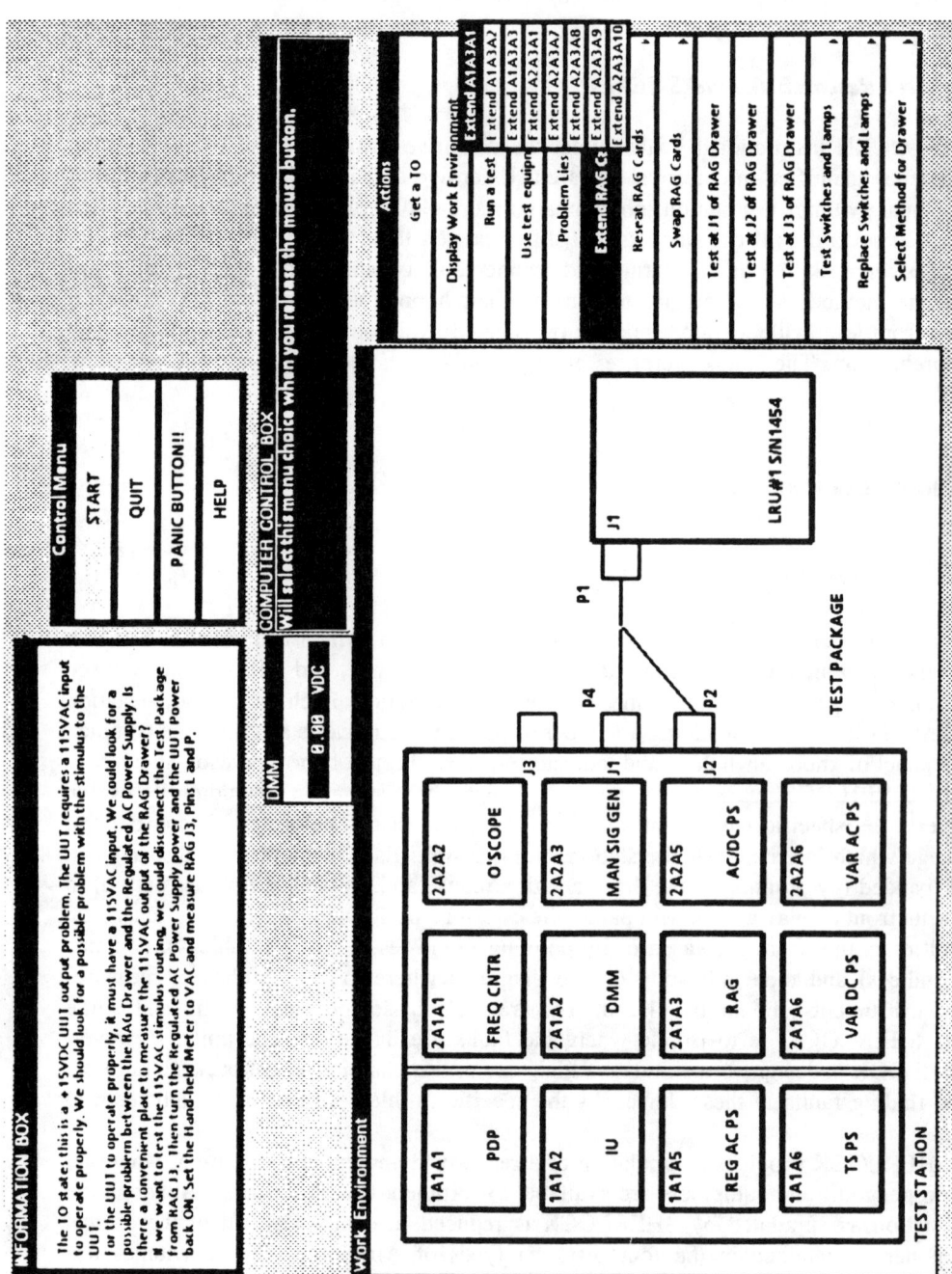

Figure 1 The SHERLOCK work environment.

SHERLOCK'S Domain Expertise

One way to develop computer-based instruction is to embody domain expertise in an expert system and to correct or instruct the student whenever his activity departs from the expert ideal. Both empirical research and expert advice suggested that such an approach would both misrepresent troubleshooting expertise and overly constrain the trainees' problem-solving activity. Often, there are multiple "best" paths or sequences for solving a troubleshooting problem. SHERLOCK's basic mode of interaction with trainees is to pose realistic troubleshooting problems and to provide expert advice on demand. Accordingly, the knowledge in SHERLOCK is organized around problems, problem solution methods, and the constraints imposed on problem solution by the structure of the work environment. Domain expertise in SHERLOCK is reflected both in its coaching and in its work environment, so skills are learned in an accurate representation of the context in which they will be used.

In order to discuss SHERLOCK's domain knowledge more completely, we need to define some terms.[3] Much of SHERLOCK's knowledge is represented in a separate structure specialized to each problem that it presents. Problem-based knowledge organization is efficient because it is based on what we call the *effective problem space*. The term *problem space* is often used to refer to a graph structure in which nodes represent partial solution states for problem and links represent actions that move the solution process from one partial solution state to another [19]. In the course of solving a problem one "traverses" the problem space via a path that represents the sequence of problem solution steps taken. For a test station diagnosis problem, there are thousands of possible moves. We did not represent all these moves. Rather, we considered only moves, good or poor, that are likely to be made by experts or trainees, plus additional moves that are infrequent but consistent with the troubleshooting approaches we are teaching. Rare and bizarre moves were omitted, since we probably would not know how to coach them anyway, except through generic techniques of reviewing the problem goal structure. It is the remaining critical subset of the problem space that we call the effective problem space.

To provide further structure, we grouped the various pieces of problem space paths into a hierarchy of larger units which we called subgoals (the overall goal is to solve the problem). Each subgoal is a piece of problem-solving procedure. Some are strategic, some tactical. Some involve heuristic plans; others involve fixed procedures. The actual SHERLOCK knowledge structure for a problem is a network in which the nodes are these subgoals rather than the more microscopic actions which they summarize. That is, most subgoals subsume a lot of mental activity; they are *abstracted* problem subgoal structures. For some problems, a number of different sequences of subgoals will produce a solution. For such problems, we can picture the subgoals as forming a lattice in which each specific solution approach is represented by a subset tree. The top node of the problem subgoal lattice represents the complete navigation of the problem space from first attack on the problem to final solution. Branches underneath any node in the lattice refer to the subgoals that must be achieved before that higher-level node can be achieved. Higher-level nodes generally reflect problem solution strategies, while lower-level nodes represent tactics and specific actions. SHERLOCK's representation of a problem, then, consists of an abstracted subgoal lattice for the problem's effective problem space (EPS), which we abbreviate as an *EPS subgoal structure*. These problem representations have fifty to sixty nodes, rather than the thousands of nodes that a "complete" problem subgoal tree for test station diagnosis would have.

The EPS subgoal structures are represented as *objects* in an object-oriented programming language. That is, for each node of the EPS subgoal structure, there is a specialist miniprogram, or object. In our *object-oriented programming* approach, a small number of generic template objects are first defined, and then each node's "personal" object is specified as an instance of the generic template. Most of the unique information in such an instance object consists of data, not program, enabling substantial use of nonprogrammers for developing such a system. In the case of SHERLOCK, the data specify hinting information, test point values, etc. Even though there are many different test measurements that a trainee might make on a component while diagnosing a test station failure, only the abstracted notion of testing the component is represented in an EPS subgoal node, along with a procedure to determine whether the appropriate testing has been done. The ability to provide test values in response to a measurement action is part of the device simulation, not the problem solution process, so it is represented separately in the model of the device.

SHERLOCK's Pedagogical Framework

We concluded from the empirical research on troubleshooting that it would be best to train troubleshooting competence holistically. With SHERLOCK, the trainee learns to do complex procedures by doing them. When he cannot handle the whole performance, he is given reminders and hints to support him through the parts he cannot handle, but at the same time he is still enacting the whole target cognitive performance. Instead of teaching separate didactic lessons, we situated our coaching of specific cognitive components in the contexts of specific problem situations in which they should be sued. This facilitates mapping new knowledge to appropriate situations. Rather than teach any one curriculum goal in isolation, each goal is addressed whenever it is relevant to overall problem-solving performance. More specifically, SHERLOCK talks only when the trainee is likely to listen—when he needs to know the information SHERLOCK is presenting. Our pedagogy parallels the cognitive apprenticeship approach in providing holistic instruction. However, instead of decomposing the troubleshooting activity into processes each of which needs to be externalized for trainees, SHERLOCK provides coaching on specific skills when trainees ask for help and demonstrate through their performance that assistance is needed for strengthening those skills.

Coaching

Our principal expert, Gary Eggan, had worked in the job as an enlisted airman and later become a civilian technical advisor to airmen doing the job. Thus, he was expert both in doing the job and in helping others learn to do it. As an articulate expert, he provided us with a rationale for problem selection and sequencing, with the coaching content, and with clear representations of the effective problem spaces for each problem. SHERLOCK permits a large number of different paths toward isolating the fault in each problem. The coaching emphasizes relevant work environment information first, following up next with broad strategy (the application of relevant "rules of thumb"), and providing specific tactics only if the trainee is still unable to reason them out on his or her own. Help is provided only on request or when the trainee's performance necessitates SHERLOCK's attention (for example, when an unsafe action is attempted).

Each of SHERLOCK's problems has an EPS subgoal structure built by our domain expert. The trainee's performance is shadowed with reference to this subgoal structure. Each EPS subgoal node has one or more expert ways to confirm or disconfirm a particular hypothesis. When the trainee asks for help, SHERLOCK knows which EPS node is relevant to his needs. The help the trainee receives depends on the appropriate actions, test outcomes, conclusions, and best next moves for the EPS node where he is, and also on an up-to-date assessment of his skills. By having both expert and novice paths planned ahead of time, it is possible to specify coaching for different levels of skill as well as for the specific nodes in the problem space.

Expert mental models of the test station, in addition to informing the coaching of "rules of thumb," are also used more broadly in the hinting process. A trainee who asks for help will often get hints that direct attention to the functional areas of the test station that are most likely to be the source of the failure. For example, one of the heuristics tutored by SHERLOCK is the history-taking procedure. History taking uses the information generated by the routine actions of the test station as it systematically exercises its components. In history taking, an expert examines the devices that were actively involved in the function which is manifesting the system failure. He also considers the functions that the system has just handled correctly. Troubleshooting hypotheses are then derived from a comparison of the successful and the failure situations to see which components are needed for the faulty function but not needed in prior successful functions. For instance, if a device active in the current testing function was also active in the prior and successful testing step, and if both functions used the device in the same way, then it probably is not faulty (of course, there is a small chance that it failed between the two steps). Otherwise, as an active component in the function that failed, it is a prime suspect.

SHERLOCK is designed to offer the least hint that can enable further problem-solving progress. Much of its coaching is designed to stimulate a trainee's thinking by asking questions rather than generating answers for what the trainee should do next. A minimal hinting approach was selected in order to encourage active construction of the problem-solving knowledge by the trainee. Thus, each time a trainee asks for help, the first hint he receives is a structured rehearsal of what he has done on that problem, so he can reflect on his plans and actions. Further, when he requests additional help he gets a trace of the hints he had already received. More elaborate hints support the trainee's problem solving much as a shop supervisor might. The content of these hints is specific to the current work environment and to the state of the trainee's knowledge. Our coaching structure was developed to provide understanding of skills in many problem contexts so that the trainee can see the conditions under which a new procedure applies. Each node can give hints about the conditions under which certain things would be true. For example,

> These data signals should cause relays K3 and K2 in the A1A1A13 card to set. One way to determine if the A1A1A13 card is operating properly is to do an ohms check on the relays being set by the stimuli code select thumbwheel switches. You could ohm out the lower contacts of K3 (between Pins 1 & 3) and the lower contacts of K2 (between Pins 55 & 57) to determine if the relays are being set.[4]

The assigned tasks, though at the limits of the trainee's capabilities, are all completed, with support from the environment available on demand.

Does SHERLOCK Have the Right Knowledge?

We began the development of SHERLOCK with the belief that it was impractical to provide knowledge of troubleshooting procedures and conceptual understanding of the entire test station in complete form, although it would be desirable to do that in an intelligent instruction system. Accordingly, we developed the approach just described. More recently, we have come to realize that what SHERLOCK really needs to know about is not the entire test station but only the circuit configurations whereby the test stations accomplish various bits of work. With a sufficient collection of circuit path models for the test station tasks, it is possible for an intelligent coach to have deep knowledge of both concepts and procedures relevant to troubleshooting and to derive specific problem-solving plans from more general knowledge on demand. Thus, more explanations can be provided, and, by introducing faults into the circuit paths, problems can be generated to meet individual training needs. The plans for SHERLOCK II are not the topic of this paper, but we want to clarify that other, and perhaps more productive, designs can be developed for intelligent coached practice environments.

Student Modeling and Adaptive Instruction

SHERLOCK provides all trainees with the same type of problems in the same sequence of graded difficulty, gradually building up the trainee's conceptual model of troubleshooting. Nonetheless, SHERLOCK is adaptive to individual differences, diagnosing the trainee's learning performance relative to the part of the problem he or she is currently attacking. Feedback is both expert driven, based on SHERLOCK's calculation of the student's performance model or learned state, and student driven, in that if a trainee asks for help, it is provided, with additional elaboration on demand. Because it is given sparingly, based on an assessment of the trainee, hinting also fades automatically as the trainee's skill develops.

SHERLOCK predicts and models a trainee's performance on each specific problem and also maintains a more enduring model of the trainee's emerging competence. Assessing the trainee's competence enables SHERLOCK to determine how much assistance is required when an impasse is reached. The level of hint provided to the trainee depends on SHERLOCK's assessment of how well the training goals that are immediately relevant have been attained. Pedagogically, SHERLOCK approximates apprenticeship—a novice trainee can independently practice goal-oriented whole performance with SHERLOCK providing individualized assistance when an impasse is reached.

The classroom cognitive apprenticeship approaches do not have SHERLOCK's approach to assessment and pedagogy. Although they evaluate a student's progress, they have no concrete rules for when to provide hints and when to fade assistance. This is pedagogically important, because SHERLOCK's hint structure allows us to examine emerging competence that might not be tapped if no assistance were given. Novices have "islands of knowledge" that can be explored only by providing the necessary bridges between these islands. For instance, a trainee may have a well-developed measurement-taking skill. But if he or she has not mastered the early steps of planning a strategy for finding a fault, the point at which the mastered skill can be demonstrated or practiced may never be reached. By supplying bridging knowledge, the hint structure enables us to assess information about the subject's knowledge that would be inaccessible in a testing environment that terminated a performance whenever it reached an impasse.

This approach taps what Vygotsky called a "zone of proximal development," emerging competence that can be assessed only by supporting it with coaching [20,21]. Vygotsky (p. 86) defined the zone of proximal development in developmentally disabled children as

> the distance between the actual development level as determined by independent problem solving and the level of potential development as determined through problem solving under adult guidance or collaboration of more capable peers.

SHERLOCK's hinting structure challenges trainees with learning opportunities that would be just beyond their reach without coaching. This permits the assessment of their skills that in turn enables appropriate coaching.

Externalization of the Trainee's Troubleshooting Process

An interesting comparison can be made between computer-based coached practice environments and apprenticeship models that have human coaches.[5] In computer-coached practice, as in traditional apprenticeship settings, a "supervisor" is available to answer the trainee's questions. The advantage of the computer is that it can provide feedback based on more detailed and accurate individual performance and competence models than humans can readily determine or remember [24]. Salomon suggests that both kinds of coach can foster the internalization of expert skills and reflective processes. A student can internalize the cognitive skills displayed by a computer coach just as one could internalize processes modeled by a human coach. Computer tools can be internalized just as advice from "more capable peers" can be. SHERLOCK was designed to take advantage of such internalization. It provides hints that externalize troubleshooting strategies.

Furthermore, SHERLOCK provides templates for cognitive skills. One source of these templates is the interface the computer presents to the trainee. The trainee interacts with the computer largely by selecting options from a menu of troubleshooting goals and actions. The menus are hierarchical, with an underlying goal structure. The trainee moves from a menu of higher-level goals to submenus of subgoals. Thus, in order to select a local goal, like extending a particular card for testing, the trainee must have already selected a higher-level goal. Higher-level subgoals represent plans to test functional units: *Test the test station, Test the test package,* or *Test the airplane unit.* Within each of these major subgoals are sub-subgoals, such as *Test the circuit cards located in the unit.* The hierarchical planning structure modeled in these menus serves as a cognitive tool for structuring the problem-solving process. Figure 1 shows a menu of actions from which a subgoal has been selected and the submenu that appears in response.

Others have attempted to use menu selection to teach structured problem-solving activity. Research has been conducted in physics to promote expert-like behavior among novices by constraining them to follow an expert-like approach. Mestre et al. developed a hierarchical computer-based problem analysis environment for mechanics problems in physics called Hierarchical Analysis Tool (HAT) [25]. HAT is a menu-driven environment that constrains users to perform a top-down analysis of classical mechanics problems. The computer poses well-defined questions, and students answer them via menu selection. The first level menu consists of general principles that could be applied to solving the problems, followed by secondary menus that consist of ancillary concepts. HAT provides no feedback or tutoring, but merely constrains the type and order of questions that need to be considered when analyzing a problem. The HAT treatment group performed better than a control group on a task that required deep structure analysis but were no more likely to apply a concept-based approach to a problem-solving task. Mestre et al. concluded that a feedback approach using HAT would be more powerful for improving problem solving. SHERLOCK incorporates hierarchical structuring and feedback and was successful. However, it is difficult to attribute SHERLOCK's success with certainty to a single instructional strategy such as hierarchical structuring, because SHERLOCK incorporates multiple instructional strategies [26, 27].

The effective problem space (EPS) subgoal structure supports another pedagogical tool. One of the purposes of the EPS, from a programming perspective, was to serve as a record-keeping device of the trainee's movements through the problem space. By monitoring the solution trace, we could summarize and display the trainee's problem-solving steps every time help was requested. From a cognitive perspective, this form of recapitulation serves as a reminder of what hypotheses the trainee generated. This reminder externalizes the trainee's thought processes and is often enough to make the trainee reflect on what parts of the problem space have been examined. Replaying the trainee's problem-solving trace is a way to encourage self-regulation of planning behavior. Repeated experience with such a tool can lead to a trainee's anticipation of the trace and eventually perhaps to self-generation of a trace.[6]

The same record-keeping capabilities are also used to generate hints for a special-purpose option we called *the panic button*. The panic button can be selected at any time during the course of problem solution. Trainees are told that the panic button should be selected when they are totally lost and do not know what step to take next. If a trainee reaches an impasse and selects the panic button, then SHERLOCK can replay the plan subgoals that the trainee has (directly or indirectly) completed. This externalizes the trainee's plans up to the point of impasse. In this manner, the panic button constrains the problem space for the trainee by explicitly telling the trainee what he has ruled out in the problem space and implicitly telling him what is left in the problem space. It helps constrain the problem space and forces concentration on plan components that have not yet been completed or proposed. This computer tool is something like a more capable peer telling the technician that he should perhaps try to "check out the unit" instead of "the test package" because his past actions have already eliminated the test package as a possible fault area. The panic button is more than a summary; it uses the trainee's prior actions to generate a statement that constrains the problem space.

Recapitulation and the panic button are both designed to foster hypothesis generation, self-regulation, and reflection. This instructional concept is similar to Schoenfeld's use of post mortems in mathematics [11]. Schoenfeld presents postmortems at the completion of a problem to draw attention to expert/novice differences at key decision points. The panic button is an interactive rather than retrospective tool in that it plays back the trainee's thought process at the point where an impasse is reached, rather than waiting until the student is done with the problem.[7] It highlights for the trainee which hypotheses were ruled out by previous actions. In addition, trainees can call up an expert's performance by asking for help and thus compare their solution to an expert model.

Evidence that a Computer-Coached Practice Environment Works

The Air Force evaluation of SHERLOCK found that subjects who spent twenty to twenty-five hours working with SHERLOCK were as proficient in troubleshooting the test station as technicians who had been on the job four years longer [4]. We performed our own evaluation of SHERLOCK's effectiveness on matched experimental and control groups of airmen at two Air Force F-15 bases [3]. The experimental group spent an average of twenty hours working through SHERLOCK's thirty-four problems. Tutoring sessions were conducted in two- to three-hour blocks that spanned an average of twelve working days. The control group subjects went about their daily activities in the manual avionics shop. Structured interviews were administered as pre- and posttests of troubleshooting to thirty-two trainees. These pre- and posttests posed problems based on actual failures encountered in the shop, rather than on SHERLOCK problems, and Air Force technical orders were used rather than SHERLOCK's modified technical orders.

The pre- and posttests of troubleshooting proficiency targeted the same functional areas as the tutor problems, but there were some surface differences. Units, test packages, test stations, and the schematics for test station drawers were different from those in our tutor. Thus, pre- and posttests measured *near transfer* from the simulated tutoring experience to on-the-job performance. There were no significant differences between the experimental and control groups at pretesting ($\chi^2(1, N = 63) = 0.00, p = 1.00$), but on the posttest performance ($\chi^2(1, N = 62) = 10.29, p < .001$) the tutored group solved significantly more problems (mean = 30) than the control group (mean = 21).

Qualitative analyses of the pre- and posttest data revealed that the experimental group was significantly different from the control group on several factors that reflected emerging competence. Pre- and posttest scoring templates were developed by Gary Eggan, our domain expert. They addressed levels of competence by classifying troubleshooting steps as *expert, good, redundant,* or *inappropriate.* An expert troubleshooting step reflected a step that an expert would make at the particular point in the problem currently being addressed, such as eliminating a device as the fault source by making one very informative measurement. A good step was one that collected meaningful information but was not necessarily expert-like, such as taking three measurements to determine what an expert might establish with one. A redundant step collected information that had been determined by a previous move. An inappropriate or bad troubleshooting step was one that was totally meaningless or irrelevant, such as measuring the resistance across a power source or measuring the voltage drop in a path that was not active when the system failed. Analyses of covariance were performed examining performance differences between the two groups on posttest criterion variables while controlling for the qualitative pretest variables. The tutored group displayed more expert-like problem solving steps (mean = 19.33) in the posttest than the control group (mean = 9.06, $F(1,27) = 28.85, p < .001$) and made fewer inappropriate or bad moves (mean = 1.89, $F(1,27) = 7.54, p < .01$) than the control group (mean = 4.19).

A qualitative analysis was performed to see how effective SHERLOCK was in influencing a technician's overall problem-solving solutions. Three problems—one early, one middle, and one late in the tutor sequence—were scored to see whether there was a change in expertise over the course of the tutored practice. Trainees were also split according to aptitude, as measured by the standard selection and placement test of the U.S. Armed Services, the Armed Services Vocational Aptitude Battery (ASVAB). An expert developed an ideal solution path for each problem and scored the trainees' solution traces using the ideal path as a template. A deviation score from the expert path was computed for each subject by using the following formula: the subject's number of solution steps minus the number of steps that overlapped with the expert path. A repeated measures analysis of variance examined the effects of aptitude (high, low) and time spent in the tutoring environment (early, middle, and late tutor problem) on type of moves (deviation scores and total steps to solution). There was a significant main effect due to time spent in the tutoring environment ($F(2,42) = 4.19, p < .02$) on the trainee's deviation scores and steps to solution. The more problems the trainee had solved using SHERLOCK, the closer the trainee's troubleshooting solution processes were to expert troubleshooting solution processes, and the fewer steps were taken to reach a solution. With practice on SHERLOCK, trainees became more expert-like and efficient. Both the high- and low-ability trainees responded in the same manner, with the low-ability individuals showing slightly more improvement than the high-ability group.

Additional data confirm that the more specific goals of SHERLOCK were achieved. The proportion of actions that involved making measurements, as opposed to swapping components, increased over tutoring sessions ($F(2,45) = 3.30, p < .05$), for example. Furthermore, the effects of improved efficiency and decreases in inappropriate steps were found whether the problems involved conceptually simple or more complex failures. Overall, our data suggest that trainees acquired expertise in all phases of troubleshooting activity, including the most difficult aspects of measurement taking, as they used SHERLOCK. Our cognitive task analysis findings have been validated by a successful tutoring environment. Tutoring trainees to test their hypotheses, to reflect on the results of their measurements, to understand the procedures and conditions under which such procedures work, resulted in positive performance gains. SHERLOCK improved troubleshooting proficiency, and it is effective for trainees of various aptitudes. In addition, SHERLOCK's student modeling indicates that improvements occurred in every major category of knowledge that was targeted for training, which is confirmed by the analyses of posttest performance.

To What Do We Attribute SHERLOCK's Success?

Since SHERLOCK has multiple instructional strategies, it is hard to attribute its success to one strategy or element. As educators we are pleased with the many indicators of its success, but as scientists we want to know why it worked and which principles it embodies are most important. Sometimes these two goals are in conflict, since ensuring positive learning outcomes may destroy the controlled experimental manipulation of instructional variables. Over the years, reviews of research on aptitude learning and instructional interactions have concurred that no one form of instruction is ideal for all learners [28, 29]. Consequently, it may be inevitable that successful training requires confounding of approaches. However, we believe that with detailed analyses of skill changes over the course of learning, and with some variation in approach from one tutor to the next, we can begin to untangle and validate the multiple instructional strategies that SHERLOCK embodies.

One way to test what is effective in SHERLOCK would be to test its underlying mechanisms in a new domain. We might term this the *cloning approach,* replicating both the instructional mechanisms and the assessment tools. An alternative way to test SHERLOCK's effectiveness would be a kind of "swapping." Having developed a successful coached practice environment, we could take out different components and see whether it still produced the same success. Whichever empirical method is chosen, we believe that the following instructional strategies are candidate contributors to successful training:

1. *Teach specific domain knowledge along with metacognitive skills and heuristics that leverage the domain knowledge.*

2. *Embed learning within a problem-solving environment that reflects the uses to which knowledge will be put.* SHERLOCK embodies activity-based learning situated in the context of the shop environment, where skills can be generalized to on-the-job learning.
3. *Provide opportunities for learning through a combination of guided practice and locally situated opportunities for observation* (i.e., tell the trainee what to do, or show him, only when he is stuck and needs a specific bit of assistance). Situate feedback as guided, on-request support. In this situated way, articulate expert problem-solving methods and foster the comparison of expert and trainee performance details.
4. *Emphasize holistic procedural training.* People learn to do complex cognitive procedures by doing them. Core mental models should be revealed and should be explicit in hints, but practice should focus on doing the thinking, not on rote repetition or playing abstractly with the background of conceptual knowledge that supports successful problem solving. When a trainee cannot handle the whole performance, he should be supported through the parts he cannot handle but should still, more or less, be enacting, with help, the whole target cognitive performance.
5. *Train the specific competence required rather than training abstractions.* Transfer comes from training concepts and procedures that are common to or can be generalized from experience in multiple situations. Abstractions must be grounded in prior concrete experience. This is the fundamental lesson about situated learning!
6. *Cognitive task analyses of the hardest thinking tasks in a job are critical to success* in designing effective thinking. These analyses identify the critical concepts and procedures situated within the job environment. Use the vocabulary of the workplace to further situate explanations and guidance.
7. *Build from the highest levels of domain expertise.* For a system to be able to shape performance toward expertise, it must reflect, in its coaching, the expertise of real workers who have learned to do the job, not just the designer's model of the work environment or the job as represented in doctrine.
8. *Enable the development of complex subgoal structures.* In SHERLOCK, we provided a tool that encouraged thinking about higher-order goals before subgoals.
9. *Adapt instruction.* Accommodate individual differences in learning by giving coaching only when it is needed. In SHERLOCK, detailed dynamic assessment, both of global competence and of local performance details, supports this individualization.

Future Enhancements

Reflective Follow-Up to Problem Solving

SHERLOCK externalizes both the trainee's thought processes and the expert's. However, it does this in a local and fragmented way. It might be helpful to also provide a postsolution opportunity for the trainee to analyze his performance in comparison with that of an expert. Our colleagues Marilyn Bunzo and Garry Eggan have designed an enhancement to SHERLOCK that combines the abstracted replay and post mortem strategies suggested by Collins et al. and Schoenfeld, respectively [7, 11]. At the end of each problem a trainee will be provided with a visual representation of an expert trace of solving the problem. This representation will illustrate the plans and actions undertaken to reach a solution. Modest animation highlights, in turn, each step that was taken. Subsequently, the student's own problem-solving trace (see Figure 2) is provided below the expert's graph. Next, a third graph appears in which the student and expert traces are overlaid so that the trainee can see where and how much his solution overlapped an expert solution.

In addition, a trainee can explore the expert problem-solving trace by pointing to any node in the graph. When pointed to, each expert node will reveal an expert's reasoning for selecting this node in the solution, as well as hints as to what the appropriate actions, outcomes, conclusions, and next steps would be for that particular node (see Figure 2). Articulating the problem solution structure makes the role of SHERLOCK as a more capable peer more explicit and allows the user to evaluate his or her own responses in comparison to an expert's.

At a finer grain size, we have a second approach for comparative analysis of the testing done on a particular circuit component. Trainees would be able to see the active circuit paths marked on a schematic diagram of a printed circuit card and would receive explanations of appropriate testing strategy that referred to the illustrated paths (see Figure 3). In this way mental models of testing functional areas can become more salient to the trainee, and testing strategies may become less mysterious.

Social Facilitation of Learning

The success models of cognitive apprenticeship in the classroom have one more element that could be incorporated into SHERLOCK's experiments on learning: the role reversal or procedure/critic element in which the trainee learns to generate and critique his own and others' work. Each of the cognitive apprenticeship approaches demonstrated to the students that learning was transitional and that perfect performance all the time was not

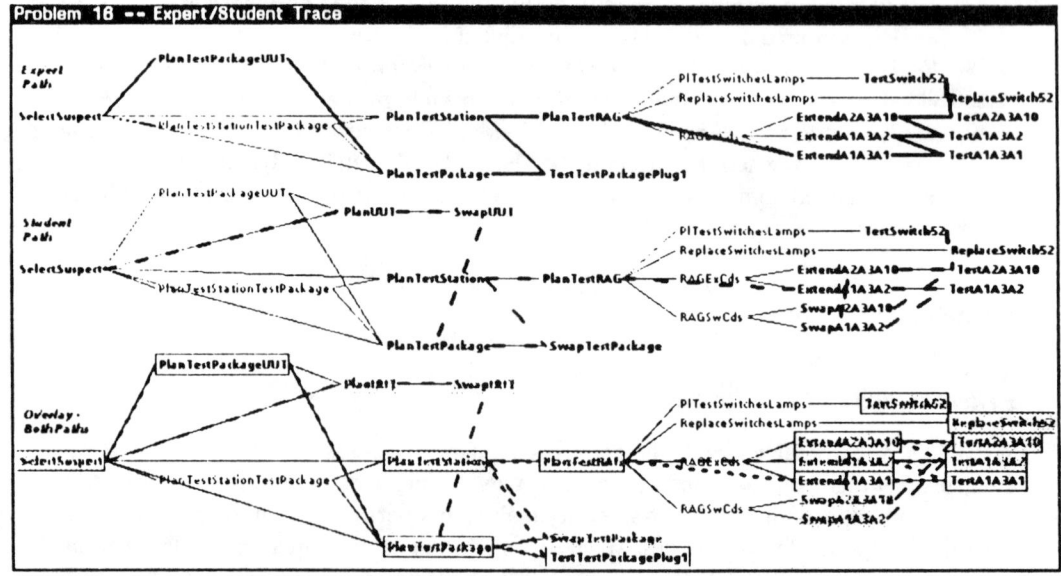

Figure 2 Comparison of expert and trainee problem solutions.

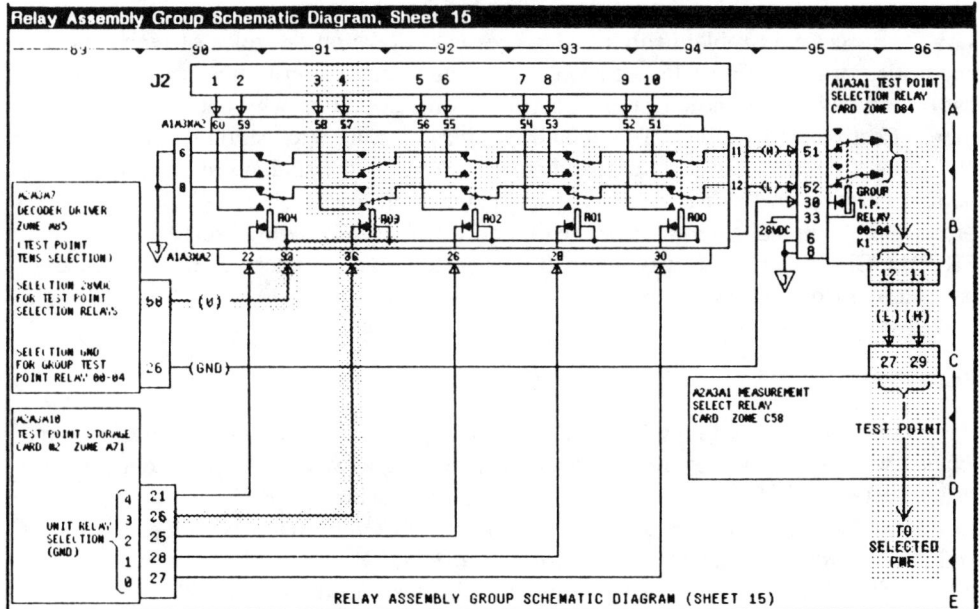

Figure 3 Display of relevant path in a printed circuit card.

the "norm." Reading comprehension was demonstrated to be transitional by discussing reading passages in a group of readers at different stages of development and having peers both produce work and critique each other's work [9]. Writing was demonstrated to be transitional in that revision was a significant component in the writing process [10]. Math problem solving was shown to be transitional by having teachers solve math problems that students generated, demonstrating that teachers can flounder toward a solution just as students can [11]. Each of these approaches has an aspect of social interaction not used in SHERLOCK. However, many of SHERLOCK's externalization mechanisms are similar to these approaches, and a controlled use of the "shop subculture" could be fostered in a new set of experiments for understanding learning transitions.

A possible extension of SHERLOCK would be to make the informal social context, in which SHERLOCK is used at the job site, a formal part of the training, so that SHERLOCK-based troubleshooting experiences are shared with peers. Learning transitions in troubleshooting might be enhanced if trainees observe other trainees at different stages in learning to troubleshoot. Experiments that use peer tutoring in the form of structured feedbacks could be informative. For example, trainees who had received the SHERLOCK treatment could tutor the control group on troubleshooting skills that they had acquired using SHERLOCK. The feedback that the tutor group would provide the control group could be structured by using SHERLOCK as a practice environment, where instead of SHERLOCK coaching, the tutored group would coach the control group. This approach would provide further learning opportunities for the tutor group by causing trainees to articulate what they had learned. Furthermore, it would clearly display transitional learning, where SHERLOCK-trainees could see other trainees at different stages in learning to troubleshoot.

At the completion of training, SHERLOCK's coaching mechanisms could be turned off completely and the trainees could be asked to coach their peers through a problem.

These teachbacks could assist trainees in developing the evaluative mechanisms necessary for successful problem solving. Trainees could take on the roles of both producer and critic, by generating efficient problem-solving traces and critiquing various solution paths, and having discussion groups that focus on elaborating why one solution path is more effective than another. This would be an interesting application of the producer/critic model described by Palincsar and Brown, in which the students are responsible for both generating their own responses and being their own critics [9]. It would also provide an additional external measure of their learning. We could observe how much of SHERLOCK's instruction was internalized and how well trainees could critique their peers in the troubleshooting process.

Tailoring to Local Needs

We previously described how SHERLOCK was designed with the assistance of avionics experts who contributed to the knowledge engineering as well as to the teaching expertise reflected in SHERLOCK. We could extend the teaching expertise aspect of SHERLOCK by having more involvement of masters at different avionics shops, so that we could have different models of expertise in SHERLOCK. Using different experts from different shops may tailor instruction more to the specific shop demands.

Technological Enhancements

Increasing the number of avionics units, test packages, and active test station drawers in SHERLOCK becomes time consuming to the problem developer. More technical documentation must be appropriated, more schematics must be developed and input into the computer, and a great deal of cross-checking of the test station setup with the TO test procedures is necessary. Creating schematics on the computer is a time-consuming activity and led us to restrict much of our instruction to a small part of the test station. With advances in interactive videodisc technology it will take less time to incorporate elaborate graphic displays. This will allow rapid development of tutors with a broader scope of devices and technical materials available in the work environment simulations, increasing the scope and realism of computer-based cognitive apprenticeship.

Acknowledgments

The work was done as part of subcontracts with HumRRO and Universal Energy Systems, who were prime contractors with the Air Force Human Resources Laboratory. Many of the tools and much of the infrastructure that permitted us to conduct this project efficiently were put in place by earlier efforts and the foresight of the Office of Naval Research, where Henry Halff, Susan Chipman, Marshall Farr, and Michael Shafto patiently invested in a future that took a number of years to unfold. SHERLOCK was a massive undertaking. It took a lot of work from a lot of people. As a real-world applied activity in a setting with real work that could not be halted, it was a test not only of ingenuity and hard work but also of courage, patience, tolerance, perseverance, and the love and respect of many players for each other. Much of the design of SHERLOCK came from joint work with Marilyn Bunzo and Gary Eggan. Joining us in the effort at one point or another were Jaya Bajpayee, Jeffrey Bonar, Lloyd Bond, James Collins, Keith Curran, Richard Eastman, Drew Gitomer, Robert Glaser, Bruce Glymour, Linda

Greenberg, Denise Lensky, Debra Logan, Maria Magone, Tom McGinnis, Valerie Shalin, Cassandra Stanley, Arlene Weiner, Richard Wolf, and Laurie Yengo. Equally important were Sherrie Gott and the many people that worked with her at the Air Force Human Resources Laboratory, Dennis Collins and other Air Force subject matter experts who helped us when we needed it most, and the officers and personnel of the 1st and 33rd Tactical Fighter Wings, at Langley and Eglin Air Force Bases, especially those in the Component Repair Squadrons who made room in their worlds for SHERLOCK and its many retainers. Arlene Weiner put considerable effort and expertise into this manuscript, restructuring much of it and making it much more coherent.

Notes

1. The term *inert knowledge* is due to Alfred North Whitehead.
2. The mouse is a common input device for graphical interfaces.
3. We thank Jill Larkin for helping us find understandable ways to discuss these aspects of SHERLOCK.
4. Note that SHERLOCK uses the jargon that trainees are known to use whenever there is a chance that more formal terms may not be as well understood.
5. There are other computer-based approaches to apprenticeship. Gott has comprehensively reviewed this literature [22]. Frederiksen, White, and Cooper have developed a computer-based simulation for teaching mental models in a domain closely related to SHERLOCK's [23].
6. A second purpose of the recapitulation hint is to compensate for processing capacity overload in early stages of training. We assume that the trainee will sometimes know what to do but simply will not be able to keep track of his progress, due to processing overload.
7. The next version of SHERLOCK will also have postproblem reflective opportunities.

References

1. A. Caramazza, M. McCloskey, and B. Green, "Naive beliefs in 'sophisticated' subjects: Misconceptions about trajectories of objects," *Cognition*, 9, 117–23, 1981.
2. L. C. McDermott, "Research on conceptual understanding in mechanics," *Physics Today*, July 1984.
3. A. M. Lesgold, S. P. Lajoie, R. Eastman, G. Eggan, D. Gitomer, R. Glaser, L. Greenberg, D. Logan, M. Magone, A. Weiner, R. Wolf, and L. Yengo, *Cognitive task analysis to enhance technical skills training and assessment,* Technical Report, University of Pittsburgh, the Learning Research and Development Center, April 1986.
4. P. Nichols, R. Pokorny, G. Jones, S. P. Gott, and W. E. Alley, *Evaluation of an avionics troubleshooting tutoring system,* Special Report, Brooks AFB, TX: Air Force Human Resources Laboratory, in press.
5. J. S. Brown, A. Collins, and P. Duguid, "Situated cognition and the culture of learning," *Educational Researcher*, 18, 32–41, 1989.
6. A. Collins, J. S. Brown, and S. E. Newman, "Cognitive apprenticeship: Teaching the craft of reading, writing, and mathematics," in L. B. Resnick (ed.), *Knowing, learning, and instruction: Essays in honor of Robert Glaser,* Hillsdale, NJ: Lawrence Erlbaum Associates, 1989.
7. A. Collins and J. S. Brown, "The computer as a tool for learning through reflection," in H. Mandl and A. M. Lesgold (eds.), *Learning issues for intelligent tutoring systems,* New York: Springer, Verlag, 1988.
8. L. B. Resnick, "Learning in school and out," *Educational Researcher*, 16, 13–20, 1973.
9. A. S. Palincsar and A. L. Brown, "Reciprocal teaching of comprehension-fostering and comprehension monitoring activities," *Cognition and Instruction*, 1, 117–75, 1984.
10. M. Scardamalia and C. Bereiter, "Fostering the development of self-regulation in children's

knowledge processing," in S. F. Chipman, J. W. Segal, and R. Glaser (eds.), *Thinking and learning skills: Research and open questions,* Hillsdale, NJ: Lawrence Erlbaum Associates, 1985.
11. A. H. Schoenfeld, *Problem solving in the mathematics curriculum: A report, recommendations and an annotated bibliography,* The Mathematical Association of America, MAA Notes, No. 1, 1983.
12. L. J. Cronbach, "The logic of experiments on discovery," in L. S. Shulman and E. R. Keislar (eds.), *Learning by discovery: A critical appraisal,* Chicago: Rand McNally, 1966.
13. J. Greeno, "A study of problem solving," in R. Glaser (ed.), *Advances in instructional psychology,* vol. 1, Hillsdale, NJ: Lawrence Erlbaum Associates, 1978.
14. S. P. Lajoie and V. Shute, "Two intelligent tutors: Principles of instructional design," presented at the 93d annual convention of the American Psychological Association, Los Angeles, CA, 1985.
15. A. L. Brown and R. A. Reeve, "Bandwidths of competence: The role of supportive contexts in learning and development," in L. S. Liben and D. H. Feldman (eds.), *Development and learning: Conflict or congruence?* Hillsdale, NJ: Lawrence Erlbaum Associates.
16. N. M. Morris and W. B. Rouse, "Review and evaluation of empirical research on troubleshooting," *Human Factors,* 27, 503–30, 1985.
17. S. P. Gott and T. S. Davis, "Introducing specific knowledge domains into basic skills instruction: From generalized power to specified knowledge," paper presented at the National Adult Education Conference, Philadelphia, PA, 1983.
18. D. H. Gitomer, *A cognitive analysis of a complex troubleshooting task,* Doctoral diss., University of Pittsburgh, 1984.
19. G. W. Ernst and A. Newell, *GPS: A case study in generality and problem solving,* New York: Academic Press, 1969.
20. L. S. Vygotsky, *Mind in society: The development of higher psychological processes,* Cambridge, MA: Harvard University Press, 1978.
21. S. P. Gott, "Apprenticeship instruction for real world cognitive tasks," In E. Z. Rothkopf (ed.), *Review of research in education,* vol. XV, Washington, DC: AERA, in press.
22. A. M. Lesgold, "The integration of instruction and assessment in technical jobs," in *Proceedings of 1987 ETS invitational conference,* 81–88, Princeton, NJ: Educational Testing Service, 1988.
23. J. R. Frederiksen, B. Y. White, and E. Cooper, "Restoring lost apprenticeship: An approach based upon intelligent tutoring," paper presented at the annual meeting of the American Education Research Association, New Orleans, LA, 1988.
24. G. Salomon, "AI in reverse: Computer tools that become cognitive," invited address at the American Educational Research Association, New Orleans, April 1988.
25. J. Mestre, R. Defresne, W. Gerace, and P. T. Hardiman, "Hierarchical problem solving as a means of promoting expertise," paper presented at the annual Cognitive Science Conference, Montreal, Canada, August 1988.
26. S. P. Lajoie, G. M. Eggan, and A. M. Lesgold, "Instructional strategies for a coached practice environment," paper presented at the annual Cognitive Science Conference, Montreal, Canada, August 1988.
27. A. M. Lesgold, S. P. Lajoie, M. Bunzo, and G. Eggan, "Sherlock: A coached practice environment for an electronics troubleshooting job," in J. Larkin, R. Chabay, and C. Scheftic (eds.), *Computer assisted instruction and intelligent tutoring systems: Establishing communication and collaboration,* Hillsdale, NJ: Lawrence Erlbaum Associates, in press.
28. L. J. Cronbach and R. E. Snow, *Aptitudes and instructional methods: A handbook for research on interactions,* New York: Irvington, 1977.
29. R. E. Snow, "Research on aptitudes: A progress report," in L. S. Shulman (ed.), *Review of research in education,* Vol. IV, Itasca, IL: Peacock, 1977.

Chapter 2

Automated Tutoring in Interactive Environments: A Task-Centered Approach

URSULA WOLZ
KATHLEEN R. MCKEOWN
GAIL E. KAISER

Columbia University, Department of Computer Science
New York, NY 10027

Abstract *Tutoring in interactive computing environments is sometimes more properly understood as* consulting. *A tutor's implied curriculum may be adaptable to the user's knowledge and experience but still not meet the user's immediate needs—to get some task done. A consultant however, can dynamically adapt to address the task at hand. The authors present a user's goal-centered approach to tutoring in interactive environments and describe how certain tutoring strategies appropriate for consulting behavior are automated. The authors have implemented their approach in GENIE, a question answering system for the Berkeley Unix Mail system. They focus on the pedagogical strategies employed by GENIE to best meet the user's immediate needs.*

Introduction

Interactive computing environments are designed to provide supportive resources for a range of users with different expertise and computational goals. Such environments may be as simple as mail systems and word processors or may encompass sophisticated data bases, design tools, or programming languages. Yet all such environments are *procedural* in that they contain an underlying set of functions or constructs with which users construct procedures to accomplish tasks. A problem arises in providing resources through which users can initially learn about the environment and then later extend their expertise. Standard curricula—even those provided on line—require users to subordinate their immediate needs to the agent embedded in the curriculum. We have developed a solution through the implementation of GENIE (GENerated Informative Explanations),

---

Ursula Wolz was supported in part by ONR grant N00014-82-K-0256, by NSF grant IST-84-51438, a grant from DARPA, and a grant from Siemens Research and Technology Laboratories.

Kathleen R. McKeown was supported by NSF grant IST-84-51438, by ONR grant N00014-82-K-0256, by a grant from Bell Communication Research, and by the Center for Advanced Technology.

Gail E. Kaiser was supported by NSF grants CCR-8858029 and CCR-8802741, by grants from AT&T, DEC, IBM, Siemens, Sun, and Xerox, by the Center for Advanced Technology and by the Center for Telecommunications Research.

an answer generating system that tutors specifically to the current needs of the user in the domain of Berkeley Unix™ Mail. Our focus is on the pedagogical behavior and domain knowledge of *consultants* rather than conventional tutors.

In our view a consultant is a "system guru" or local expert in a domain to whom others come for information and advice. This is in contrast to a conventional tutor who initiates personalized, often remedial instruction within the constraints of a prescribed curriculum. Automatic consulting provides a valuable test bed for theories about personalized tutoring because of the unique relationship between the consultant and the user. The knowledge conveyed in a consulting session is determined by the current task or goal of the user and the user's questions about that task. In most tutoring, both on and off the computer, the teaching agenda is predetermined by an implied curriculum chosen by the tutor. This is not appropriate for a consultant, because users have a multitude of different needs, backgrounds, and deficiencies in what they know. Furthermore, most tasks can be done in more than one way, and the core functions can be used in more than one task. A curriculum may cover all the tasks and functions but will rarely capture the richness of the relationships between them or the many possibilities for performing any given task.

Although we focus on task-centered environments, exploratory learning environments that encompass a curriculum can benefit from consulting. In task-centered settings, consulting is based on tasks initiated by the user. In a learning-centered setting, a tutoring agent initiates the task, but the student may still require consultation in order to accomplish it.

We take a user's goal-centered approach to consulting, in which help given is a direct function of users' goals, their knowledge, and the current context. Specific tutoring strategies determine the form and content of an answer to a user's questions. A user model represents the plans that the consultant thinks the user has for achieving goals. An expert domain model provides a representation for a consultant's knowledge of computational goals, plans to accomplish them, and functions of the environment that can instantiate plans.

We abandon the classification of aspects of the environment according to level of expertise and the categorization of users as "novice," "intermediate," or "expert." Instead, information about a user's exposure to the environment influences the pedagogical agenda of the consultant. Decisions about how to answer a user's question are based on an analysis of the discrepancies between the expert domain knowledge and the user model, as well as the current situation. In this respect GENIE bases its answer on "local" user expertise, by analyzing whether entities are present or absent in the user model. However, we intentionally do not include indices for rating entities on a continuum such as "expert" or "novice," since these labels tend to hide important epistemological reasons for making a choice.

Consulting can be characterized as a three-stage process comprised of question understanding, problem analysis, and answer generation. We concentrate on the latter two components: analysis, through an algorithm called the Plan Analyst; and generation, through an algorithm called the Explainer. GENIE attempts to answer a question by doing a two-phase search of the knowledge base, first trying to construct a coherent relationship between the user's question and computational goal in an attempt to find the most appropriate information for a response; then trying to construct a coherent textual explanation that takes into account what the user already knows.

In this paper, our focus is on the tutoring strategies GENIE employs in various settings. These include:

- *Introducing*—presenting functions and plans that the user has not encountered before
- *Reminding*—briefly describing functions and plans that the user has been exposed to but may have forgotten
- *Clarifying distinctions*—explaining distinctions and options about functions and plans
- *Elucidating misconceptions*—clearing up misunderstandings that have developed about functions and plans to which user has been exposed.

We also show how the content of the answer GENIE provides depends on the type of question asked and on an analysis of discrepancies between the user model and the expert domain knowledge.

The next section presents our perspective on tutoring and consulting in interactive environments. Subsequent sections introduce the tutoring strategies we use, describe the prerequisite knowledge for choosing a tutoring strategy, and summarize under what conditions different strategies are chosen. Then we present GENIE, our experimental consultant system, and give examples of how the strategies are used. The remaining sections summarize related work and present conclusions and directions for future work.

Tutoring in Interactive Environments

Interactive computing environments are *procedural environments* in which learners develop skills rather than learn facts and associations between facts. In these environments, sophistication and complexity, and consequently power, are built upon simple interfaces. Good environments are often characterized as *customizable*—in which users can adapt the tools to their own personal needs, and *extensible*—in which users can build new tools from those that already exist.

Expertise in a procedural environment is not simply a matter of knowing what functions allow simple tasks to be performed. One must know how functions interrelate or interfere with each other when attempting to achieve more complex goals. In other words, one must know *plans* for accomplishing goals, where plans may be decomposed into subgoals or directly executed by functions. A particular plan is directly related to a particular goal, but the components of a plan, the steps or subgoals, may occur in many different plans. Furthermore, in most computing environments there is often more than one plan to achieve a goal.

Learning skills and doing real tasks are intimately woven together. While initial learning may require extensive supervision, once the key concepts have been learned, users are expected to initiate their own goals and solicit expertise from others only when necessary. Within such an environment, lending expertise is often thought of as "consulting" rather than tutoring.

It is important to articulate some of the differences between consulting behavior and tutoring. In more classic settings a tutor has an agenda based on an externally prescribed curriculum and diagnosed student deficiencies. The tutor assumes some external force motivates the student to learn, or the tutor must find incentives to motivate the student. A consultant, on the other hand, has a limited agenda based on showing the user how to accomplish a task. Motivation for learning is a matter of being able to do the task independently of the consultant. Brown and VanLehn [1] introduced the notion of *felicity conditions* under which new learning is most likely to take place. Within a procedural setting, two situations are extremely motivating. The user attempted to do something and it didn't work. The user finds a task tedious and suspects there is a more efficient way to do it.

Since the user's current task initiates the tutoring dialogue, in the student's mind, whatever learning takes place is secondary to actually accomplishing the task. A prescribed curriculum may not always be appropriate for a particular task. In a traditional curriculum it is often assumed that a set of skills must be learned in a particular order, and to a specified level of mastery. Within the context of consulting, this is inappropriate, because some skills may not be necessary for a given task. Furthermore, users' knowledge of the functions available within the environment may vary considerably depending on the tasks they do regularly and how much they attempt to customize the environment to their needs.

We do not claim that consulting behavior is appropriate to all kinds of tutoring—only that interactive environments provide a potentially rich metaphor for creating procedural learning environments. Even within these environments, there is a place for more traditional tutoring. During the initial stages of using the environment, it seems imperative that the basic concepts and functionalities of the environment are introduced. Here a didactic approach would be more expedient. Later, however, when users take off in different directions, consulting behavior is more appropriate, providing truly individualized instruction.

Tutoring Strategies

It is clear that automated consulting in interactive environments requires complex interactions between sophisticated processes and knowledge bases. We have chosen to carve a niche within this larger domain that has not been explored as fully as others—namely, how best to choose what to say and how to say it. In order to prevent further clouding of the pertinent issues, we have chosen not to concern ourselves with *when* to say it. At the present time we view consulting as *passive*—information is presented only after it is explicitly solicited through a user's question. Obviously a feedback loop through which the user can rephrase or expand on the question is necessary. We discuss this briefly below.

Our narrow view of tutoring is therefore concerned with understanding a user's question about how to do something, deciding what skills are necessary to do it, and choosing the appropriate way to present the skills to the user. We focus here on the last of these. Others [2, 3] have addressed understanding the question, and we have described elsewhere how to determine what skills to present [4, 5].

We have identified four necessary tutoring strategies in computing environments through an analysis of various types of help. They specify the kind of information that is typically included. Two of the strategies we posit, *introducing* and *reminding*, are found as part of written reference material, whether on line or off. They can be used to provide help about a function or goal the user has, either by introducing details about a function or plan that can accomplish the goal or by simply reminding the user of forgotten functions and plans. If a user already has some information about a function or plan, but it is incomplete or incorrect, then one of the two other strategies, *clarifying distinctions* or *elucidating misconceptions,* is more appropriate. These strategies are corroborated primarily by informal observations of human consulting situations.

Our analysis of on- and off-line help for Lisp, UNIX, Pascal, BASIC, Logo, and a number of word processing programs reveals that the reference material tends to fall into three categories:

- *Reference manuals* that provide details and definitions of the environment. The material is either alphabetically ordered or grouped according to the function of the constructs.
- *Support manuals* that provide more explanation about how to use the functions. These are organized according to the function of the constructs.
- *Tutorials and textbooks* that introduce the concepts behind the functions. These tend to be much more explanatory and less definitional than reference or support material. Although they may be organized according to the function of the constructs, there is a greater emphasis on how constructs are combined.

Tutorials and, to some degree, support manuals are intended to *introduce* new material, while the concise definitions in reference manuals can efficiently *remind* users of how functions work. All three kinds of material may help users clarify details or clear up misunderstandings of *functions,* but the user must possess strategies for locating the relevant information. *Clarifying distinctions* or *elucidating misconceptions* about plans for goals occurs only in a limited way in tutorials and textbooks. The user's goal may not occur in a written text. Even if it does, texts tend to introduce the simplest techniques for accomplishing a task. Although the information necessary for learning a better way may exist, the user is responsible for finding that information and must often piece it together from various points in the book.

Evidence for two of our strategies, *introduce* and *remind,* is also provided by Magers [6] and Borenstein [7], who have drawn a distinction between information that is *definitional* or *instructional.* Definitional information is more appropriate for reminding someone about something that has previously been learned, while instructional information is more appropriate for introducing new information. These types differ not only in their format and level of detail, but also in their emphasis and the degree to which related information is included. We therefore choose to *remind* or *introduce* depending on the user's knowledge and goals. We further refine the distinction of Magers and Borenstein, however, by including the possibility of *elucidating misconceptions* or *clarifying distinctions.*

We also extend our strategies by using them differently to satisfy distinct tutoring needs: the need for tutoring that is in *direct response* to the question and tutoring that is intended as *enrichment.* The former prevails in order to satisfy principles of informativeness: answer the question that was asked. But it is also possible to present new skills to the user opportunistically. For example if the user asks whether a particular plan will accomplish a particular goal, the consultant must respond informatively that it does or does not. However, if the consultant knows of a better way to accomplish the goal, the opportunity should be taken to mention it. Each strategy can be used both responsively and as enrichment.

In the following sections we describe each strategy in more detail and show examples of their use in reference material. Steps within a strategy that occur in direct response but not as enrichment are marked with an asterisk, "*."

Introducing

Introducing is used to provide generic details about a function or plan, assuming no previous background about it. In the reference materials, we found certain kinds of information consistently being presented in the support manuals, tutorials, and textbooks. We have identified the kind of information used for introduction and provided an ordering on the information that is often, but not always, used in the texts.

1. Informally, the process of introducing a plan consists of:
   a. Stating the goal*
   b. If the plan maps to a function, introducing the function; otherwise:
   c. Summarizing the subgoals for the plan
   d. For each subgoal either introducing or reminding about a plan for the subgoal, depending on whether the user model does or does not contain a plan for the subgoal
   e. If it is a top-level plan, reviewing the steps in the plan through an example
   f. Relating each step in the example to a subgoal.*
2. Introducing a function consists of:
   a. If not in the context of introducing a plan, stating the goal
   b. Presenting the syntax
   c. Describing the parameters
   d. Describing any preconditions that must exist for it to work
   e. Describing the effects (which is not the same as stating the goal)
   f. If not the context of introducing a plan for a subgoal, giving an example.*

This strategy can be seen in the example shown in Figure 1 taken from a textbook on Lisp [8]. In the first paragraph, the goal is stated ("define procedures such as COUNT-ATOMS") and a summary of the steps (MAPCAR and APPLY) is provided. The remaining paragraphs *introduce* the first subgoal (following step 1d). The second paragraph equates the subgoal "iterate" with the function "MAPCAR," and the remainder of the section recursively uses the *introduce* strategy for the function (step 1b). Following the second strategy for introducing functions, the third paragraph presents the syntax

Dealing with Lists May Call for Iteration Using MAPCAR

A somewhat more elegant way to define procedures like COUNT-ATOMS is by means of the primitives MAPCAR and APPLY, two new procedures that are very useful and very special in the way they handle their arguments.

Iterate is a technical term meaning to repeat. It can be used to iterate when the same procedure is to be performed over and over again on a whole list of things. Suppose, for example, that it is to record which numbers in a list are odd. From the list (1 2 3), we expect to get (T NIL T).

To accomplish such transformations with MAPCAR, you supply the name of the procedure together with a list of things to be handed to the procedure one after the other. Consider the following, for example, where the primitive 'ODDP has the obvious definition:

```
(MAPCAR 'ODDP     ;Procedure to work with.
        '(1 2 3)) ;Arguments to be fed to the
                   procedure.
(T NIL T) (1 2 3))
```

MAPCAR is said to cause iteration of ODDP, because the MAPCAR causes ODDP to be used over and over again.

There is no restriction to procedures of one parameter, but if the procedure does have more than one parameter, there must be a corresponding number of lists of things from which to extract arguments that can be fed to the procedure.

As shown in the following example, MAPCAR works like an assembly machine, taking one element from each list of arguments and assembling them for the procedure. . . . . .

Figure 1 An example of *introducing*.

of the function ("supply the name of the procedure together with a list of things to be handed to the procedure one after the other") and describes the parameters ("a list of things"). Preconditions are provided out of order in the last paragraph, and the effects are provided in the immediately preceding sentence ("MAPCAR is said to cause iteration of ODDP, because the MAPCAR causes ODDP to be used over and over again"). Since this is not the top-level goal, we skip step 1e, finishing introduction of the function, and pop back to introduction of the subgoal "iterate" continuing with step 1e, reviewing through an example. There is only one step in the plan for this goal, MAPCAR, and an example is provided (again, out of order) through ODDP. Step 1f is not needed, because there is only one step in the plan.

The strategy *introduce* as shown here provides a high-level characterization of the type of information provided in the naturally occurring example. This use of strategies is quite similar to the use of schemas in works by McKeown [9], Paris [10], and McCoy [11].

Reminding

Reminding is used to present the bare minimum of information about a function or plan under the assumption that the reader has some knowledge about it from previous experience. Manuals most often use this strategy.

1. Reminding about a goal consists of
   a. Stating the goal
   b. If the plan maps to a function, reminding about the function; otherwise:
   c. Summarizing the subgoals for the plan
   d. If this is the top-level plan, reviewing the steps in the plan through an example.
2. Reminding about a function consists of:
   a. If not in the context of introducing a plan for a subgoal, stating the goal
   b. Presenting the syntax
   c. Describing the parameters
   d. If not in the context of introducing a plan for a subgoal, giving an example
   e. If not in the context of introducing a plan for a subgoal, relating the function to other pertinent information.

An example of reminding is shown in Figure 2 taken from a Lisp manual [12]. The first sentence provides a general description of the action (step 2a). The syntax and parameters are then presented (steps 2b and 2c), followed by an example (step 2d). Additional information about side effects is also provided (step 2e).

Clarifying Distinctions

Clarifying is used to compare two functions or two plans for a goal. This is done occasionally in tutorials and textbooks but is essential for face-to-face consulting and questions, when the user often queries about the difference between two plans or specifically asks for a better way to achieve a goal.

To clarify a plan, we assume that between two alternative plans, one of them is clearly better. Here functions can simply be considered degenerate plans. Which one is better can vary depending on context. Clarifying a plan consists of:

MAPCAR   Function

The mapcar function applies a function to each element of each list in a series of lists.

(mapcar *function list* &rest *more-lists*)

Examples

```
(mapcar #'1+ '(1 2 3))                    => (2 3 4)
(mapcar #'list '(smith) '(sue janie mary))  => ((SMITH SUE))
```

Comments

The iteration of mapcar applies to the elements of the shortest list in the series of lists; excess elements in longer lists are ignored. mapcar returns a list whose elements are the values returned by calls to function.

See also, Steele: 7.8.4 - p. 128.

Figure 2 An example of *reminding*.

1. Stating that a best plan (B) exists for the goal
2. Summarizing an alternative plan (P)*
3. Summarizing B
4. Describing the relationship between B and P
5. For steps in B that differ from P, introducing or reminding about plans for subgoals based on whether those goals exist in the user model, and describing the relationship between this step and the corresponding step in P
6. Summarizing the steps of B through an example.

Examples from texts do not correspond exactly to our strategy, as there is generally no context for identifying which of a set of alternatives is the better plan. In some cases, a context may be hypothesized, and one of the alternatives is labeled as better for this case. An example in which context is not hypothesized is shown in Figure 3, taken from a Lisp support manual [13]. The goal is mapping over lists and is described in paragraphs 1 and 2. Several alternative plans (single-step plans that map directly to functions) are then given (step 2 of the strategy). Although it is not explicitly identified as a best plan, because of the detail given, we can assume the author intends to identify mapcar as a best, or at least default, plan. A more detailed summary of mapcar follows in paragraph 3 (step 3), followed by an example (step 5). Following this, each alternative function is compared with mapcar and differences noted (step 4).

Elucidating Misconceptions

Elucidating is used to clear up misconceptions and will be used most often when dealing with an individual's problems. Because most texts do not specifically address an individual reader, elucidating is found even less often in texts than clarifying. However, some texts (particularly tutorials) sometimes will posit a possible misconception that can occur. Figure 4 identifies five kinds of invalidities that can be diagnosed based on work by Joshi, Webber, and Weischedel [14].

Elucidating misconceptions in plans (or functions) consists of:

1. Stating that the plan does not work for the goal
2. Summarizing the plan, identifying the problem
3. If the problem is missing preconditions, either state that no plan exists to satisfy

7.8.4 Mapping

Mapping is a type of iteration in which a function is successively applied to pieces of one or more sequences. The result of the iteration is a sequence containing the respective results of the function applications. There are several options for the way in which the pieces of the list are chosen and for what is done with the results returned by the applications of the function.

The function map may be used to map over any kind of sequence. The following functions operate only on lists.

mapcar *function list* &rest *more-lists*	[function]
maplist *function list* &rest *more-lists*	[function]
mapc *function list* &rest *more-lists*	[function]
mapl *function list* &rest *more-lists*	[function]
mapcan *function list* &rest *more-lists*	[function]
mapcon *function list* &rest *more-lists*	[function]

For each of these mapping functions, the first argument is a function and the rest must be lists. The function must take as many arguments as there are lists. mapcar operates on successive elements of the lists. First the function is applied to the car of each list, then to the cadr of each list, and so on. (Ideally all the lists are the same length; if not, the iteration terminates when the shortest list runs out, and excess elements in other lists are ignored.) The value returned by mapcar is a list of the results of the successive calls to the function. For example:

```
(mapcar abs '(3 -4 2 -5 -6)) => (3 4 2 5 6)
(mapcar cons '( a b c ) '(1 2 3)) => ((a . 1) (b . 2) (c . 3))
```

maplist is like mapcar except that the function is applied to the list and successive cdr's of that list rather than to successive elements of the list. For example:

```
(maplist '(lambda (x) (cons 'foo x))
         '(a b c d))
    => ((foo a b c d) (foo b c d) (foo c d) (foo d))
(maplist '(lambda (x) (if (member (car x) (cdr x)) 0 1)))
         '(a b a c d b c))
    => (0 0 1 0 1 1 1)
   ;An entry is 1 if the corresponding element of the input
   ;  list was the last instance of that element in the input list.
```

mapl and mapc are like maplist and mapcar respectively, except that they do not accumulate the results of calling the function.

Compatibility note: In all LISP systems since LISP 1.5, mapl has been called map. In the chapter on sequences it is explained why this was a bad choice. Here the name map is used for the far more useful generic sequences mapper, in closer accordance to the computer science literature, especially the growing body of papers on functional programming.

Figure 3 An example of *clarifying*.

- A step in the plan has missing preconditions.
- A step in the plan is missing.
- A step in the plan is extraneous.
- A step in the plan that has missing preconditions is related to a step that is missing.
- A step that is missing is related to a step that is extraneous.

Figure 4 Types of invalidities found in plans.

them, or introduce or remind about a plan to satisfy them, depending on whether the plan exists in the user model

4. If the problem is a missing step, introduce or remind about it depending on whether it exists in the user model
5. If there is an extraneous step, identify it, and describe why it is extraneous—what effects does it have that are redundant with some other step
6. If missing preconditions are related to a missing step, identify the relationship between the two, and introduce or remind about the missing step depending on whether it exists in the user model
7. If a missing step is related to an extraneous one, clarify the difference between them.

Two examples taken from a Logo tutorial are shown in Figure 5 [15]. In both cases the first sentence identifies that there may be a problem with the plan "typing POTS." The first example illustrates a missing precondition—the workspace must contain procedures (step 2). The second sentence suggests a plan for viewing procedures ("retype the ones you want", step 3). The second example illustrates a missing step—namely, to pause when the screen is full. The discussion of Ctrl-NumLock provides the missing skill (step 4).

Knowledge that Affects the Choice of Strategy

Strategies dictate the type of information to include in a response, but they will be used and combined differently depending on circumstances. Four kinds of knowledge influence how a consultant decides what to say and how to say it:

- *The Expert Model:* The consultant's expertise is the primary influence on the choice of what should be presented.
- *The User Model:* The consultant's beliefs about what the user knows affect both what should be presented and the form in which to present it.
- *The Current Context:* The context in which the question was asked provides a basis for both content and form.
- *The Question Intent:* The intention of the user's question influences the focus of the response.

Representing Expertise

Expertise includes knowing which plan is most appropriate in a given situation, and therefore requires knowing the relationship between alternative plans for accomplishing a goal. For example, there are at least two ways in most mail systems to send a message

Printing Out Procedures—POTS Command
  Possible Bugs:
    1. If you don't see anything when you type POTS, you have no procedures in your workspace. You can review the previous chapters and retype the ones you want.
    2. If you have many procedures in your workspace, you may find that the titles go by too quickly on the screen and disappear as more titles come on. If this happens, repeat the command POTS and press Ctrl-NumLock as soon as you see the title or titles you're looking for. Ctrl-NumLock will pause whatever is on the screen. Press any key to see the rest of the titles.

Figure 5 An example of *elucidating*.

to a set of people. One can type each address in turn when prompted for the receiver of the message. One can also create an *alias,* which is a named list of addresses that can be reused, and type the alias name at the prompt. The first method is most appropriate when the set occurs only in this instance or is very small and easy to remember. The second method is more appropriate if over a period of time many messages will be sent to this set of people.

Procedural knowledge can be characterized as a *web* of interrelated goals for doings tasks, plans for accomplishing those goals, steps within plans that are either goals themselves (subgoals), or functions that describe the actions available in the environment. Choosing the functions that will execute the plan for a goal is a matter of navigating the web, making decisions about what plans, subgoals, and, ultimately, functions to use.

The Expert Model and the User Model

The web is the basis of both our Expert Model and User Model. Presumably the former is considerably richer than the latter. The Expert Model is traversed in order to locate relationships between goals, plans, and functions. Encoded within it is the knowledge to present choices between plans. For our purposes, we view the expert model as a static declarative structure that is searched in order to locate information. Extending, updating, and modifying the Expert Model is discussed below.

The User Model is also a web of relationships between goals, plans, and functions. It is used to decide both what to present and how to present it. In the example above, if a consultant knows that a user has a very sparse knowledge of sending mail, it may decide to describe how to type the list of addresses rather than create an alias, since the latter will require introducing yet another unknown function. On the other hand, if the consultant thinks the user has a number of inefficient plans[1] for this goal, it might introduce aliasing. The User Model also affects the choice of strategy in answering the user's question. The form of an answer will depend on whether the consultant thinks the user does or does not already know something, or whether it thinks the user has a misconception. Constructing and maintaining a User Model is also discussed below.

From this perspective, expertise and complexity of functions can be characterized by the richness of the web, rather than as simple spectra as described by Chin [16]. Although spectra give the illusion of quantifiable criteria, the methods used to develop and validate them are often superficial at best. In our model, functions that would be classified as "hard" or "advanced" can be better characterized as requiring an understanding of complex interactions in order to navigate the web. Those that are "simple" or "basic" have more straightforward paths. Similarly, classifications such as "novice," "intermediate," and "expert" are hard to quantify. It our approach, they are unnecessary because users are judged by what they know about the current task, rather than how much they know about the entire environment. It is perfectly plausible for a user to have an extensive web of knowledge (local expertise) about a portion of the environment, and almost no expertise about others. Like a spider web, density and strength of paths occur where they are most useful in accomplishing specific tasks.

Current Context

Although the User Model provides a means for choosing the most appropriate plan for a goal, the current context is also relevant. For example, if the user wants to reply to a message sent to a group of users rather than initiate correspondence to the group, his or her current collection of old mail may contain a message that was addressed to all members of the group. Many mail systems contain a function for replying only to the individual who sent the message and a similar function for automatically "carbon copying" the reply to all who received the original message. The choice of how to reply to the group may be dependent on whether the collection of old mail contains such a message.

The context also plays an important part in the content of the answer. Rather than using "canned" examples, the explanation of what to do can be based on actual objects in the environment. In an electronic mail environment the objects include messages, users, and named collections of both.

Question Intent

The last kind of knowledge that influences tutoring in a procedural environment is the intent of the user's question. This is often termed the *discourse goal,* since it is the goal of constructing some expression to elicit some information. It is distinguished from the computational goal of trying to accomplish some task in the environment. In order to decrease confusion between the two, we refer to the computational goal about which help is sought as the "goal," and the discourse goal as the question "intent."

Within a procedural environment it is possible to reason about, and consequently ask questions about, the relationship between goals and plans, goals and functions, plans and functions, goals and other goals, plans and other plans, functions and other functions. The range of question intentions reduces to those in Figure 6. The utterance identifies a goal or plan (or function) or both, and implies an assumption about their validity. Its form also implies an expected answer. Therefore, in order to reply informatively, a consultant system must provide the information users *expect,* namely a goal, plan, function, or relationship. It must also confirm or deny their assumptions about the validity of the goal, plan, or function mentioned in the question.

Choosing a Tutoring Strategy

The knowledge described in the previous section provides four basic parameters for choosing a tutoring strategy. These are:

- A *question intent—QI,* that provides an expected discourse focus of a goal, plan, and/or function.
- A *computational goal—G,* which is either identified by the user in the question, or is inferred by the consultant from the plan or function specified in the question.
- A *User Model plan—U,* for the computational goal, which is the plan that the consultant thinks the user has used in the past for accomplishing the goal.
- A *best plan—B,* for the computational goal, which is the plan that the consultant thinks the user has used in the past for accomplishing the goal.
- A *related plan—R,* for the computational goal which is chosen by the consultant from the Expert Model when B = U or B = S. It is introduced as enrichment.

Each subsection below describes how these parameters affect the choice of strategy and whether the strategy is used responsively or as enrichment. When the distinction is

Question Intent	Question Identifies	Question Assumes	Question Expects in Reply	Intuitively
1. What plan/function is required to satisfy the goal?	Goal	Goal is possible	Plan or Function	How do I do it?
2. What goal is satisfied by this function or plan?	Function or Plan	Function or Plan satisfies some Goal	A goal	What does it do?
3. Does this plan satisfy this goal?	Function or Plan and Goal	Plan or Function satisfies Goal	Confirmation of assumption or explanation	Does it do it?
4. This plan doesn't work, for this goal, what's wrong?	Function or Plan and Goal	Plan or Function does not satisfy Goal	Confirmation of assumption or explanation	What's wrong with it?
5. Is there a better way for this goal?	Function or Plan and Goal	Plan or Function is not best way to satisfy goal	Confirmation of assumption or explanation	What's a better way?

Figure 6 Information embedded in the question intention.

not important, we refer to both plans and functions as plans. The following notation is used for the logical relationships between goals and plans.

Notation: P_E = P exists $P_{\sim E}$ = P does not exist
P_V = P is valid $P_{\sim V}$ = P is not valid
P_f = P after invalidity is fixed
Q_n refers to the question intents in Figure 6.
& = AND, | = OR

Introducing

A plan is introduced responsively when the consultant thinks the user has no previous knowledge of it or has a faulty plan for the goal. An exception exists when the question intent specifically asks for a better way, in which case the clarify strategy is used. More formally:

Introduce Responsively:

Intro(B): $S_{\sim E}$ & ($U_{\sim E}$ | $U_{\sim V}$)
Intro(S): ($U_{\sim E}$ | $U_{\sim V}$ | [U_V <> S_V])

$$\& \;[(S_V \;\&\; [Q_2 \mid Q_3 \mid Q_4]) \mid [Q_5 \;\&\; (B = S_V)]]$$

For example, in the domain of Berkeley Unix Mail, if the user knows nothing about replying to a message and asks, "How do I answer a message" (i.e., specifies a goal), the function *Reply* is introduced. Conversely, if the user asks, "what does *Reply* do?" (i.e., specifies a function), the best plan in the current context for the goal *reply to a message* is introduced.

Introducing as enrichment occurs when either the stated plan or the User Model plan is the same as the best plan, and the consultant has found a related plan. More formally:

$$\text{Introduce As Enrichment: Intro(R): } [S_{\sim E} \mid S_V = B] \;\&\; (U_V = B)$$

For example, if the user asks, "To respond to the sender of a message, do I type *Reply*?" and the consultant believes this is the best plan, and the only one the user knows, then the consultant would first confirm the statement by reminding the user about the command. Only then would the consultant proceed to introduce a related plan for replying to a group of users, for example, all of those who received the original message.

Reminding

A plan is reminded responsively when the user does not state a plan and the consultant thinks the user already knows the best plan. Reminding is also used when the stated plan is a plan that the consultant thinks the user already knows. More formally:

Remind Responsively:

$$\text{Remind(U): } S_{\sim E} \;\&\; U_V$$
$$\text{Remind(S): } (S_V = U_V) \;\&\;$$
$$[Q_2 \mid Q_3 \mid Q_4 \mid (Q_5 \& (S_V = B))\,]$$

For example, if the consultant has seen the user reply to messages in the past and the user asks, "How do I answer a message?" (no S is stated), then the user just needs to be reminded about the command; a long introduction is not necessary. Similarly, if S is stated: "Does *reply* let me answer a message?" then a terse affirmative answer is appropriate.

We have not found a reason to use reminding as enrichment. In fact, it seems rather pedantic.

Clarifying Distinctions

A plan is clarified responsively when the intent of the question is specifically for a better plan for the goal. A plan must be stated, and clarification occurs if it is not the best plan. More formally:

$$\text{Clarify Responsively: Clarify(S): } Q_5 \;\&\; (S_V <> B)$$

For example, the user asks, "Normally, to send mail to a group of users, I just type all the addresses at the TO: prompt, is there a better way?" The consultant may decide that it is time to introduce *aliasing,* and will clarify the difference between using a new plan that includes the *alias* function and the user's stated plan.[2]

Clarifying as enrichment occurs whenever the question intent is not specifically asking for a better way, but the plan stated is not the best plan. More formally:

Clarify As Enrichment:

$\text{Clarify(U)}: S_{\sim E}$ & $(U_V <> B)$
$\text{Clarify(S)}: (\sim Q_5$ & $[S_V <> B])$ |
$\quad (S_{\sim v}$ & $[S_f <> B]$ & $[Q_2 | Q_3]$
$\quad$ & $[U_{\sim E} | U_{\sim v} | (U_V <> B)])$

For example, the user asks, "Can I type more than one address to the TO: prompt" and the context suggests that the best way would be to create an alias. First since the plan is the focus of the question, the consultant must responsively introduce or remind about the plan depending on what it thinks the user knows. Only then would it say, "by the way . . ." and clarify aliasing as enrichment.

Elucidating Misconceptions

A plan is elucidated responsively when the user states a plan for a goal that the consultant does not think is valid. Five invalidities were presented in Figure 4. More formally:

Elucidate Responsively: $\text{Elucid(S)}: S_{\sim v}$ & $\sim Q_5$

For example if the user asks, "To send mail to a group of users, do I type *alias* at the TO: prompt?" the consultant notices that a precondition to using the *alias* command is that one be at the *Mail>* prompt. The consultant provides a solution: return to the *Mail>* prompt, create the alias, then begin to compose the message.

Elucidating for enrichment occurs when the user is looking for a better way and the plan stated is invalid. It also occurs when the consultant thinks the user has a plan for the goal that is not valid. Since the plan in the User Model is never the focus of discourse (unless it is equal to the stated plan, in which case the stated plan is still the focus of the discourse), it can never be elucidated in response. However, under some circumstances it may seem opportune to address what the consultant thinks the user knows. This must proceed in a delicate manner, because it is based on knowledge the consultant believes rather than knows.

More formally:

Elucidate As Enrichment:

$\text{Elucid(S)}: S_{\sim v}$ & Q_5
$\text{Elucid(U)}: U_{\sim v}$ & $(S_E | [S_V = B])$

The Use of Tutoring Strategies in GENIE

In order to answer a user's question, a consultant must *understand* the question, *analyze* the problem phrased in the question to find the best solution, and *generate* an answer that is informative. GENIE consists of three processes, an Understander, a Plan Analyst, and an Explainer, that operate on three knowledge bases, a World Model, an Expert Model, and a User Model. Since the focus of this paper is on tutoring strategies, this section shows how the form and content of an answer is developed by the Explainer. The other processes and the knowledge bases are also described briefly.

The Explainer uses the rules discussed in the preceding section to chose a strategy in response and possibly as enrichment. Then the schemata presented in the section on tutoring strategies are used to flush out an answer. We have focused on issues of deep structure generation—that is, what to say rather than the surface level wording through which to say it. Consequently, this discussion describes the directives produced by the Explainer, but does not show in detail how actual English text is produced. Issues pertaining to surface generation are addressed below.

Four example questions show how seven different answers occur depending on the user's knowledge and the context in which the question was asked. Through the first question we show how the form of an answer varies depending on the tutoring strategy chosen, while the skill chosen remains the same. The second shows how the answer varies with the skill that is described, while the tutoring strategy remains the same. Finally, through two more questions we show how the intent of the question itself influences both the form and content of the answer.

Understanding and Analysis in GENIE

The Understander must be able to parse a user's question and identify the components listed in Figure 6. It must identify the goal, plan, or function to which the user is referring. It must determine what the user assumes; for example, that the identified plan does or does not satisfy the identified goal. It must also determine what the user expects as an answer: either a goal, plan, or function, or a relationship between them. From this it must identify the *intent* of the question. Since research by Wilensky et al. [2] and Pollack [3] has been devoted to question understanding, building of robust understander in GENIE is currently a low priority. At the present time, we produce the stated goal, plan, or function, and the user assumptions, expectations, and question intent by hand, and provide these as input to GENIE.

When the Plan Analyst is given a plan or function and a goal, it uses the Expert Model to find a "match" between them. A match is defined as confirmation that a plan or function satisfies a goal. When it is given a function and goal, the Plan Analyst confirms or denies that they match. Given a plan and a goal, it tries to find a match. If it is unsuccessful it uses the cases in Figure 4 to identify the mismatch. When the Plan Analyst is given only a plan, it attempts to locate the goal satisfied by the plan. If no actual match emerges from a list of candidates, then the goal with the least complex mismatch is chosen, but is marked as a mismatch along with its causes.

Given a goal, the Plan Analyst searches the Expert Model for the "best" plan for the goal using two sets of heuristics. First it tries to choose steps in plan using "world knowledge" such as efficiency and temporality based on the current context which we call the @i(World Model). For example, users normally prefer plans that accomplish goals now rather than later, or that require fewer rather than more steps. As discussed below in the context of a mail environment, the choice may be affected by whether a message, address, or alias already exists. The second set of heuristics compares the candidate plans to knowledge in the User Model. The plan whose subsets occur most frequently in the user model is considered the *best plan*.

Knowledge Bases in GENIE

The Expert Model is a network of the computational goals that can be satisfied in the computing environment. Figure 7 shows the structure of this frame-based knowledge representation. Computational goals contain links to alternative plans for satisfying the goal. A plan can be linked to a subgoal or an ordered sequence of subgoals that describe how it can be executed, or to a function that executes it directly. Encoded within a computational goal are links that describe the relationship between plans.

Functions describe the operators of the environment. Their representation includes information about the correct syntax of the function, any preconditions and effects, and the actions associated with parameters. Preconditions define a state that must be true before a function can be correctly executed. They may also contain a link to a goal that could satisfy it. Effects encode the actions of functions when applied to the World Model. Currently the World Model is represented as a simple add/delete list that describes possible states in the mail environment. Therefore, effects are encoded as directives to add or delete a state from the World Model.

The User Model has the same representation as the Expert Model. It contains a history of what the user has done in past sessions in terms of what goals have been accomplished and what plans and functions were used to accomplish them. Its goals may contain plans that were attempted but didn't work or plans that do not exist in the Expert Model.

How the User Model Affects Choice of Strategy

We first present an example where the best plan (B) chosen by the Plan Analyst is described by the Explainer through three different tutoring strategies depending on what the user knows. Consider the question, "How can I answer a message?" The question identifies a goal (*reply to a message*) and has a question intent (QI) to receive a plan in response. Assume that the World Model contains a message that the user is currently reading that was sent only to the user, not to some group that included the user. Figure 8 is a graphic representation of the Expert Model required to answer this question. The World Model dictates that the Plan Analyst choose as the best plan B:

    Goal: reply.to.message (satisfied by)
    P2: (through steps)

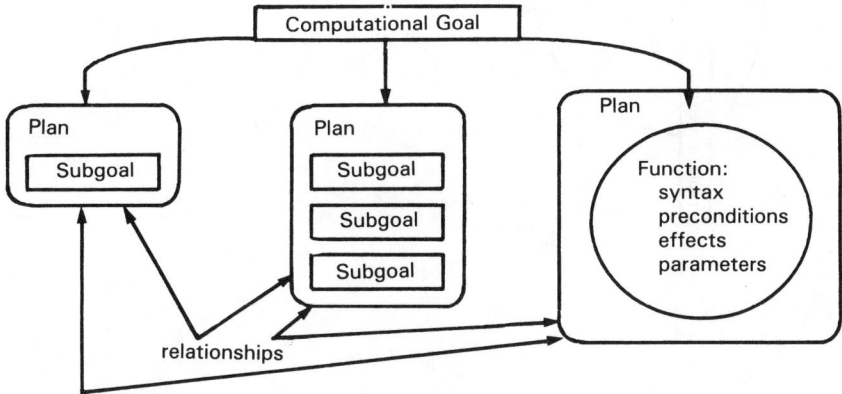

Figure 7 Computational goal of the Expert Model.

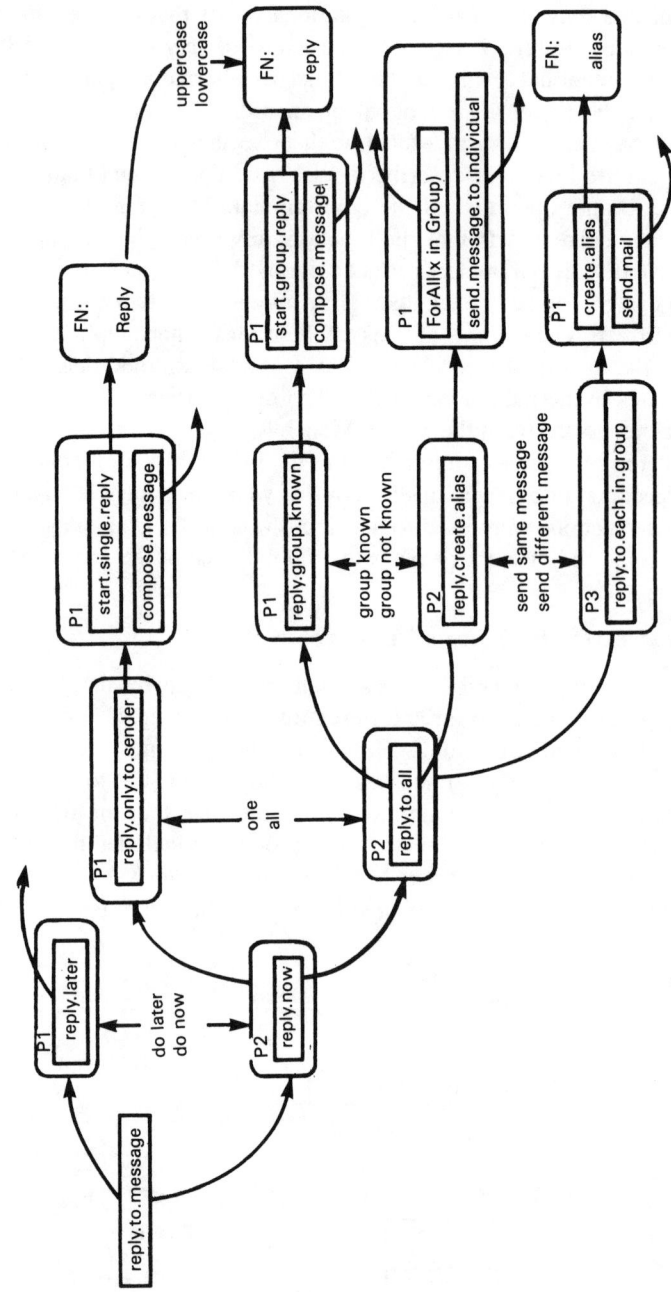

Figure 8 GENIE's expert model knowledge of replying to a message.

```
            Goal: reply.now (satisfied by)
              P1: (through steps)
                Goal: reply.only.to.sender (satisfied by)
                  P2: (through steps)
                    Goal: start.single/reply (satisfied by) . . .
                    Goal: compose.message (satisfied by) . . .
```

Three different tutoring strategies are invoked depending on what is in the User Model. In all three cases we assume that the user has a plan for *compose.message*.

In the first case the User Model contains no plan for the goal, that is, GENIE has no knowledge that the user has a plan like B to reply to a message. The Explainer chooses to *introduce responsively* because no plan U exists in the User Model for the goal. Figure 9 shows both a trace of the directives that are assembled for producing text and the resulting text.

In the second case the User Model contains a plan U that is identical to B, that is, GENIE believes the user has replied to messages correctly in the past. The Explainer

```
 Introducing: reply.to.message
   STATEGOAL: reply.to.message
    SUMMARIZE: reply.to.message.P2
     Introducing: reply.now
      STATEGOAL: reply.now
       SUMMARIZE: reply.now.P1
        Introducing: reply.only.to.sender
         STATEGOAL: reply.only.to.sender
          SUMMARIZE: reply.only.to.sender.P1
           Introducing: start.single.reply
            STATEGOAL: start.single.reply
             IntroducingFunction: Reply
              PRESENT-SYNTAX [Reply]
              DESCRIBE-PRECONDITIONS:
                 (mode = reading.messages)
              DESCRIBE-EFFECTS:
                Delete (mode = reading.messages)
                Add (addressee = current.sender)
                Add (interrupt.mode = compose.message)
              Reminding: compose.message

         EXAMPLE: (FN: Reply, Compose.message)
```

In order to reply to a message it is assumed you want to reply right away. In order to reply right away it is assumed you want to reply only to the sender. To do this, you must indicate you wish to reply and compose a message. You can indicate you wish to reply by using the command 'Reply'.

The syntax is: Reply

To use this command you must be in read mode. The command removes you from read mode, makes the address of your message the current sender and temporarily puts you in write mode.

To compose a message just type your message and end with <esc>.
For example,
    Type the command: Reply
    Then just type your message and end with <esc>.

Figure 9 *Reply to a message* when user model contains no plan for the goal.

chooses to *remind responsively* because U exists and is equal to B. Figure 10 presents a portion of the trace.

In the third case, the User Model contains a plan that does not exist in the Expert Model, but that is valid:

> Goal: reply.now (satisfied by)
> P: (through steps)
>   Goal: save.message (satisifed by) . . .
>   Goal: leave.read.mode (satisfied by) . . .
>   Goal: send.message (satisfied by) . . .

The Explainer chooses to *clarify responsively* because the user has a valid plan for the goal, but that plan isn't best. Figure 11 presents a portion of the trace.

How Content Varies within a Strategy

We now show how the Plan Analyst's choice of a plan for a goal varies the content but not the form of the answer. Consider the question, "To reply to a group of users I reply to each individually—is there a better way?" The question identifies a goal—*reply to all* and a plan (S)—*Forall (x in group) send.message.to.individual*. It has a question intent (QI) to get the best plan. Assume that the User Model contains the goals *compose.message* and *send.mail*, but does not contain *create.alias* or *start.group.reply*.

For the first case, assume the World Model contains a message that was addressed to a group. The Plan Analyst determines that *reply.group.known* is the best plan because the group is known and it is less work that the stated plan S. Since QI = get a better plan B than S, and S < > B, the Explainer chooses to *clarify responsively*. Figure 12 presents a portion of the trace.

In the second case, assume the World Model contains a single message sent only to the user. The Plan Analyst determines that the best plan is *reply.create.alias* because no group message exists and it is less work than the user's plan. Since QI get a better plan B than S, and S < > B, the Explainer still chooses to *clarify responsively*. Figure 13 presents a portion of the trace.

How Question Intent Influences Form and Content

The two questions in the previous sections generated five different answers even though they were asking essentially the same thing—"How to reply." We believe this illustrates the fundamental difference between intelligent tutoring systems that use generated answers and those that use canned text. The former provide flexibility in answers that the latter cannot. We illustrate this point further with two more questions that also ask about *reply.to.message*. Both questions require *elucidating misconceptions,* but the form and content differ because the stated plans fail in different ways.

In the first case, consider the question, "I'm trying to reply to a group of users using

```
Reminding: reply.to.message
   STATEGOAL: reply.to.message
   SUMMARIZE: P2
   EXAMPLE: (FN: Reply, Compose.message)
```

Figure 10 *Reply to a message* when user model contains a plan for the goal.

```
Reminding: reply.to.message
  STATEGOAL: reply.to.message
  SUMMARIZE: U
  EXAMPLE: (FN: Reply, Compose.message)
Clarifying: reply.to.message
  SUMMARIZE: B
  COMPARE_STEPS(send.message,reply.only.to.sender)
    SHOW-RELATIONSHIP: (initiate addressee) (sender is addressee)
    Introduce: Reply
```

Figure 11 *Reply to a message* when user model does not contain best plan.

reply but I only seem to be able to reply to the sender of the message. Why?'' The stated goal is *reply.to.all* and the plan implied in the question is (FN: *reply.compose.message*). Assume that the message in the World Model was only sent to the user, and the user knows about the *alias* function. The fault lies in the fact that there are no other group members in the message header, causing a precondition of the function *reply* to be violated. The Plan Analyst finds the fault and suggests using the *reply.create.alias* plan. Given that S is invalid, the Explainer chooses to *elucidate responsively* and expands on a missing precondition. Figure 14 presents a portion of the trace.

In the second case consider the question, "I'm using *reply* to reply to the sender of a message, but seem to send mail to everyone else to whom the message was addressed. Why?" The stated goal in this case is *reply.only.to.sender* and the implied plan is (*FN: reply.compose.message*) with the subgoal *start.single.reply* being satisfied by the function *reply*. Assume that the message in the World Model was sent to a group of users. Given the knowledge in Figure 8, the Plan analyst notices that *FN: reply* seems to be an extra step while *FN: Reply* is missing a step.[3] From the figure these steps are related both at the level of the function—by the case of the first letter and two levels up in the distinction between *reply.to.all* and *reply.only.to.sender*. Since S is again invalid, the Explainer chooses to *elucidate responsively*, expanding on an extra step related to a missing one. Figure 15 provides a portion of the trace.

Related Work

Previous research in both intelligent tutoring systems and natural language processing have influenced the design of GENIE. The representation of our Expert and User Models is based on work by Goldstein [17], Genesereth [18], and Clancey [19]. Goldstein introduced the notion of a network, a *genetic graph* of knowledge that represents stages of development of a student's understanding of a domain. Rather than overwhelm the beginner with an optimal method, the genetic graph provides a basis for choosing

```
Clarifying: reply.to.group
  STATE-BETTER-WAY-EXISTS: reply.to.group
  SUMMARIZE: U
  SUMMARIZE: B
  SHOW-RELATIONSHIP: (group known) (group not known)
  COMPARE-STEPS: (start.group.reply, ForAll(x in group))
    Introducing: start.group.reply . . .
  COMPARE-STEPS: (compose.message, send.message.to.individual)
    Reminding: compose.message . . .
  EXAMPLE: B
```

Figure 12 Clarifying that the best plan is *reply group known*.

```
Clarifying: reply.to.group
  STATE-BETTER-WAY-EXISTS: reply.to.group
  SUMMARIZE: U
  SUMMARIZE: B
  SHOW-RELATIONSHIP: (send same message) (send different message)
  COMPARE-STEPS: (reply.create.alias, ForAll(x in group))
    Introducing: start.group.reply . . .
  COMPARE-STEPS: (send.mail, send.message.to.individual)
    Reminding: send.mail . . .
  EXAMPLE: B
```

Figure 13. Clarifying that the best plan is *reply create alias*.

what to say. Its structure, however, is based on degrees of student development, which, as we argued earlier, are very hard to quantify and evaluate. Since our web is task-connected as well as knowledge-connected, we should have less difficulty approaching a problem from a novel perspective. Genesereth showed through his *dependency graph* of goals and plans that this is the case. However, since he was concerned with diagnostics, he did not attempt to encode semantic information about the relationships between plans, which we see as crucial to explanation. Clancey first articulated the need for the separation of domain expertise from tutoring knowledge. Although he implicitly captures goal/plan knowledge in the rules, unlike GENIE, his tutors cannot reason abstractly about relations between goals and plans.

More recent work, especially that of Bonar and Lesgold, has focused on developing systems where the major emphasis in the relationship between Expert and User models is on articulating the curricular aspects of knowledge [20, 21]. Bonar introduced the notion of "bite sized architecture," which is the set of features collected around a teachable concept. These features can include domain knowledge such as part/whole relationships, and teaching knowledge, such as what knowledge is prerequisite to what other knowledge. Lesgold's theory encompasses three layers of knowledge; at the lowest level is the knowledge layer, then the curriculum layer, then the aptitude layer. Both of these approaches and that of other researchers such as Anderson et al. [22] have been concerned with representing a teaching agenda.

We differ in that we take as a premise that the teaching agenda must be constructed dynamically within the current context. Our starting point is an epistemological rather than pedagogical analysis of the domain. In other words, we began by analyzing *what you can do* in Berkeley Mail, and then looked at how you might talk about it. This is in contrast to looking at *how to teach someone* about Berkeley Mail, and then looking at what aspects of the domain are important to teach. The difference is subtle but has a profound effect on the requirements of a User Model, which is described below.

Research by Brown and Burton [23] and by Sleeman and Smith [24] illustrated the difficulty in determining what the student knows in unguided settings. The difficulty lies in diagnosing what the student doesn't know based on expected behavior. The prob-

```
Elucidating: reply.to.all
  STATE-BROKEN-PLAN: reply.group.known.P1
    ProblemIn: MissingPrecondition (exists group)
      IN start.group.reply
    Fix: reply. create.alias.P1
      Reminding:reply.create.alias
```

Figure 14. Elucidating when a precondition is missing.

```
    Elucidating: reply.only.to.sender
  STATE-BROKEN-PLAN: reply.only.to.sender.P1
    ProblemIn: ExtraMatchesMissingStep
           Extra: start.group.reply
         Missing: start.single.reply
  DescribeMatch: Extra: FN: Reply (uppercase)
               Missing: FN: reply (lowercase)

            Extra: reply.to.all (all)
          Missing: reply.only.to.sender (one)
      Introducing: reply.only.to.sender
```

Figure 15. Elucidating when a missing step matches an extra step.

lem lies in expecting a set of behaviors in the first place, either to confirm that a component of the teaching agenda has been satisfied, or to fire a new remedial agenda if the behavior is deemed incorrect. The root of the problem is that the choice of agenda is determined by comparing an expected solution path with a possibly unreliable or incomplete representation of the solution path taken by the student. Since our aim is to provide information for the task at hand, our decision on what to say is not strongly dependent on analyzing the user's behavior in relation to the expected behavior of a teaching agenda.

Our user model is merely a reference point for choosing the form and content of the consulting dialogue, but is rarely the focus. In only one instance, elucidating as enrichment, does GENIE expect to find a "buggy plan" in the User Model. This case is a by product of our representation rather than its focus and therefore does not play a significant role in how GENIE works. However, this does not mean we can completely ignore the problem of how the user model is built and maintained, and more is said about this below.

Research on User Modeling and Natural Language Generation falls into two categories: either stereotypes are used to represent large categories of users and answers are geared to a category, or explicit beliefs and goals of the user are represented and reasoned about to determine an answer. Chin's work [16] on the UC system [2] best exemplifies the stereotype approach and is most relevant to our system. Chin's system provides help within the Unix environment based on a dual set of stereotypes. Users are classified as novice, intermediate, or expert, functions are classified as easy, two levels of intermediate, and hard. The system uses rules stating how much detail to provide about the different classes of functions depending on which class the user falls into. Our approach is in direct opposition to Chin's: its is based on the assumptions that knowledge cannot be quantified in discrete chunks and users are unlikely to learn and progress from one neat division to another. Rather, they are likely to learn functions based on tasks they have performed, and this will vary substantially, depending on the tasks.

A second body of work, by Allen and Perrault [25], Pollack [3], and Appelt [26], uses detailed information about user beliefs and plans in combination with a formal reasoning system to determine what to include in an answer. Because they have relied on such detailed informal models in combination with a theorem prover, they have tended to operate in limited, well constrained domains, producing shorter responses than those at which we aim. Pollack is an exception, although she has focused more on the representation required to produce helpful responses for plan invalidities and less on the generation of the responses themselves. Like Pollack, we also represent details about beliefs, plans, and goals. However, we combine the representation of beliefs with the use of tutoring strategies rather than a theorem prover, to allow us to produce more complex responses as part of a less well constrained domain.

Summary and Conclusions

The great hope of educational computing, that computer technology can intelligently deliver unprecedented individualized instruction, remains to be fulfilled. This paper describes a step in that direction, through the development of GENIE, an automated consultant for interactive environments. We have demonstrated how the form and content of an answer are chosen based on the current needs and knowledge of the user. Our knowledge representations and the algorithms and heuristics used to traverse them allow us to abandon spectra of user expertise and functional difficulty. We therefore circumvent the need for a broad sequential curriculum and have much greater flexibility in when, what, and how to present skills to users.

Our demonstration system GENIE is by no means as sensitive as a skilled human tutor. At the present time, it cannot handle certain aspects of context, one cannot ask questions in any natural way, the knowledge bases must be updated by hand, and the answer that is generated could stand stylistic improvement. These are all primarily implementation issues, and in the paragraphs that follow we address each of them in turn.

One major goal of this work was to answer questions within the current context. We provide the context in symbolic form as input to GENIE. Our assumption is that it can be derived by a subsystem that observes users' actions. In order to answer "what if" questions where the user presents a hypothetical context, a different subsystem would be required that allows users to construct a context in some friendly way such as natural language. Furthermore, in human discourse a question is rarely answered fully through a single utterance. Often a user's initial question spawns other questions that may be combinations of question types and require a mixture of answering strategies. Here too, a different subsystem, one that monitors a discourse history, would contribute to the construction of the current context.

At the present time we make the simplifying assumption that the user either knows or does not know about things in the domain. By definition, if something exists in the User Model, even if it is wrong, then it is known. If it does not exist, then it is not known. In practice this assumption is inadequate, first because knowledge is not binary, and second because tutoring can occur across a broader continuum. Rather than fall for the seduction of simple categories of levels of mastery, we would like to evaluate and record knowledge in a less quantified manner. This in turn will affect how GENIE's heuristics and strategies must be changed.

We also do not presently do any sort of diagnostics. However, since the user is engaged in using an environment, it should be possible to build a system to monitor his or her actions. For example, our Marvel software development environment [27] monitors all commands entered by the programmer. Inferences could therefore be drawn on whether particular goals have been attempted and successfully accomplished and through what plans or functions. Selker [28] and Quilici [29], among others, have demonstrated how such plan recognition is possible and useful for developing a User Model.

We have thus far avoided consideration of a "front end" for asking questions. Our justification has been that our interests lie in generation rather than understanding of language. We are now at a stage where we have a clearer picture of what information we must get from the user and are in a better position to design an understanding component for demonstration purposes. We have begun work on a simple Augmented Transition Network [30] parser that will be able to extract the information described in Figure 6.

Clearly, all possible goals and plans for those goals cannot be known when a computing environment is set loose on users. Furthermore, if the environment is extensible, even the relationships between functions cannot be fixed. We view the problem of extending, updating, and modifying both the Expert and User Models as one of knowledge acquisition and learning and therefore not within the immediate scope of this research. Work on machine learning by Lebowitz [31] has influenced the design of our knowledge representations, and we are confident that they can be updated automatically in the future.

Finally, the current version of GENIE relies on textual templates to produce actual text. We have discovered that the prose generated is stylistically stilted, referentially awarded, often redundant, and therefore inadequate. We are currently exploring the use of a functional unification grammar [32, 33] that should allow us to produce more graceful text.

To summarize, GENIE is an automated consultant that attempts to meet the contextual needs of the user by basing its answers on the user's current task and those tasks the user has successfully accomplished in the past. Our current plans are to develop a front end for demonstration and testing purposes, to make use of a functional unification grammar to produce better text, and to expand both the analysis and explanation facilities to handle a wider variety of consulting situations. Through these projects we hope to shed more light on the complex tutoring that occurs during consulting in interactive environments.

Acknowledgments

We would like to thank Elaine Kant and Steve Feiner for helpful comments on earlier versions of this paper. Implementation of components of GENIE has been and continues to be done by Lourdes Andre, David Robinowitz, and Michael Tanenblatt.

Notes

1. In some contexts our first plan is less efficient than the second. Another less efficient plan is to retype the message laboriously to each addressee.
2. Note that asking, "What's the best way?" is not the same as asking, "What's a better way?" The former is actually a form of "How do I do it?"
3. For the Unix uninitiated, character case matters, so *Reply* and *reply* are indeed two different functions. We admit this is grossly user unfriendly and should not occur in the first place, but the example is too good to resist.

References

1. J. S. Brown and K. VanLehn. "Repair theory: A generative theory of bugs in procedural skills," *Cognitive Science,* 44 379–415, October 1980.
2. R. Wilensky, Y. Arens, and D. Chin, "Talking to UNIX in English: An overview of UC," *Communications of the ACM,* 276 574–93, June 1984.
3. M. Pollack, *Inferring domain plans in question-answering,* Ph.D. diss., Moore School, University of Pennsylvania, 1986.

4. U. Wolz, "Analyzing user plans to produce informative responses by a programmer's consultant," Technical Report CUCS-218-85, Department of Computer Science, Columbia University, 1985.
5. U. Wolz and G. E. Kaiser, "A discourse-based consultant for interactive environments," in *Proceedings of the Fourth IEEE Conference on Artificial Intelligence Applications,* Institute for Electrical and Electronic Engineers, San Diego, CA, March 1988, 28-33.
6. C. S. Magers, "An experimental evaluation of on-line HELP for non-programmers," in *CHI'83 Conference Proceedings,* ACM Special Interest Group on Computer and Human Interaction and the Human Factors Society, Boston, MA, December 1983, 277-81.
7. N. S. Borenstein, *The design and evaluation of on-line help systems,* Ph.D. diss. Carnegie Mellon University, 1985.
8. P. A. Winston and B. K. Horn, *Lisp, second edition,* Reading, MA: Addison Wesley, 1984.
9. K. R. McKeown, *Text generation: Using discourse strategies and focus constraints to generate natural language text,* Cambridge, England: Cambridge University Press, 1985.
10. C. L. Paris, *The use of explicit user models in text generation: Tailoring to a user's level of expertise,* Ph.D. diss., Columbia University, 1987.
11. K. F. McCoy, "The ROMPER system: Responding to object-related misconceptions using perspective," in *Proceeding of the 24th Annual Meeting of the ACL,* Association of Computational Linguistics, New York, NY, June 1986, 97-105.
12. Hewlett-Packard Company, *LISP language reference (M-Z),* Fort Collins, CO, 1986.
13. G. L. Seele, *Common Lisp, the language,* Bedford, MA: Digital Press, 1984.
14. A. Joshi, B. Webber, and R. Weischedel, "Living up to expectations: Computing expert responses," in *Proceedings of AAAI-84,* American Association of Artificial Intelligence, Austin, TX, August 1984, 169-75.
15. International Business Machines Corporation, *Logo: Programming with turtle graphics,* Boca Raton, FL, 1983.
16. D. N. Chin, *Intelligent agents as a basis for natural language interfaces,* Ph.D. diss. University of California, Berkeley, 1988.
17. I. R. Goldstein, "The genetic graph: A representation for the evolution of procedural knowledge," in D. Sleeman and J. S. Brown (eds.), *Intelligent tutoring systems,* London: Academic press, 1982, 51-78.
18. M. R. Genesereth, "The role of plans in intelligent tutoring systems," in D. Sleeman and J. S. Brown (eds.), *Intelligent tutoring systems,* London: Academic Press, 1982, 137-55.
19. W. J. Clancey, "Tutoring rules for guiding a case method dialogue," in D. Sleeman and J. S. Brown (eds.), *Intelligent tutoring systems,* London: Academic Press, 1982, 201-25.
20. J. G. Bonar, R. Cunnigham, and J. Schultz, "An object-oriented architecture for intelligent tutoring," in *Proceedings of OOPSLA '86,* Association for Computing Machinery, Portland, OR, September 1986, 269-76.
21. A. M. Lesgold, "Toward a theory of curriculum for use in designing intelligent instructional systems," in H. Mandl and A. Lesgold (eds.), *Learning issues for intelligent tutoring systems,* New York: Springer-Verlag, 1987.
22. J. R. Anderson, C. F. Boyle, and B. J. Reiser, "Intelligent tutoring systems," *Science,* 228:4698, 456-62, April 1985.
23. J. S. Brown and R. R. Burton, "Diagnostic models for procedural bugs in mathematics," *Cognitive Science,* 2:2, 155-92, April 1978.
24. D. H. Sleeman and M. J. Smith, "Modelling student's problem solving," *Artificial Intelligence,* 16:2, 171-87, May 1981.
25. J. F. Allen and C. R. Perrault, "Analyzing intention in utterances," *Artificial Intelligence,* 15:3, 143-78, December 1980.
26. D. E. Appelt, *Planning natural language utterances,* Cambridge, England: Cambridge University Press, 1985.
27. G. E. Kaiser, P. H. Feiler, and S. S. Popovich, "Intelligent assistance for software development and maintenance," *IEEE Software,* 5:0, 40-49, May 1988.

28. E. Selker, "The COGnitive adaptive computer (COACH) interface," Technical Report RC14130, T. J. Watson Research Center, IBM, 1988.
29. A. E. Quilici, G. Dyer, and M. Flowers, "Understanding and advice giving in AQUA," Technical Report UCLA-AI-85-19, UCLA Artificial Intelligence Laboratory, 1985.
30. W. Woods, "An experimental parsing system for transition network grammars," in *Natural language processing,* New York: Algorithmics Press, 1973.
31. M. Lebowitz, "An experiment in intelligent information systems: RESEARCHER," in *Intelligent information systems: Progress and prospects,* London: Ellis Horwood, 1986, 127–50.
32. M. Kay, "Functional grammar," in *Proceedings of the 5th meeting of the Berkeley Linguistics Society,* Berkeley Linguistics Society, 1979.
33. K. R. McKeown and C. L. Paris, "Functional unification grammar revisited," in *Proceedings of the 25th Annual Meeting of the Association for Computational Linguistics,* Association for Computational Linguistics, Stanford, CA, July, 1987, 97–103.

Chapter 3

Explanation, Analogy, and Transfer in an Intelligent Tutoring System Paradigm

PETER PIROLLI

Graduate School of Education
University of California, Berkeley
Berkeley, CA 94720

> **Abstract** *The successive refinement of cognitive models and intelligent tutoring systems (ITS) provides a novel research paradigm for cognitive scientists interested in modeling human learning. Intelligent tutoring systems can be viewed as unique scientific instruments that yield new and richer measures of human behavior than the traditional tests and assessment devices. Traces of ongoing student–ITS interactions provide the best data available for refining computational models of the dynamic flow of cognition and learning. This chapter describes the beginnings of our explorations of this new paradigm for improved understanding of learning.*

Intelligent tutoring systems provide a novel investigative paradigm for cognitive scientists interested in the nature of learning. Such systems can be viewed as scientific instruments that provide richer and more informative measures of dynamic behavior than the simple outcome measures traditionally used in research on learning and instruction. Traces of ongoing student–tutor interactions are precisely the sort of data appropriate for refining computational models of the moment-to-moment flow of cognition and learning that are favored in cognitive science [1]. Furthermore, intelligent tutoring systems make dynamic decisions about the flow of instructional activity and can be used to rigorously define complex experimental manipulations that would be difficult to achieve by other means. These refinements in instrumentation and experimental control are a promissory note for improved understanding of learning, and in this chapter I describe how we are beginning to cash in on this note.

For several years, I have been working on formal models of how people learn to program recursive functions. The LISP Tutor [2] developed at Carnegie-Mellon University has been an integral part of this research. The general instructional situation of interest in this research is a fairly common one, in which people read some instruction or listen to a lecture and then work through some exercise problems. The general aim of this research is to develop a unified model of performance and learning in this instructional situation. Such a model must address a complex array of phenomena, including how people interpret texts and examples, how they use these interpretations to perform novel tasks, how people acquire task-specific knowledge, how such knowledge transfers across situations, and how people reflect on, assess, and modify their own performance. At the core of this theoretical endeavor are computational models that simulate the moment-by-moment psychological processes of our subjects. In addition to simulations of individuals, we develop mathematical models that address more general learning and performance effects. Such a complex model should provide a powerful set of tools for the analysis of instructional designs.

Figure 1 presents a general framework for structuring our analyses. The boxes in Fig. 1 indicate different kinds of knowledge content, and the arrows indicate different kinds of cognitive processes or mechanisms. Learners are presented with instruction and actively construct interpretations of texts and examples for themselves. These interpretations and self-explanations vary with learners' prior knowledge. The results of this processing are essentially representations of what the texts are about and what the examples illustrate. We consider these representations to be declarative in nature and may be stored in memory for future use.

In typical school-like lessons, learners are presented with problems to exercise their knowledge. For familiar parts of the problems, the learner applies efficient domain-specific skills wherever possible. At problem-solving impasses, where no previously acquired skills are applicable, the learner resorts to weak-method problem solving. These weak methods operate on the declarative knowledge acquired from texts and examples. These problem-solving experiences are summarized and stored as new

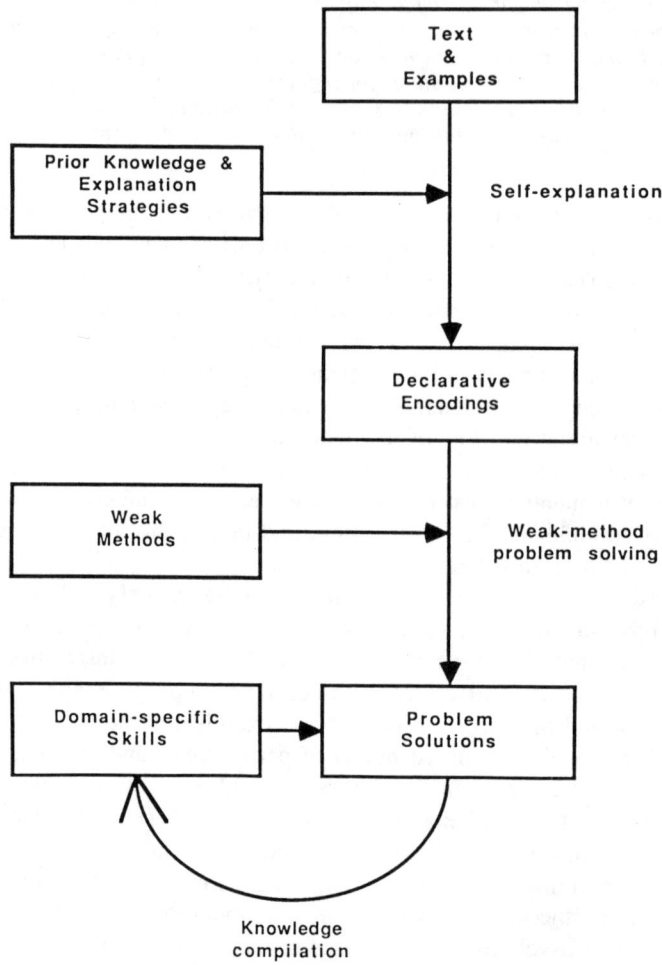

Figure 1. The transformation of examples and text into problem solutions and domain-specific skills.

domain-specific skills by skill acquisition mechanisms. In much of our research, we have assumed that knowledge compilation [3] underlies such skill acquisition, although alternatives such as chunking [4] might also account for the observed phenomena.

In general, investigation of this learning situation has started at the end of this process and worked toward the beginning: Models of problem-solving performance for a variety of domains were developed first [e.g., 5, 6], followed by models of knowledge compilation and chunking [e.g., 3, 4, 7]. More recently work has furthered our understanding of the kinds of problem solving and learning that occur in the initial acquisition of skill. In part, this history of developments reflects a research strategy proposed by Newell and Simon [8]: by first understanding the details of problem-solving performances, one is in a better position to understand learning.

The presentation in this chapter loosely parallels the history of these developments, starting with models of performance and working back toward learning that affects the initial acquisition of skill. The chapter first reviews analyses of data collected in a study using the LISP Tutor, focusing on evidence for the transfer of cognitive skills and the important role of instructional examples. The discussion then turns to an overview of computational models that attempt to address these phenomena.

Performance and Skill Acquisition in Programming: The GRAPES Ideal Model

My colleagues and I initially investigated the acquisition of skills for programming recursive functions [9, 10], which resulted in a set of production system models, implemented in GRAPES [5], that simulated novice skill development and expert skill. This work served as the basis for the development of instruction on recursion in the LISP Tutor [2, 11]. To construct the LISP Tutor lessons on recursion, we developed a set of production system models called *ideal models* that capture an effective set of skills for programming a wide variety of recursive functions in LISP. We also identified *bugs* that students often have when learning recursion. Table 1 presents a small subset of production rules used in an ideal model for programming recursive functions and Fig. 2 presents the goal tree represented by these production rules. In Fig. 2, the boxes represent programming goals, and the descending links joined by an arc represent the decomposition of a goal into subgoals by a particular production rule from Table 1.

The ideal models of programming recursion mainly involve the successive refinement of program specifications into code. However, the models also capture the abstraction of program specifications from examples of the application problem. Figure 3 presents a goal tree implemented by a set of productions that generate examples of the recursive cases of a function and then summarizes (abstracts) the specifications that characterize the examples. Another set of productions uses these specifications to actually generate the required code for the recursive cases of the function.

The LISP Tutor

Instruction with the LISP Tutor takes the following form. The LISP curriculum is divided into a set of lessons, very similar to the standard lists of chapter topics found in LISP programming textbooks. For each lesson, students read some text introducing some new programming feature or technique—specifically, a chapter from the text by Anderson, Corbett, and Reiser [12]—and then work through a set of programming problems with the LISP Tutor. Figure 4 depicts the LISP Tutor interface. Students

Table 1
A Subset of Production Rules from the GRAPES Ideal Model
for Tail Recursion

P1:	IF	the goal is to write a function
	THEN	write "defun"
		and set subgoals to:
		Specify the name of the function
		Specify the function arguments
		Code the body of the function
P2:	IF	the goal is to code the body of a function
		and the function recurses on a number
	THEN	write "cond"
		and set subgoals to
		Code the terminating cases
		Code the recursive cases
P3:	IF	the goal is to code a terminating case
		and the terminating case has been determined
	THEN	set subgoals to:
		Code a condition to test for the terminating case
		Code an action for the terminating case
P4:	IF	the goal is to code the recursive cases
		and the recursive cases have been determined
	THEN	set subgoals to code each case
P5:	IF	the goal is to code recursive cases
	THEN	set subgoals to
		Characterize the recursive cases
		Code the recursive cases

generally work on their program in the middle window, and the LISP Tutor provides feedback and help in the top window.

The LISP Tutor instructs by using a *model tracing methodology,* which involves comparing a student's programming behavior to the behavior of the LISP Tutor's internal ideal and buggy models. To anthropomorphize a little, the LISP Tutor basically watches what the student is doing as he or she programs, and compares this behavior to what it knows about what good students (ideal models) or poor students (buggy models) might do. If the student's behavior corresponds to that of an ideal model then the LISP Tutor usually does not interfere (except in cases where some major program planning may be forthcoming, or the LISP Tutor needs to determine which of several goals the student is attempting). If, on the other hand, the student's behavior matches that of a buggy model, then the LISP Tutor interrupts with feedback, hints, and help that should get the student back on the correct solution path and overcome the student's misconception. When a student does something that does not match anything in the system's ideal or buggy models, the system attempts to at least provide enough information that the student can figure out the next correct step. If the student flounders for some time, the LISP Tutor will eventually show the student what to do next.

Explanation, Analogy, and Transfer

The LISP Tutor provides a means for automatically collecting and classifying detailed behavioral data. Basically, each cycle of interaction with the LISP Tutor can be characterized as an opportunity for a correct production rule to apply (i.e., one from an ideal model) or for a bug to occur. The student can exhibit a correct solution step, make an error and get feedback, ask for help, or make three successive errors and be shown the correct step.

The Transfer of Procedural Knowledge: A Production System Analysis

Given that a student has solved one or more programming tasks, can we predict his or her performance on the next task? In a recent experiment, we had 20 subjects work their way through a series of lessons using the LISP Tutor. In the recursion lesson, there were two groups of subjects, with each group receiving a different sequencing of recursion problems. For one group, the problems were blocked such that recursive programs that

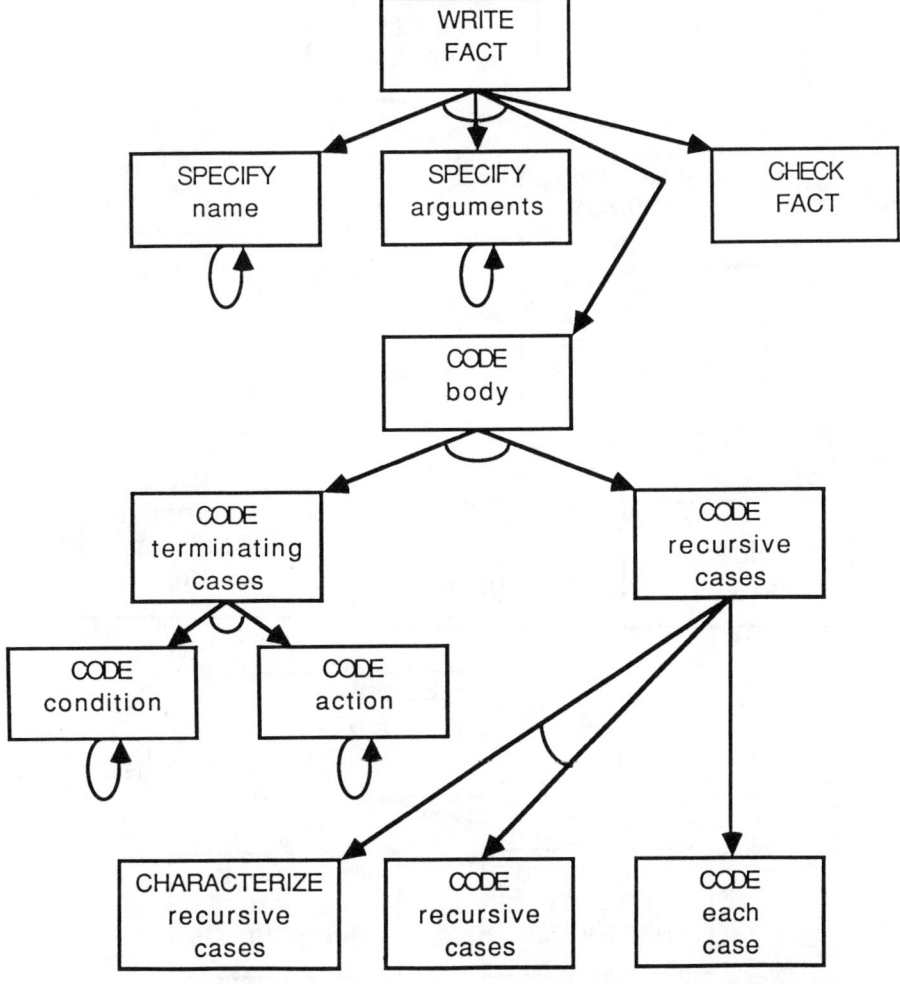

Figure 2. The goal tree represented by the productions in Table 1.

recurse on integer inputs were coded before recursive programs that recurse on list inputs. For the other group, the two types of programs were intermixed (i.e., odd numbered problems dealt with integers and even numbered problems dealt with lists). Orthogonal to this manipulation of problem sequencing, we divided subjects into two groups that received different example programs as part of their instruction. One group received an example of a recursive program that operated on list inputs, and the other group received an example program that operated on integer inputs.

One of the goals of this study was to obtain data relevant to the analysis of transfer of skill across problems and the relation of instructional examples to learning. Figures 5 and 6 present errors per problem across sequences of problems for each group. On first examination, the data in Figs. 5 and 6 appear to be quite varied and unsystematic. However, a production system analysis can be used to reveal a compelling underlying order to these data.

Our analysis of these data derives from recent work on production system models of

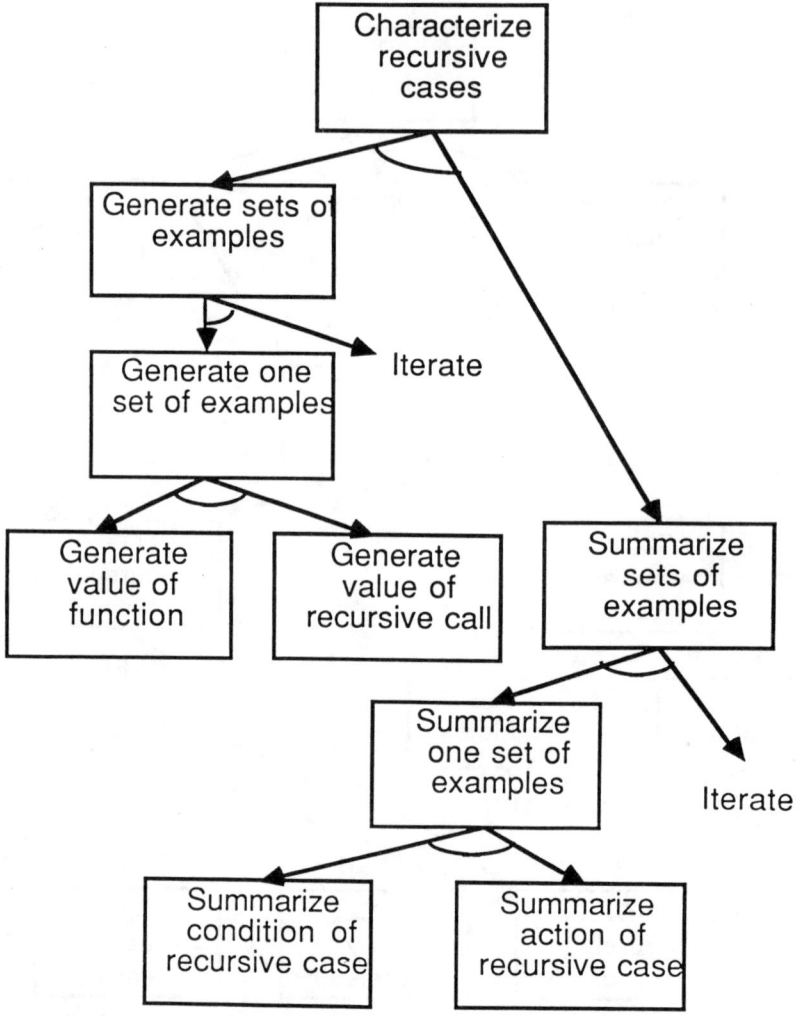

Figure 3. A goal tree for the abstract of program specifications for recursive functions.

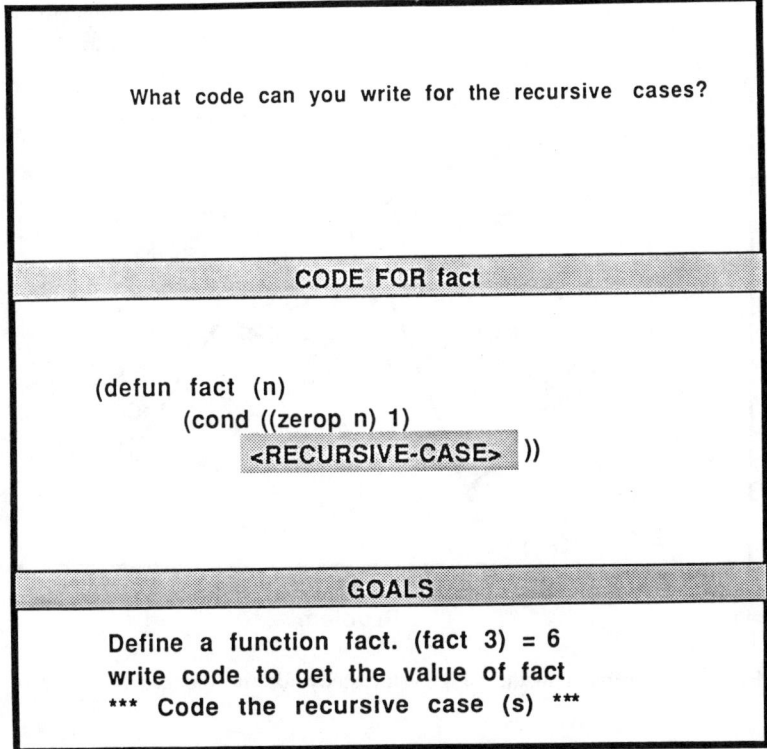

Figure 4. The LISP Tutor interface.

transfer [13, 14], which are in the spirit of Thorndyke's identical elements theory of transfer. It has long been recognized that the main problem with Thorndyke's formulation of the identical elements theory of transfer has been his vagueness in defining the elements of transfer. Singley and Anderson [13] have recently presented a detailed revision of the identical elements theory in terms of the ACT* production system theory. In its bare-bones form, the ACT* theory of transfer states that productions are the elements of transfer. Each production represents a unit of cognitive skill at any level of generality. Singley and Anderson analyze a wide variety of data, including transfer in text-editing, LISP programming, and calculus. Their analyses corroborate their assumption that productions have the following features: (a) productions are learned independently and transfer independently; (b) they are compiled in an all-or-none manner; (c) productions accrue strength with practice, and the reliability and speed of performance is proportional to strength; and (d) productions may characterize abstract components of a skill and consequently are not necessarily tied to features of the external representation of the problem.

Ignoring the effects of practice for the moment, we examined each programming problem and used the ideal model production system in the LISP Tutor to predict which skills transfer from previous learning and which parts of the programming task will be novel. More precisely, we determined which ideal model productions were expected to be evoked in solving a particular problem, and these productions were divided into one set that were evoked on previous problems (old productions) and a set that were being evoked for the first time (new productions). The data in Figs. 5 and 6 are replotted in

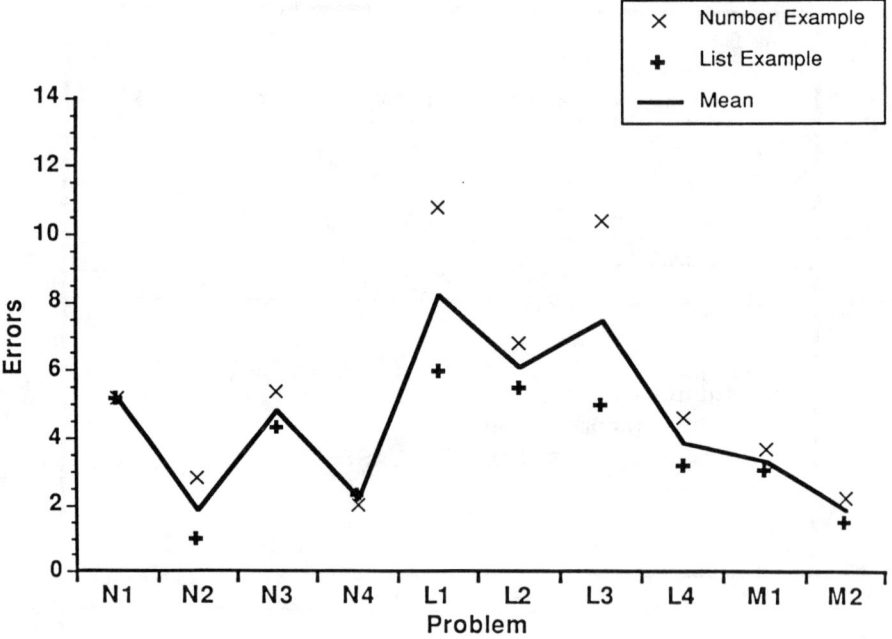

Figure 5. Errors per problem in the LISP Tutor study for the blocked sequence of recursion problems.

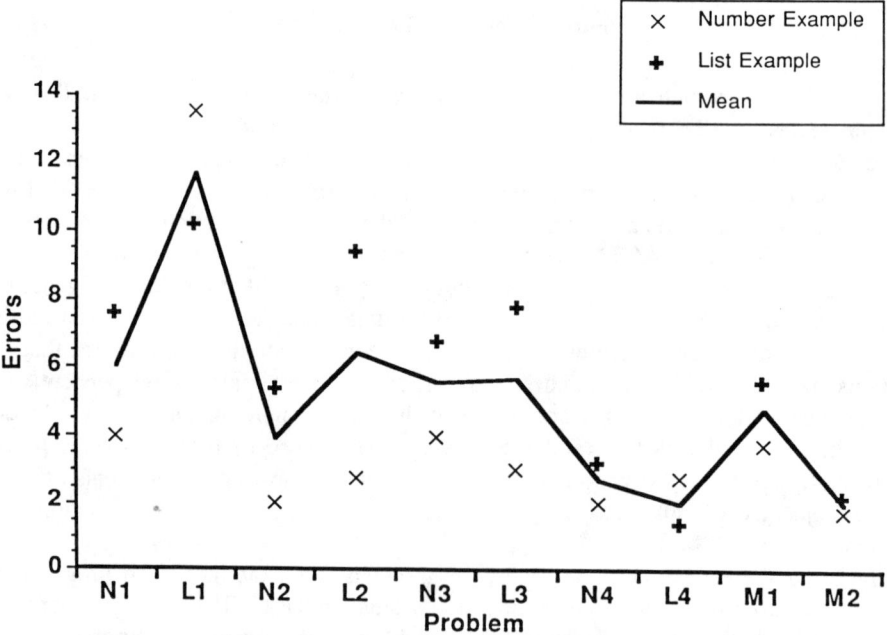

Figure 6. Errors per problem in the LISP Tutor study for the interleaved sequence of recursion problems.

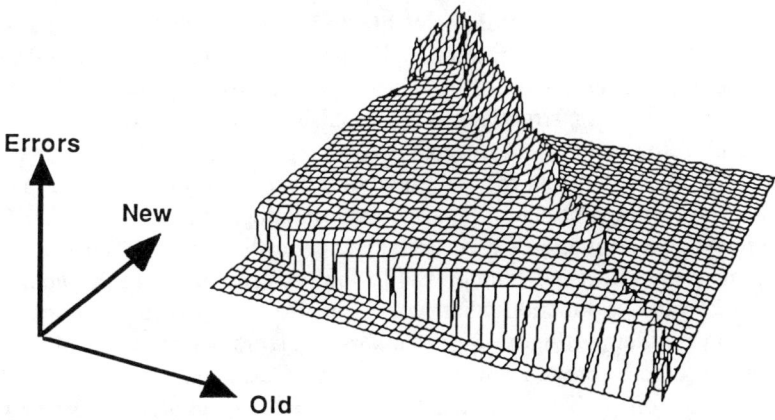

Figure 7. Errors per problem in the LISP Tutor study as a function of the number of new and old productions per problem. New productions per problem range from 0 to 6. Old productions per problem range from 0 to 37.

Fig. 7 as a three-dimensional perspective plot in which the surface is the mean errors per problem as a function of the number of new and old productions evoked in each problem. (Portions of the surface that do not rise above the plane are not covered by any empirical data.) The axis for new productions ranges from 0 to 6 and the axis for old productions ranges from 0 to 37. Figure 7 shows that errors per problem increase with the number of old productions evoked in a problem, but an even steeper increase is associated with increases in the number of new productions.

A more precise quantitative fit of these transfer data are presented at the end of the next section. For now, however, it is clear that much of the variation in problem-solving performance in Figs. 5 and 6 can be captured by a careful task analysis using an ideal model production system, and by identifying the amounts of compiled knowledge that can apply to a problem and the amounts of weak problem solving that will need to be done at novel goals. Complexity is added to such production system analyses of transfer by considering the role of declarative knowledge [13]. New productions are compiled as summarizations of the operation of weak methods operating over declarative structures. The same declarative knowledge may serve as the basis for the compilation of many different productions. Variations in the declarative sources of new productions can influence the number and generality of skills that are learned and the efficiency of learning. The work of Kieras and Bovair [15], which showed that acquiring a mental model of device improved performance on novel operating procedures, is one example of how declarative knowledge can improve the efficiency of learning. The next section discusses how another source of declarative information, instructional examples, can also have a substantial impact on the efficiency of learning.

Analogy from Examples

Recent research suggests that examples presented in instruction play an important role in the early development of skill. LeFevre and Dixon [16] found that their subjects largely ignored written instructions and preferred to use examples to guide their performance. Anderson [17] found that 81% of the high-frequency errors made by students learning from a variety of intelligent tutoring systems could be attributed to the particular examples that had been presented. Similarly, VanLehn [18] found that 86% of the systematic errors in his data on arithmetic problem solving were a result of example-driven learning. Likewise, our earlier studies of learning recursion [10, 11] revealed the importance of examples presented in instruction on the acquisition of novel skills.

The LISP Tutor experiment discussed earlier presented subjects with one of two examples. Half of the subjects received textbook instruction on recursion that included an example of a recursive function that operated on integer inputs. The other half of the subjects received an example of a recursive function that operated on list inputs. One way to examine the effects of examples is to look at error rates on problems requiring the coding of recursive functions that work with list inputs (list problems) or integer inputs (number problems). Figure 8 presents mean errors per problem on list and number recursion problems broken down by the type of example available during instruction. Performance on problems similar to the available example is superior to performance on problems different than the example, and the interaction in Fig. 8 is significant $t(144) = 2.00, p < .05$. However, this measure of similarity is based on a rather simple metric of whether or not the same kind of input is operated on by two programs.

A more precise analysis could again be based on a production system analysis. Two programs could be said to be similar to the extent to which they invoke the same skills. A theoretical estimate of the similarity of one program to another can be calculated by determining the proportion of the total productions in the one program that is shared with other. Using this method, we determined the similarity of each programming problem to the examples available to subjects. Since we expected that it is the novel parts of a programming problem—those invoking new productions—that are affected by the availability of a relevant example [19], we multiplied these similarity values by the proportion of new productions invoked in each problem. The resulting values form a predictor variable, $A(i)$, indicating the degree to which each new production on problem i will involve knowledge that could be derived from the available example program.

A multivariate linear regression on the data in Figs. 5 and 6 was performed involving variables for the number of new productions per problem, old production per problem, and the example influence, $A(i)$. To capture practice effects, the regression also included problem trials as a separate variable. The resulting best-fit linear function, $E(i)$,

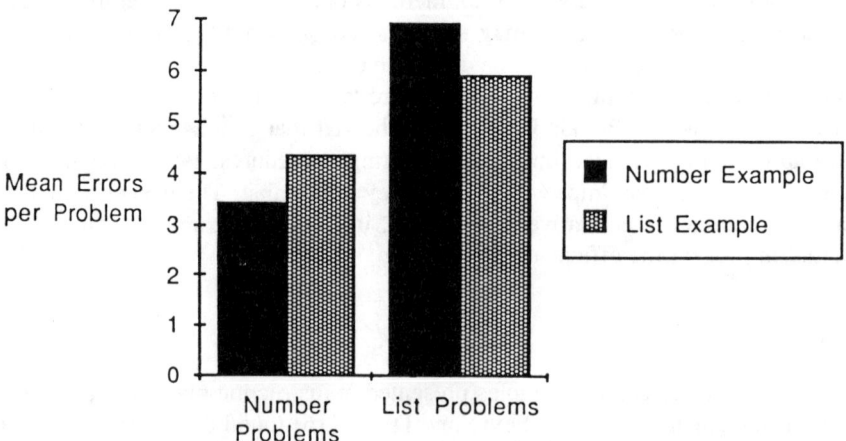

Figure 8. Mean error per problem on list and number recursion problems in the LISP Tutor study.

for predicting errors on problem i is:

$$E(i) = 1.39\, P_{new}(i) + .25\, P_{old}(i) - 1.91 \log(t) - 5.00\, A(i)P_{new}(i)$$

where $P_{new}(i)$ is the number of new productions on problem i, $P_{old}(i)$ is the number of old productions on problem i, and t is the problem trial in the sequence of the problem. This regression is highly significant, $F(4, 36) = 60.71$, $p < .0001$, accounts for substantial variance, $R^2 = .87$, and each of the predictor variables is significant. The estimated parameters also make a great deal of sense. As suggested by Fig. 7, the parameter estimates show that new productions are more likely to cause errors than old productions. The regression also suggests that errors decrease with log trials, indicating an overall practice effect. Finally, errors on new productions are reduced by the availability of declarative knowledge about the example.

The X Model

The GRAPES ideal production system model for programming recursion provided a rigorous task analysis that permitted excellent fits of the learning data from the LISP Tutor experiment. To perform that analysis, however, we made some additional assumptions about the effects of examples that are not directly incorporated in the ideal models. The important role of examples in learning to program was the main motivation for our development of an extension of GRAPES, called X, that includes mechanisms for problem solving by analogy to examples. A summary of the X system and some simulations is presented in Pirolli [20]. The current X model is essentially an instantiation of the PUPS [21] production system learning mechanisms in the GRAPES language.

Like several other proposals for problem solving by analogy [22–26], the X analogy mechanisms supply a method for problem solving when domain-specific methods are lacking or inadequate. The general notion is that the learner has some declarative knowledge of how the structure, S_e, of an example achieves various functions, F_e, under certain preconditions, C_e, and is faced with achieving goals, G_t, under conditions, C_t, in a target problem. The task in analogical problem solving is to come up with a target solution, S_t, by solving the analogy $S_e:F_e::S_t:G_t$, subject to the constraint that the mapping of S_e onto S_t transforms C_e, into a set of preconditions that are in C_t or satisfiable in the target solution.

Problem solving proceeds by selecting productions that match goals and working memory information. These productions lay out declarative representations of a solution structure. Both planning knowledge and knowledge of the function and structures of domain objects are used in problem solving [cf. 25]. The function of a unit representing a plan is the goal that it achieves, and the structure is the partial ordering of subplans or actions that it specifies. The function of a unit representing an object is the purpose it may serve (as occurs in cases in which objects are structured to make a device or as part of an action in a plan). Objects may have several functions associated with them, and their structures consist of properties or other objects with which they are configured.

Table 2 presents part of an X encoding of an example function that computes exponents m^n and part of an encoding of a target problem involving the calculation of the factorial function, $n!$. The encoding states that the LISP definition of power has a structure containing "defun," the name of the function, the arguments, and the body of the function. The last three parts of the structure for power in turn have their own encod-

Table 2
Examples of X Representations

The X Representation of the POWER Program

Power-definition
    Functionality: defines(power-function args power-result)
    Preconditions: implemented-in(power-function LISP)
    Structure: steps(defun power-name args body)

The LISP Definition of POWER

```
(DEFUN POWER (M N)
  (COND
    ((ZEROP N) 1)
    (T (TIMES N (POWER M (SUB1 N)]
```

The X Representation of the Factorial Problem

Fact-definition
    Functionality: defines(fact-function x-arg fact-result)
    Preconditions: implemented-in(fact-function LISP)

---

ings, which are not presented here. Note that the factorial problem has an encoding of the program's functionality, but its structure is absent.

When a goal is set to code the factorial function, a set of production rules searches through a network of instantiated plans and domain objects that comprise the current state of a problem solution to select goals that will serve as the next focus of planning. This explicit representation of goal selection procedures is one significant respect in which X differs from GRAPES and PUPS, which used fixed goal selection strategies. These goal selection strategies can be acquired and modified by the same learning mechanisms governing the acquisition of programming plans, and consequently may serve as the vehicle for modeling the transition from simple programming strategies to more sophisticated software design strategies. It appears that the system could learn to select goals either on the current state of instantiated plans, the current configuration of objects making up a design solution, or some combination of both.

In X, in solving novel problems, analogical problem solving is invoked when a goal is activated and no production matches (Fig. 9). In theory, we assume that an associative retrieval, through spreading activation mechanisms [7, 27], retrieves an encoding of a plan unit derived from an example (or prior experience) that achieves a similar goal in a similar problem context [cf. 28]. Alternatively, a configuration of objects (e.g., a piece of LISP code) that achieves a similar goal could be selected. In the actual X implementation, the effects of spreading activation are approximated by determining a match between encodings of example solutions and a target problem, and using a specificity principle to select the best match.

For instance, if the factorial problem in Table 2 were the current goal in X, and the power program was available as an example, X would match the functionality of the factorial to the functionality of power, as well as the functionality of any other available example units. This process is called *function matching*. The result of the match of functionalities in X would be the correspondence set {*power–function/fact–function, args/x–arg,power–result/fact–result*}, where *a/t* means *a* corresponds to *t*. Such correspondences are later used by other analogy processes to map information about the example onto information about the target problem. A particular example unit is selected for further analogical processing by a specificity principle in which correspondence sets having greater numbers of correspondences and more identical correspondences (e.g., *a/a*) are given higher specificity. Since the goal of analogical problem solving is to map an existing solution structure onto a target problem, one constraint on the selection of an example unit is that it must contain an instantiated structure slot. Another constraint is that preconditions on the example unit are not violated when they are mapped onto the target problem.

Having performed a function match and selected a relevant example, the system then performs a structure mapping process. This involves mapping an analog structure onto a new target structure, using the available correspondence set to perform substitutions of target elements for example elements. However, there is no guarantee that the correspondence set will be elaborate enough to permit such a mapping. When an element of an example structure is encountered that has no correspondence to a target element, a process called *function elaboration* is invoked. There are a number of ways that function elaboration can be carried out to map a particular structural element e_a of an analog onto a new structural element e_t in a target. First, the functionality of e_a may match the functionality of some existing target unit e_t. The correspondence set can be elaborated with this correspondence plus the correspondences resulting from the function match of e_a and e_t. Second, a new target unit can be created and assigned a functionality mapped from e_a, and this may recursively invoke further function elaboration. Third, additional correspondences can be found by elaborating the match of an analog unit to a target unit. This is achieved by recursively matching the functions of elements already placed into correspondence. This may lead to an elaborate set of correspondences that permits e_a to be mapped onto a new e_t. When the structure mapping is complete, new goals are activated and the problem-solving process begins anew. Note that the entire example solution is not used and that different examples can be used to solve different goals. Also

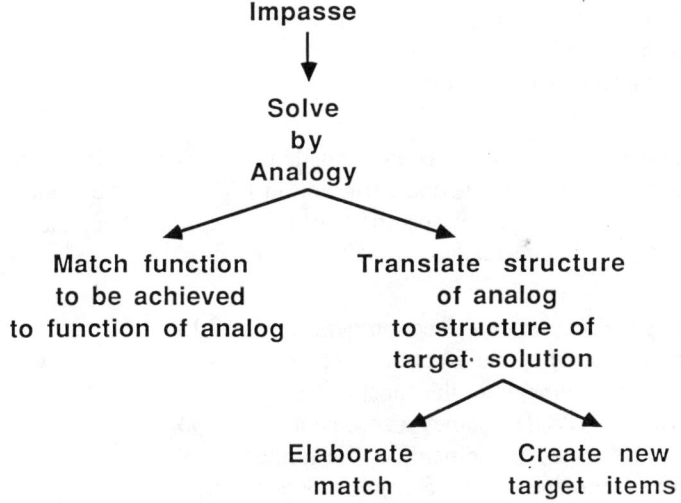

Figure 9. The functional flow of processing in the X analogy process.

note that the particular selection and mapping of analogs is determined by the particular goal and problem feature context, which contrasts with models of analogy that are based on purely structural features [29].

One major outcome of analogical problem solving is the induction of new production rules through knowledge compilation mechanisms. The interaction of analogical problem solving and knowledge compilation produces new production rules that generalize over information present in the declarative units representing example and target problems and their solutions. Interestingly, we did not alter the basic knowledge compilation mechanisms of GRAPES to achieve this induction of generalized procedural knowledge. Thus, the interaction of analogy and knowledge compilation offers an alternative procedure for the generalization of cognitive skills in ACT* [30].

To see how general skills are acquired from analogy, consider that analogical problem solving in X consists of two things: (a) matching preconditions and functionalities of the analog and target, and (b) creating target conditions, functionalities, and structures based on the analog. Knowledge compilation creates new productions with conditions that specify the target functions and conditions that were matched in the analogy process and that have actions that specify the structures, functions, and conditions that were created in the target by analogy. The conditions created by compilation retain the components of the target that matched exactly to the analog and variablizes over target information that mismatched. Thus, a compilation of the analogical problem solving involved in achieving the goal unit *fact-definition* based on the example *power-definition* produces the production:

```
L1:   IF   the goal is to achieve
           =definition
               functionality: defines(=function =arguments =result)
               precondition: implemented-in(=function LISP)
           and
           =name
               functionality: name-of(=function)
      THEN the structure of =definition is
               steps(defun =name =arguments =body)
           and the functionality of =body is
               implements(=function)
```

where the items preceded by the equal sign denote variables. Production L1 applies when the goal is to achieve a function definition in LISP when the name of the function has been decided on. The action specified by L1 lays out a template for the code to define the function. A more readable rendition of L1 might be:

```
L1':  IF   the goal is to write a function definition in LISP
           and the name of the function is given
           and the arguments to the function are given
      THEN write "(DEFUN ⟨name⟩ ⟨arguments⟩ ⟨body⟩)"
           where ⟨name⟩ is the name of the function
           and ⟨arguments⟩ are the arguments to the function
           and set a subgoal to code the ⟨body⟩ of the function
           which implements a process that achieves the function
```

Production L1 would thus apply in situations similar to those that invoked analogical problem solving in the first place.

The final learning mechanism of note in X is the strengthening mechanisms inherited from GRAPES. Each time a production rule is applied in solving a problem, it yields an increase in a strength parameter associated with the production. Strengthening is associated with improvements in the efficiency and accuracy of production application [7] and is assumed to account for practice effects, such as those found in the LISP Tutor experiment discussed earlier.

Simulations in X of subjects learning to program recursion account for the differences we observed in the effectiveness and efficiency of learning [11, 20]. Subjects who had encoded examples in a relatively literal manner showed poor performance and transfer, whereas subjects who had elaborated the examples developed more general skills and showed better performance. The production system theory of transfer of Singley and Anderson [13] is thus complicated by the fact that declarative encodings can have a dramatic impact on the acquisition of novel skills, over and above the transfer of productions from prior tasks. Furthermore, the structure and semantics of productions derive from the declarative knowledge from which they were compiled. The productions compiled in X simulations of subjects using literal analogies show less transfer across problems than simulations of analogies that involved representations of more abstract features of the examples. Unfortunately, this begs the question of how such differences in encodings of examples arise in the first place, and points to a major gap in our model of analogical problem solving.

The Role of Self-Explanations in Learning to Solve Problems

Chi, Bassok, Lewis, Reiman, and Glaser [31] report a study that begins to address the issue of individual differences in understanding and encoding examples. Specifically, Chi et al. [31] analyzed the verbal protocols of students as they studied examples in a physics text and then solved physics problems. Chi et al. divided these students into groups of good and poor students based on a post-hoc median split on their performance in solving physics problems. Good and poor students showed differences in the number and kind of elaborations that they generated while processing examples. Good students generated more elaborative ideas about the worked-out examples, and more specifically, generated more statements that were explanations, as opposed to simple monitoring statements ("I see how they did it") or other statements such as paraphrases. Furthermore, proportionally more of the explanations provided by good students (98%) were justifications of the actions presented in an example than were observed for poor students (65%). These justifications were statements that (a) refined or expanded the conditions of an action in the example, (b) explicated or inferred additional implications of an action, (c) imposed a goal or purpose for an action, or (d) imposed meaning on quantitative expressions. Additionally, good students preferred to generate these explanations when first processing an example, rather than while looking back at the example to solve a problem.

Chi et al. propose that the main reason self-explanations are so important is because of the way examples are usually presented in texts (the examples used in their study came directly from a standard physics text). Example solutions usually present a set of actions, with little explanation of the rationale for those actions. It is left to the student to infer when those actions might be applicable, or what their relation is to underlying principles or plans. One might expect that some students are fortunate enough to learn general strategies for coping with such examples.

It is notable that texts in programming may be even worse at presenting example solutions than physical science or mathematics texts. At least in these latter cases, examples of worked-out solutions are presented. Programming texts frequently present nothing more than the final code that serves as a solution to a programming problem. The task facing a student in the situation of programming is more arduous because he or she also has to infer most of the intermediate cognitive actions that were involved in producing a program.

Our modeling activities are now turning to addressing the processes and knowledge involved in self-explanations in the domain of programming. David Rockower, a student, and I worked on a model that constructs explanations of example recursive functions in which we assumed that self-explanation is itself a problem-solving process. In this model, the goal is to show how an example program achieves its stated purpose. As in explanation-based learning [26, 32, 33], this can also be viewed as a proof process, in which prior knowledge about the structure and function of program components is brought to bear in the interpretation of a program, along with new information derived from the surrounding text.

Table 3 presents facts about a recursive function example that has been used in one of our simulations of the self-explanation process. At this stage of our research, we have been less concerned with modeling protocol data than with getting plausible computer models

Table 3
Inputs to the Explanation Model

Facts About the Example Program

Function1 checks for science books in a list.
When it finds one, it adds it to the output list.
When it does not find one it continues.

The Example Program (in Pseudo Code)

Function1 (list) =

    IF   (list = { })
 THEN { }

    IF   First (list) is a science book
 THEN Insert (first) list into (Function1 (rest list))

  ELSE (Function1 rest (list))

Two Explanation Rules

    IF   Each case in the body of the function returns what it should
 THEN The function works

    IF   The function should not add anything to its output under condition C
  AND It should continue processing its input
  AND Function (⟨args⟩) is the recursive call
 THEN The action should be
            function (⟨args⟩)
            in condition C

---

that will serve as a springboard for empirical investigations. The example comes from previous work that examined learning in the SIMPLE language [34]. The example, Function1, takes a single-level list as input. It checks each element of this list to determine if the element is an atom that tests as a "science book," using a built-in predicate of the SIMPLE language. When Function1 finds a science book, it adds to the output list; when the item is not a science book it continues processing the remainder of the input list.

Pseudocode for Function1 is also presented in Table 3. The code contains three conditional clauses: (1) When the input list is empty, an empty output list is returned; (2) when the first item in the input list is a science book, it is inserted into the result of a recursive call to Function1 called on the rest of the input list; (3) when the first item in the input list is not a science book, the result of the recursive call to Function1 is returned.

Two of the inferential explanation rules used in our simulation are also presented in Table 3. The first rule infers that a function works if each of its conditional clauses work. The second rule infers that a recursive call is appropriate if the function should continue its processing of an input list. Our simulation[1], in analyzing how the Function1 example works, generates an explanation structure that is partially depicted in Fig. 10. These explanation structures are proof trees in which the facts at the bottom of the figures support the facts at the top through some rule of inference. The first rule in Table 3 generates the explanation structure at the top of Fig. 10, and the second rule in Table 3 is used in the generation of the structure at the bottom of Fig. 10.

Summary

Returning to Fig. 1, I have described our work on a model that learns and solves problems in the following manner. In encountering a novel domain, the model is provided with texts and example solutions that serve as instruction. The model uses relevant background knowledge and strategies to generate declarative encodings of these examples. In solving problems in the novel domain, the model uses domain-specific skills, represented as productions, when applicable. At problem-solving impasses, the model resorts to analogical problem-solving processes that operate over representations of the target problem and available examples. Knowledge compilation processes summarize these experiences into new domain-specific productions. With repeated practice, productions acquire strength, leading to more efficient and accurate skills. Our ultimate goal is to provide precise qualitative and quantitative characterizations of knowledge acquisition and transfer in this paradigm, and the role of individual differences in background knowledge.

This research is similar in spirit to a few other studies in which intelligent tutoring systems have played a major role in the investigation of human cognition and learning. Anderson, Conrad, and Corbett [36] performed extensive exploratory and confirmatory analyses of data collected by the LISP Tutor, focusing on the assumptions and predictions about skill acquisition made by the ACT* theory [7]. As implied by the results discussed in this chapter, Anderson et al. found that the production rule representation of ACT* revealed systematic practice and transfer effects in the LISP Tutor data. In another subject-matter domain, Shute and Glaser [37] found that certain measures in a student model in an economic tutor could be treated as indicators for different discovery strategies associated with varying degrees of learning effectiveness. Such a finding could be important in designing new tutors that focus on communicating general learning

---

[1] The current simulation is implemented in the MRS logic programming language [35].

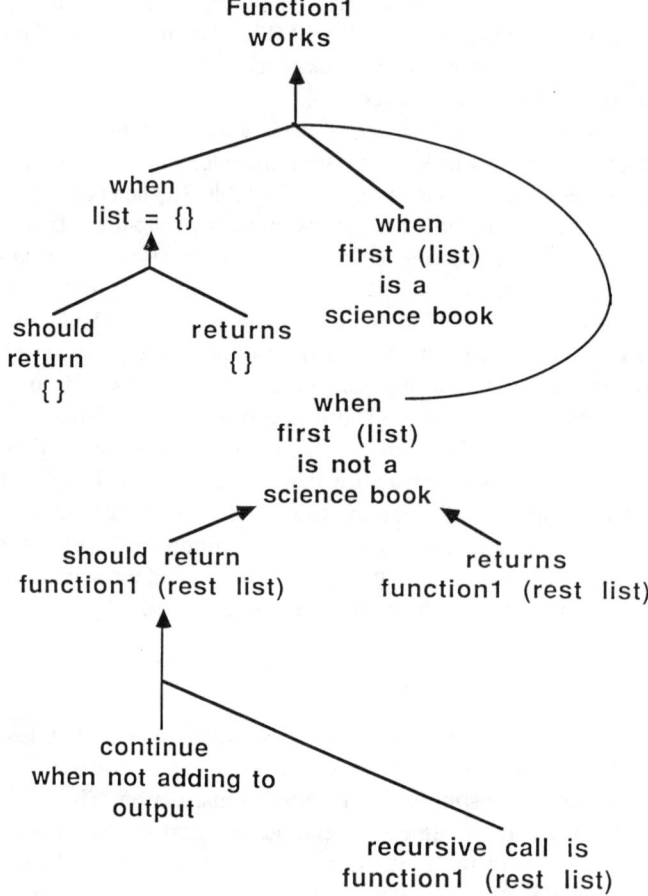

Figure 10. Part of the explanation structure generated for the inputs in Table 3.

strategies. Similarly, our continuing research examining how learners explain examples to themselves may lead to the design of systems that provide more effective individualized explanations or that shape learners to improve their own self-explanation abilities.

Learning to program is certainly a complex and difficult skill. No student who has taken such a course would argue otherwise. Constructing models of learning and performance in such a complex domain is a nontrivial matter, but we propose that our methodology, involving the successive refinement of cognitive models and intelligent tutoring systems, will prove to be a fruitful one in the study of complex phenomena in psychology and education.

Acknowledgments

Portions of this paper were presented at the Annual Meeting of the American Educational Research Association, April 5, 1988. Portions of this research were funded by a National Academy of Education Spencer Fellowship, the University of California, Berkeley, and the Institute for Cognitive Studies. I would like to thank Kate Bielaczyc,

Beatrice Lauman, Margaret Recker, and David Rockower for their assistance in conducting various parts of this research.

References

1. J. R. Anderson, "Methodologies for studying human knowledge," *Behavioral and Brain Sciences,* 10:467–505, 1987.
2. B. J. Reiser, J. R. Anderson, and R. G. Farrell, "Dynamic student modelling in an intelligent tutor for Lisp programming," in *Proceedings of the Ninth International Joint Conference on Artificial Intelligence,* Los Altos, CA: Morgan-Kaufman, 1985, 8–14.
3. J. R. Anderson, "Skill acquisition: The compilation of weak-method problem solutions," *Psychological Review,* 94:192–210, 1987.
4. P. Rosenbloom, and A. Newell, "Learning by chunking: A production system model of practice," in D. Klahr, P. Langley, and R. Neches (eds.), *Production system models of learning and development,* Cambridge, MA: MIT Press, 1987.
5. J. R. Anderson, R. Farrell, and R. Sauers, "Learning to program in LISP," *Cognitive Science,* 8:87–129, 1984.
6. J. H. Larkin, J. McDermott, D. P. Simon, and H. A. Simon, "Expert and novice performance in solving physics problems," *Science,* 208:1335–1342, 1980.
7. J. R. Anderson, *The architecture of cognition,* Cambridge, MA: Harvard University Press, 1983.
8. A. Newell and H. A. Simon, *Human problem solving,* Englewood Cliffs, NJ: Prentice-Hall, 1972.
9. J. R. Anderson, P. L. Pirolli, and R. Farrell, "Learning to program recursive functions," in M. Chi, R. Glaser, and M. Farr (eds.), *The nature of expertise,* Hillsdale, NJ: Lawrence Erlbaum, 1989, 153–183.
10. P. L. Pirolli and J. R. Anderson, "The role of learning from examples in the acquisition of recursive programming skills," *Canadian Journal of Psychology,* 39:240–272, 1985.
11. P. Pirolli, "A cognitive model and computer tutor for programming recursion," *Human-Computer Interaction,* 2:319–355, 1986.
12. J. R. Anderson, A. T. Corbett, and B. J. Reiser, *Essential LISP,* Reading, MA: Addison-Wesley, 1987.
13. M. K. Singley and J. R. Anderson, *Transfer of cognitive skill,* Cambridge, MA: Harvard University Press, 1989.
14. D. E. Kieras and S. Bovair, "The acquisition of procedures from text: A production system analysis of transfer of training," *Journal of Memory and Language,* 25:507–524, 1986.
15. D. E. Kieras and S. Bovair, "The role of a mental model in learning to operate a device," *Cognitive Science,* 8:255–273, 1984.
16. J. LeFevre and P. Dixon, "Do written instructions need examples?" *Cognition and Instruction,* 3:1–30, 1986.
17. J. R. Anderson, "The analogical origins of errors in problem solving," in D. Klahr and K. Kotovsky (eds.), *Complex information processing,* Hillsdale, NJ: Erlbaum, 1989, 343–371.
18. K. VanLehn, "Arithmetic procedures are induced from examples," in J. Hiebert (ed.), *Conceptual and procedural knowledge: The case of mathematics,* Hillsdale, NJ: Lawrence Erlbaum, 1986.
19. P. Pirolli, "Effects of examples and their explanations in a lesson on recursion: A production system analysis," *Cognition and Instruction* (in press).
20. P. Pirolli, "A model of purpose-driven analogy and skill acquisition in programming," in *Proceedings of the Cognitive Science Society Conference,* Seattle, WA, 1987.
21. J. R. Anderson and R. Thompson, "Use of analogy in a production system architecture," unpublished manuscript, Department of Psychology, Carnegie-Mellon University, Pittsburgh, PA, 1986.
22. J. G. Carbonell, "Derivational analogy: A theory of reconstructive problem solving and

expertise acquisition," in R. S. Michalski, J. G. Carbonell, and T. M. Mitchell (eds.), *Machine learning, volume II,* Los Altos, CA: Morgan-Kaufman, 1986, 371–392.
23. M. L. Gick and K. J. Holyoak, "Analogical problem solving," *Cognitive Psychology,* 12:306–355, 1980.
24. M. L. Gick and K. J. Holyoak, "Schema induction and analogical transfer," *Cognitive Psychology,* 15:1–38, 1983.
25. K. J. Holyoak and P. Thagard, "Analogical mapping by constraint satisfaction," *Cognitive Science,* 13:295–355, 1989.
26. C. Lewis, *Why and how to learn why: Analysis-based generalization of procedures* (Tech. Rep. CS-CU-347-86), Boulder, CO: University of Colorado, Department of Computer Science, 1986.
27. J. R. Anderson and P. L. Pirolli, "Spread of activation," *Journal of Experimental Psychology: Learning, Memory, and Cognition,* 10:791–798, 1985.
28. B. H. Ross, "Remindings and their effects in learning a cognitive skill," *Cognitive Psychology,* 16:371–416, 1984.
29. D. Gentner, "Structure-mapping: A theoretical framework for analogy," *Cognitive Science,* 7:155–170, 1983.
30. J. R. Anderson, "Knowledge compilation: The general learning mechanism," in R. S. Michalski, J. G. Carbonell, and T. M. Mitchell (eds.), *Machine learning, volume II,* Los Altos, CA: Morgan Kaufman, 1986.
31. M. T. H. Chi, M. Bassok, M. W. Lewis, P. Reiman, and R. Glaser, "Self-explanations: How students study and use examples in learning to solve problems," *Cognitive Science,* 13:145–182, 1989.
32. G. DeJong and R. Mooney, "Explanation-based learning: An alternative view," *Machine Learning,* 1:145–176, 1986.
33. T. M. Mitchell, R. M. Kellar, and S. T. Kedar-Cabelli, "Explanation-based generalization: A unifying view," *Machine Learning,* 1:47–80, 1986.
34. J. Shrager and P. L. Pirolli, *SIMPLE: A simple language for research in programmer psychology* [Computer program], Pittsburgh: Carnegie-Mellon University, Department of Psychology, 1983.
35. S. Russell, *The compleat guide to MRS* (Tech. Rep. KSL-85-12), Stanford, CA: Stanford University Knowledge Systems Laboratory, 1985.
36. J. R. Anderson, F. G. Conrad, and A. T. Corbett, "Skill acquisition and the Lisp Tutor," *Cognitive Science,* 13:467–505, 1989.
37. V. J. Shute and R. Glaser, "Large-scale evaluation of an intelligent tutoring system: Smithtown," *Interactive Learning Environments,* 1:51–76, 1990.

II
Tools and Environments

Chapter 4

Control for Intelligent Tutoring Systems: A Comparison of Blackboard Architectures and Discourse Management Networks

WILLIAM R. MURRAY

Artificial Intelligence Center
FMC Corporation
1205 Coleman Avenue, Box 580
Santa Clara, CA 95052

Abstract *An intelligent tutoring system must have some control mechanism for deciding what instructional action to perform next. Traditional CAI systems have inflexible control mechanisms that procedurally encode tutorial strategies. Two alternative architectures proposed recently to address these problems are discourse management networks and blackboards. This paper compares these two architectures and argues that an intelligent tutoring system controlled by a blackboard architecture can deliver significantly more effective and flexible instruction than one controlled by a discourse management network.*

Introduction

The research discussed in this paper addresses the problem of control for intelligent tutoring systems. At any point during instruction a tutoring system must select from many possible instructional actions. For example, topics can be introduced, reviewed, motivated, summarized, explained in depth, related to earlier topics, etc. Similarly, student knowledge can be assessed by true/false or multiple choice tests, direct questioning, self-assessment, and many other means. The problem of control is to select the most effective action, given the current lesson objectives, student model, tutorial strategy, subject matter, and resources available.

An important decision in designing an intelligent tutoring system is the choice of architecture for this control mechanism. Traditional CAI systems have inflexible control mechanisms that procedurally encode tutorial strategies, and which prevent them from being readily extended to handle new domains or tutorial strategies. Two alternatives proposed recently to address these problems are discourse management networks and blackboard architectures. This paper compares the advantages and disadvantages of using these two architectures for control in intelligent tutoring systems. Based on this comparison, we argue that blackboards address important limitations of discourse management networks that limit their ability to deliver customized, globally coherent, planned instruction and to support more flexible mixed-initiative instruction.

The framework for comparison is the ability of each architecture to support the dynamic planning of instruction. *Dynamic instructional planning* is a planner-based approach to control for intelligent tutoring systems in which appropriate plan generation[1] and revision occur *during* the tutorial session and in response to the changing tutorial situation. A dynamic instructional planner reasons about alternative lesson objectives and alternative instructional plans to realize the tutorial strategies it selects. We argue that such planning is necessary to build effective, robust, and flexible tutoring systems and that blackboard architectures provide better support for this planning than discourse management networks. Before comparing the two architectures, we first justify the relevance and importance of dynamic instructional planning to intelligent tutoring systems.

Importance of Dynamic Instructional Planning in Intelligent Tutoring Systems

Why should tutoring systems incorporate dynamic instructional planning? Two points must be supported here: first, that a tutoring system should have a plan; and second, that it should be able to plan. It should have a plan to properly manage its time and generate globally coherent instruction. Proper time management prevents spending too much time on relatively unimportant topics or packing too many topics or exercises into one lesson. Global coherence means that topics are logically connected to support instructional objectives and that topics are sequenced and presented in a manner sensitive to the student's perceived knowledge, the tutor's instructional objectives, the student's interest and motivation, and the time available for lessons. Transitions between topics are smooth and the tutor is consistent; that is, it does not contradict previous or future instruction. One sign of incoherent instruction is recurring student confusion about the tutor and its intentions, resulting in distractions from the subject matter being taught.

Another advantage of having a plan is the ability to introduce material in a layered fashion, first providing an overview of the material and lesson plan, and then introducing successive layers of detail that build on and refine earlier material. A plan also allows the tutor to recognize opportunities to motivate and lead into future instruction in response to unexpected student questions, or to defer answering the questions because they will be addressed later in the lesson. The tutor can use a plan to select a sequence of problem-solving cases that cover a set of topics, and to focus its conversation on the central topics for each case, deferring or abbreviating discussion of other topics best addressed in subsequent cases. Finally, a plan assists in explaining and motivating current instruction based on its relationship to future instruction.

One alternative to dynamic instructional planning is for the tutor simply to interpret a highly detailed plan provided by a human curriculum author; this approach is used in traditional computer-assisted instruction (CAI). Although *theoretically* the plan could provide for any eventuality that might arise, *practically* the combinatorics of different student responses and tutorial states require that student initiative be curtailed, fine-grained student modeling avoided, and a significant amount of time spent preparing lesson plans. One goal of intelligent tutoring systems is to explicitly represent the tuto-

---

[1] For the purposes of this paper we allow plan selection to be considered a trivial form of plan generation, otherwise discourse management networks would not support dynamic instructional planning merely by their inability to assemble a plan during instruction.

rial strategies that are encoded procedurally in such a system, allowing for more rapid development of tutors for new instructional domains, and experimentation with alternate strategies.[2] By providing intelligent tutoring systems with a planning capability, their reliance on human-generated plans is reduced while their ability to generate the high-quality expository instruction of well-crafted CAI systems is increased.

A tutor that can plan is also better able to handle a mixed-initiative dialog than a tutor that cannot plan. Replanning can be used to handle topic transitions brought about by student questions and to revise lesson plans or omit or cover topics to satisfy student requests. Again, the combinatorics of the different possible student requests, questions, and tutorial states argue for representing the knowledge for handling student initiative in planning knowledge rather than as different plan contingencies in a pre-stored plan. This planning knowledge allows the tutor to plan topic transitions, to replan in response to the amount of time remaining in the lesson, and to decide how to return to the current topic or whether to abandon it altogether.

Planning knowledge is used not only to customize the lesson plan in response to student requests during the lesson, but also before the lesson begins. A tutor that can plan can generate an initial lesson plan customized to the student's interest, assessed capabilities, and background. The lesson plan can be revised as the tutor's student model changes or in response to student requests. Instead of the tutor's assuming a fixed set of instructional objectives for all students, the student can request instruction on particular areas of a subject. As before, it is impractical to anticipate the number of different student interests, objectives, and backgrounds in teaching complex subjects. It is more economical to represent lesson planning knowledge than to provide lesson plans for all the different possible combinations. The representation is more economical in the sense that the same knowledge is represented explicitly in a concise form and not replicated implicitly in the branches of a single highly conditionalized lesson plan, or as a very large number of lesson plans unique to different situations.

These arguments for the tutor's being able to plan rest on a subtle point that is worth reemphasizing. Simple plan-following tutors (CAI) can always replicate the behavior of a dynamic instructional planner for any particular situation. However, the dynamic instructional planner is better able to handle the combinatorial explosion of different tutorial situations when mixed-initiative instruction is allowed, and where lesson plans are tailored to different student models and to varying amounts of time for each lesson. The CAI system limits mixed-initiative dialog and lesson customization, and omits or simplifies the student model to reduce the number of different tutorial situations. The dynamic instructional planner imposes fewer limitations, because it can apply its planning knowledge to the different tutorial situations that arise.

The economic representation of planning knowledge in a dynamic instructional planner also contributes to its greater ease of use than that of a CAI system. The CAI system procedurally encodes planning knowledge and implicit assumptions about the tutorial state at each branch of its prestored instructional plan. The planning knowledge in a dynamic instructional planner makes fewer assumptions about the tutorial situation and can thus be more readily applied in the construction of new tutoring systems.

---

[2] Intelligent tutoring systems also differ from CAI systems by their emphasis on sophisticated student modeling, the tailoring of instruction to the student model, and the ability to solve problems, model expertise, and answer questions about the domain they are teaching.

Types of Instructional Planning

We may distinguish three levels of instructional planning:

- *Curriculum planning*—planning an extended sequence of lessons for a subject
- *Lesson planning*—determining the subject matter to present in a single lesson, and the order of presentation
- *Discourse planning*—planning communicative actions between the tutor and student within a lesson.

These levels cannot in practice be so cleanly separated. Frequently, therefore, discourse planning and lesson planning are intertwined, as are lesson planning and curriculum planning. Typically, a human instructor's lesson plan will include not only a sequence of topics but also some common discourse procedures such as collecting homework or having students solve and explain problems on the board. This three-level distinction is useful here, because discourse management networks support primarily discourse planning[3] while the blackboard architecture supports all three kinds.

The remainder of this paper compares discourse management networks and blackboards with respect to their support for dynamic instructional planning. The next section formalizes the discourse management network architecture, provides two examples, and then discusses this architecture's support for dynamic instructional planning. Similarly, the following section formalizes the blackboard architecture, provides two examples, and then discusses support for dynamic instructional planning. Next, a specific application, control of a Lisp tutor, is considered to illustrate the advantages of the blackboard architecture. The last section summarizes these arguments.

Discourse Management Networks

Discourse management is the selection by the tutor of the next discourse action or intended sequence of actions. In this architecture, the tutor does not reason about the results of discourse actions or future tutorial situations; the only planning performed is plan selection or skeletal planning. However, planning is not a prerequisite for flexible behavior [1], and tutors built with this architecture allow sophisticated control of tutorial dialogs.

A discourse management network (DMN) is a kind of procedural network similar to an augmented transition network. Nodes represent tutorial procedures or actions corresponding to tutorial states. Arcs represent state transitions. The tutor is in one state at any time. The state determines what actions are performed. After performing the actions or procedures, control follows one of the arcs leaving from the state. The choice of arc, if there is more than one, depends on the predicates on the arcs.

Architectural Formalization

Figure 1 represents the key features of the discourse management network architecture. Circles represent tutorial states named $S_0 \ldots S_n$ and $S_{default}$. Arrows represent possible transitions to other states, not all of which are shown. Assume the tutor is originally in state S_0. Then A_0 represents the actions that will be performed in state S_0. The heavy line

---

[3] Planning in the sense of choosing a course of action, not in the sense of reasoning about the results of actions and projecting future discourse situations.

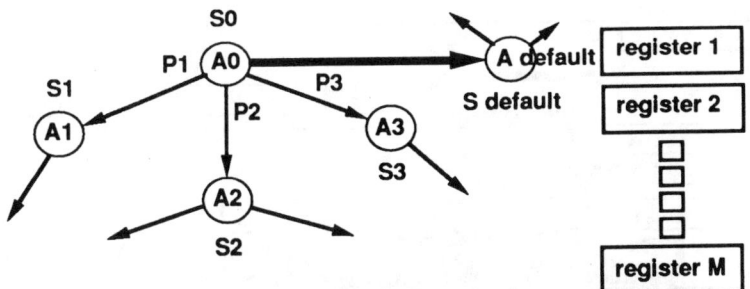

Figure 1 The discourse management network architecture.

represents a *default transition* to the state $S_{default}$. Other lines from S_0 represent other possible transitions that can be taken. Each predicate $P_1 \ldots P_n$ is considered in turn, and the first true predicate causes control to transfer to that state. If no predicate $P_1 \ldots P_n$ is true, then the default transition is taken. The discourse management network has no memory other than the current state and the registers $R_1 \ldots R_m$. These registers can be set to scalar or symbolic values by any of the actions and accessed in any of the predicates. Actions can only set registers or perform communicative actions (e.g., give a test). Actions cannot change control in any other way other than that described above. For example, an action cannot change the current state or the arcs emanating from a state.

As formalized here, discourse management networks are a kind of finite state machine with registers used to reduce the number of states. Usually discourse management networks are coupled with other control mechanisms (e.g., an agenda) and an external memory to provide a topic selection mechanism. Two examples of discourse management networks are discussed below, along with extensions they make to the abstract model above.

Two Exemplars

MENO-TUTOR [2] uses a discourse management network to control a Socratic question-and-answer dialog with a student that alternately probes the student's knowledge and refines his knowledge by teaching new information or correcting misconceptions. Figure 2 [3] shows the state transition diagram of MENO-TUTOR. Numbered states indicate a possible sequence of states that might occur in a dialog. It differs from the abstract model above in its use of hierarchical abstraction: tutorial states at the bottom of the diagram are refinements of the tutorial states above them. In this diagram only possible transitions from more abstract to less abstract states are shown; default transitions are not shown. The actions of a particular state result in the generation of a single utterance by a surface natural language generator.

MENO-TUTOR also differs from the idealized model in its use of meta-rules to represent state transitions other than to default states. Rather than explicitly representing these arcs and their predicates, the meta-rules operate as demons that can override default transitions. This movement to a new state from a specified prior state happens only when a meta-rule's trigger condition is true. Since the meta-rule can be replaced by arcs to the new state with predicates using the meta-rule's trigger condition, the meta-rules provide only a more compact representation but no added functionality compared to the abstract model above.

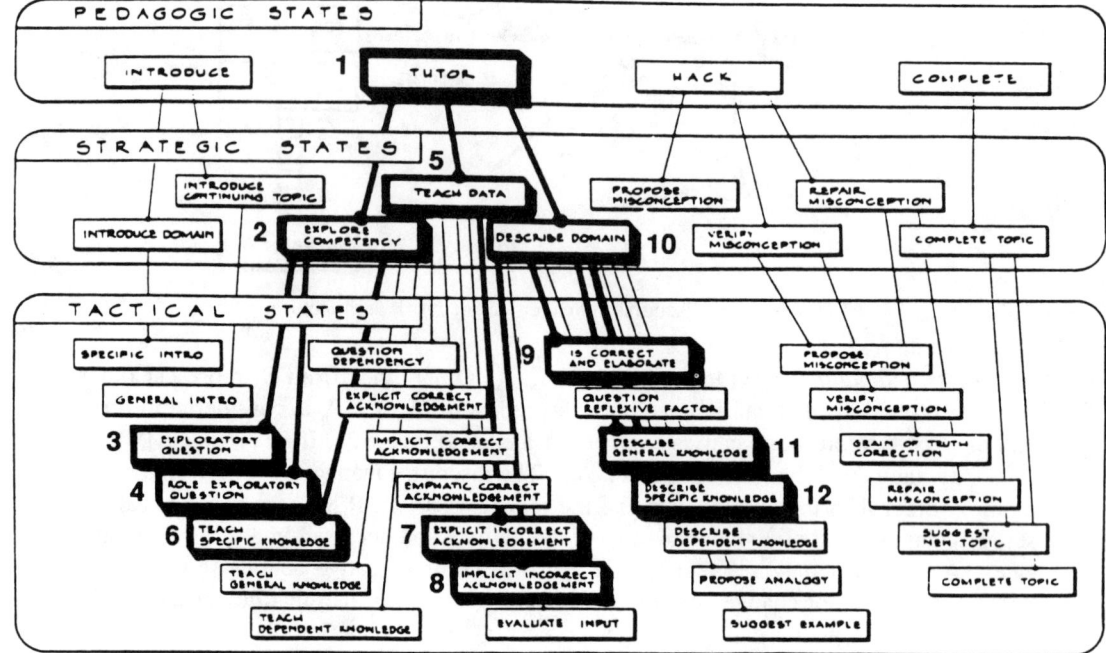

Figure 2 MENO-TUTOR's discourse management network.

MENO-TUTOR was reverse-engineered to produce Socratic question-and-answer dialogs similar to those obtained from human protocols. One such dialog that MENO-TUTOR could produce, if coupled to a surface language generator, is shown below (these examples are from [3]). The Pascal program being referred to is shown in Figure 3.

Tutor: Do you know that GRADES, in line 9, is a control variable for the WHILE loop?
Student: Yes
Tutor: Does that mean that GRADES should be
(a) Initiated before the loop?
(b) Added into SUM within the loop?
(c) Modified within the loop?
(d) Both (a) and (c).
Student: C
Tutor: You seem to understand about the control variable of the WHILE loop. Now let's talk about some other command statements.
What do you think READ (GRADES) in line 8 does with the value of GRADES?
(a) Reads all values of GRADES at one time.
(b) Reads in a single value of GRADES.
(c) Assigns several new input values to GRADES.
Student: A
Tutor: Then do you think that GRADES := GRADES + 1 updates the next of those values?

```
1  PROGRAM LESSON1(INPUT, OUTPUT);
2    VAR
3      SUM, GRADES, STUDENTS:INTEGER;
4      MEDIAN:REAL;
5    BEGIN
6      SUM:=0;
7      STUDENTS:=0;
8      READ (GRADES);
9      WHILE GRADES>0 DO
10       BEGIN
11         SUM:=SUM + GRADES;
12         STUDENTS:=STUDENTS+1;
13         GRADES:=GRADES+1
                 should be READ(GRADES);
14       END;
15     MEDIAN:=SUM/STUDENTS;
16     WRITELN
17     ('THE MEDIAN GRADE IS', MEDIAN:8:3)
18  END
```

Figure 3 Student program in MENO-TUTOR dialog.

Student: Yes

Tutor: There is your misconception. Adding one to a variable is a common way to step through a range of values. But it cannot apply here because your READ instruction in line 8 retrieves only a single value. You should replace line 13 with a second call to READ (GRADES).

GUIDON [4] extends the utility of the discourse management network formalism by introducing quite sophisticated tutorial actions. Instead of simple actions, such as suggesting a new topic, a packet of tutorial rules (t-rules) can be invoked. Such a collection of t-rules is called a discourse procedure, and t-rules themselves can call other discourse procedures. The result is that most of the selection of discourse actions occurs as a result of the operation of the t-rules within a state, and the discourse management structure is more of a convenient means of organizing discourse procedures, rather than the primary control resulting in the selection of discourse actions.

The domain of discourse for which GUIDON was developed is a MYCIN consultation. A medical student using GUIDON attempts to learn or refine his understanding of the knowledge represented by MYCIN's diagnostic rules. Discourse topics include MYCIN rules, goals, clinical data, and student hypotheses. Discourse procedures are organized around these topics as shown in Figure 4 ([5, p. 59]). Selection of the next discourse procedure is also performed within each state; thus, the predicates on the arcs are implicit in the t-rules that determine the next discourse procedure to select.

The point here is not that GUIDON's enhancements alter the DMN architecture or that GUIDON does not handle discourse in a sophisticated way (it does), but that GUIDON's power comes from its use of these discourse procedures and t-rules and *not* from the underlying discourse network architecture. The discourse management network architecture—the states, arcs, and predicates—is not where most of the decision making is performed regarding what discourse action to take. Rather this is performed by the numerous t-rules in each state.

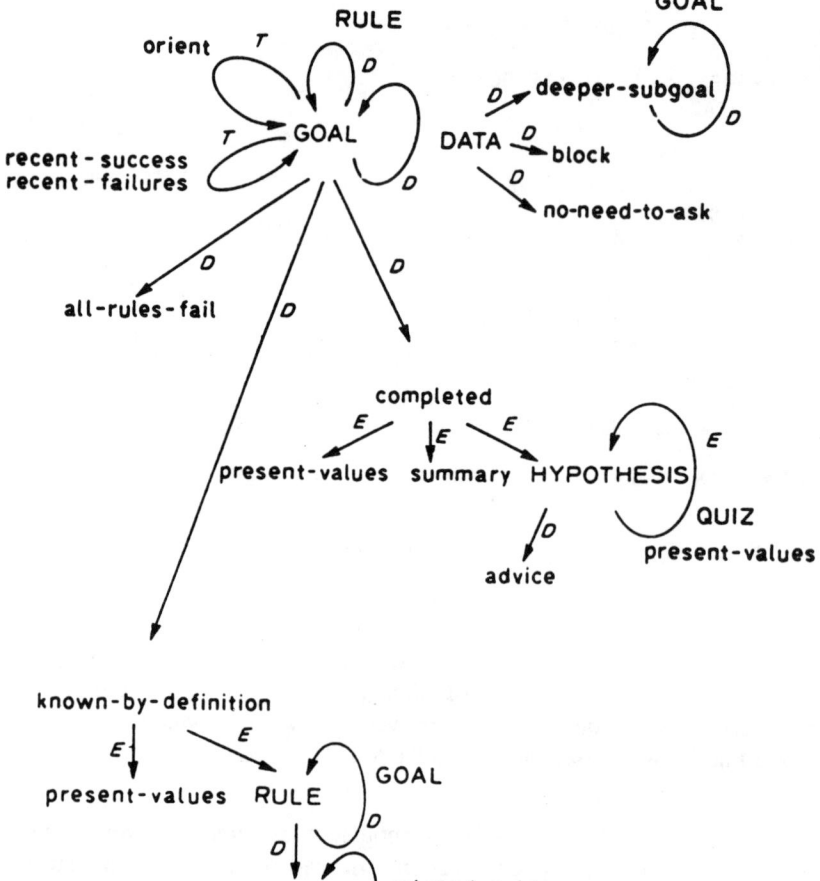

Figure 4 GUIDON's discourse management network.

Support for Dynamic Instructional Planning

The main feature that discourse management networks lack to support dynamic instructional planning is an explicit plan representation that can be generated and interpreted *during* instruction. The lack of such a representation hampers

- *Incremental planning,* because there is no explicit representation of partially specified subplans or operations, or representation of constraints on those parts of the plan that have not been elaborated to the level of primitives.
- *Plan repair or improvement by revising existing plans,* because there is no plan data structure that can be modified and then subsequently interpreted.
- *Gracefully suspending, resuming, or abandoning plans,* because suspension and resumption of *instructional* plans requires more than placing a plan on a stack and then popping it off later. Instead a coherent line of discourse must be maintained in suspending or abandoning a plan and moving to an entirely new or resumed plan. Topic transitions may need to be planned or an explanation generated for why the current line of instruction is now changing.

Sophisticated dialog management in GUIDON and MENO-TUTOR is still maintained, essentially by dynamically selecting from prestored discourse plans.[4] This dynamic selection allows reactive discourse that responds to student questions and requests, but still does not provide the flexibility and global coherence that dynamic instructional planning could provide. Two examples from GUIDON illustrate limitations that can be addressed by dynamic instructional planning. First, note that GUIDON's discussions can be verbose and lacking in global direction, because GUIDON does no planning before the dialog begins. Instead, GUIDON considers only one level of MYCIN's AND-OR goal tree at any time when controlling its dialog. Clancey [5, pp. 231–32] discussed two deficiencies that became apparent when GUIDON was applied to the nonmedical domain of SACON (structural analysis):

> First, better overviews are required. A SACON dialogue is tedious because GUIDON methodically guides the dialogue in a depth-first way, only motivating the discussion one step at a time. To convey a global sense of purpose, the program must reason about the relation between subgoals (e.g., distinguish between definitions and hierarchical subtype) and summarize subtrees.
>
> Second, the constraint on time for the session prevents treating each "context" (e.g., culture, loading, organism) with equal emphasis; some contexts must be omitted or mentioned in passing. While GUIDON has a number of methods for reducing the numbers of rules that are mentioned and has methods for mentioning them economically, it never looks down into the solution, beyond one level of the tree, to decide what topics to focus on.

GUIDON is also limited in its ability to tolerate student interruption because it lacks dynamic instructional planning. Although students can interrupt, students cannot receive new case data during the interruption, because the current (suspended) discourse procedure or tutorial rule may need to be abandoned or modified to account for the student's new knowledge. GUIDON cannot do this and so attempts to avoid situations where this problem might arise:

> Exchange of initiative in the tutorial involves complex interactions with the knowledge models and discourse procedures, given that student interruptions can disrupt a plan at any time. We minimize these interactions in GUIDON by taking the initiative for just short sequences; long lectures or explanations are more prone to student interruptions that overturn the teacher's plans. When it makes a difference, as during hypothesis evaluation, the program preserves the current dialogue context, and hence the logic behind its initiative, by not giving new case data during a student interruption. In a more flexible tutor, modeling and tutoring would occur in tandem. Specifically, in presentations of any considerable length, like hypothesis evaluation, it is desirable for the tutor to periodically stop, listen, and reevaluate its present course. ([5], pp. 89–90)

It is precisely this ability to replan dynamically that would allow dynamic instructional planning to provide greater flexibility in handling student interrupts than that provided by GUIDON.

---

[4] These prestored discourse plans are presented in GUIDON as discourse procedures. In MENO-TUTOR, they are sequences of states connected by default transitions.

The Blackboard Architecture

This section considers the blackboard architecture: its architectural formalization, two examples of planners implemented in that architecture, and then its support for dynamic instructional planning.

Architectural Formalization

The distinguishing features of a blackboard architecture are

(1) *Hierarchically structured global database.* Solution elements to the problem being solved are posted here. The hierarchical structure of the blackboard facilitates problem abstraction and the efficient triggering of knowledge sources.
(2) *Independent knowledge sources.* These are rules triggered by changes to the blackboard whose actions contribute to an evolving solution by adding or modifying solution elements on the blackboard. Knowledge sources communicate only by adding or changing the contents of the blackboard.
(3) *Agenda control.* An agenda of possible actions to perform is maintained. Activation records created by the triggering of knowledge sources are added to the agenda. A scheduler selects the next action to execute from the agenda.

Figure 5 depicts the standard blackboard interpretation cycle. Knowledge sources are triggered by changes to the blackboard. This causes new activation records to be added to the agenda, indicating possible actions corresponding to the knowledge sources. The scheduler selects the next action to perform. Its execution results in additional changes to the blackboard, possibly triggering new knowledge sources.

Two Exemplars

The BLACKBOARD INSTRUCTIONAL PLANNER [6] (abbreviated here as BB-IP) best illustrates the direct application of a blackboard architecture to construct a dynamic instructional planner. Conceptually, the problem being solved—a lesson plan[5] being incrementally constructed—is represented on a *domain blackboard.* A second blackboard, called the *control blackboard,* represents decisions about how to construct and modify

---
[5] Part of the lesson plan is a current discourse plan.

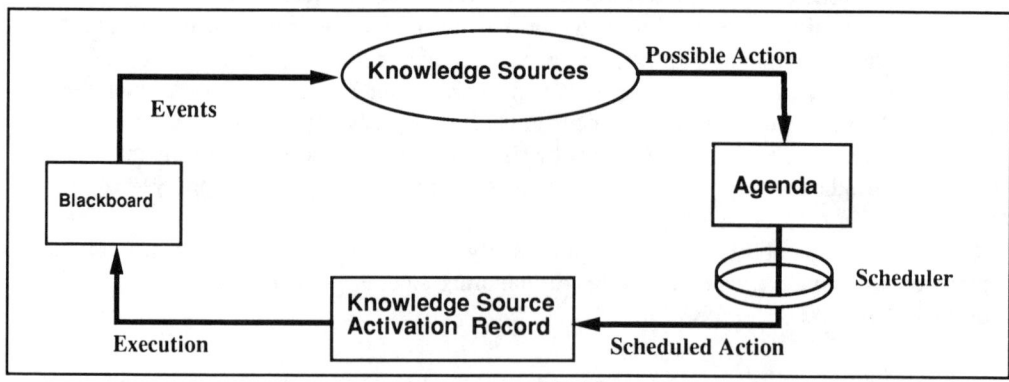

Figure 5 Blackboard interpretation cycle.

this plan. The lesson plan itself represents decisions about what instructional actions to take now and in the future. These blackboards are shown in Figure 6. There are three levels in the domain blackboard. The highest level represents a partial ordering of lesson objectives, concepts, or skills to teach. The second level represents tutorial discourse plans. Each plan consists of one or more steps that are in turn represented on the third level.

Each step influences the scheduler to select instructional actions from one particular category of instructional actions. For example, an instructional plan could first bias the scheduler to select instructional actions to motivate a topic, then actions to provide an overview of the topic, then actions to present the topic in detail, and finally actions to assess the student's understanding of the actions. There are several types of actions to present or assess a topic, including presentation of graphics or specific kinds of tests, such as a true-false test. The categories of instructional actions are shown in Figure 7.

Plan steps can direct the scheduler to favor actions from any one of these categories. The next step in a plan is selected once the current step's goal is satisfied. The goal might be that all such actions on the agenda have been executed or that the tutor's belief that the student is likely to know a lesson objective has exceeded some threshold. The control blackboard sets up default preferences to favor particular kinds of tutorial discourse plans (e.g., those that always assess a student's knowledge of a topic before discussing it). These initial preferences are altered during the tutorial session by the action of knowledge sources that monitor time remaining, the efficiency of instructional plans, student questions, and student requests.

IDE-INTERPRETER [7] is a second example of an instructional planner implemented in a blackboard architecture. The planner is not built over a general-purpose blackboard system, as BB-IP is built over BB1; instead, the planner's architecture contains the three key components of the blackboard architecture. There is an explicit representation of the lesson plan at various levels of abstraction using a tree data structure (first key component). Changes to the tree trigger rules that correspond to the knowledge sources of the blackboard architecture (second key component). The actions of these knowledge sources are to further refine parts of the lesson plan. Each rule is encapsulated in a record as a possible task to be performed on an agenda (third key component—agenda control). A scheduler, consisting of a collection of heuristics, selects the next goal expansion task.

Terminal nodes in the tree correspond to instructional units that are directly executable self-contained instructional procedures. Typically these are associated with the presentation or assessment of a particular topic. Thus this planner emphasizes lesson planning over discourse planning, because most specific discourse actions are procedurally encoded in the instructional units and not reasoned about by IDE-INTERPRETER. BB-IP differs in its equal emphasis on both lesson and discourse planning, as illustrated by its reasoning about more fine-grained discourse actions along with actions that refine or sequence lesson objectives.

Support for Dynamic Instructional Planning

Each of the three key architectural features of the blackboard architecture supports dynamic instructional planning. The hierarchically structured global data base supports an explicit hierarchically structured plan representation. This in turn supports the following functionality that facilitates dynamic instructional planning:

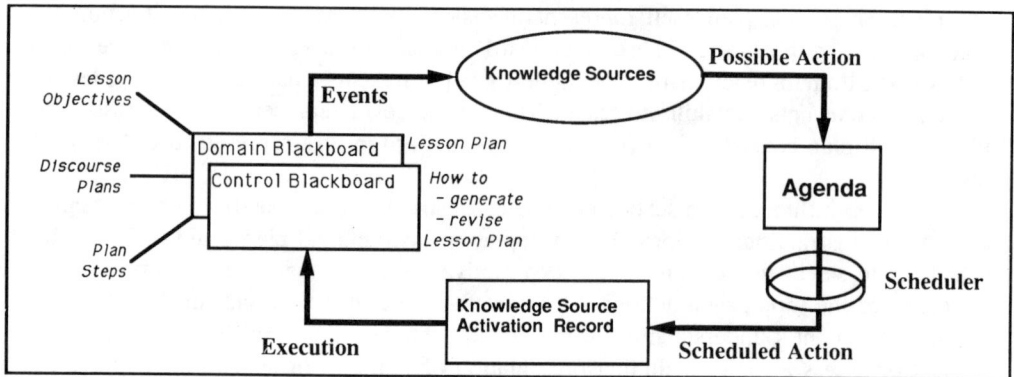

Figure 6 Domain and control blackboards of BB-IP.

- *Planning at multiple levels of abstraction.* This plan abstraction reduces the search space for an effective instructional plan and facilitates the separation of those aspects of the instructional plan related to curriculum planning, lesson planning, and discourse planning.
- *Incremental planning.* The explicit plan representation allows the representation of those parts of the plan that have not yet been refined and any constraints on partially refined plans. This allows the planner to avoid premature commitment while retaining constraints that ensure plan consistency.
- *Plan repair and plan improvement.* The current plan need not be discarded entirely if some part of it fails; instead, it can be patched to retain the remainder of the plan. If the plan is not failing, but an unexpected opportunity to improve it arises—perhaps a student question leading into planned instruction, then the plan can also be revised accordingly. Without the ability to modify only parts of the plan, either the entire plan must be discarded and another selected, or the curriculum designer must attempt to anticipate all possible plan failures or opportunities for improvement—and again the combinatorics argue against the practicality of this approach.

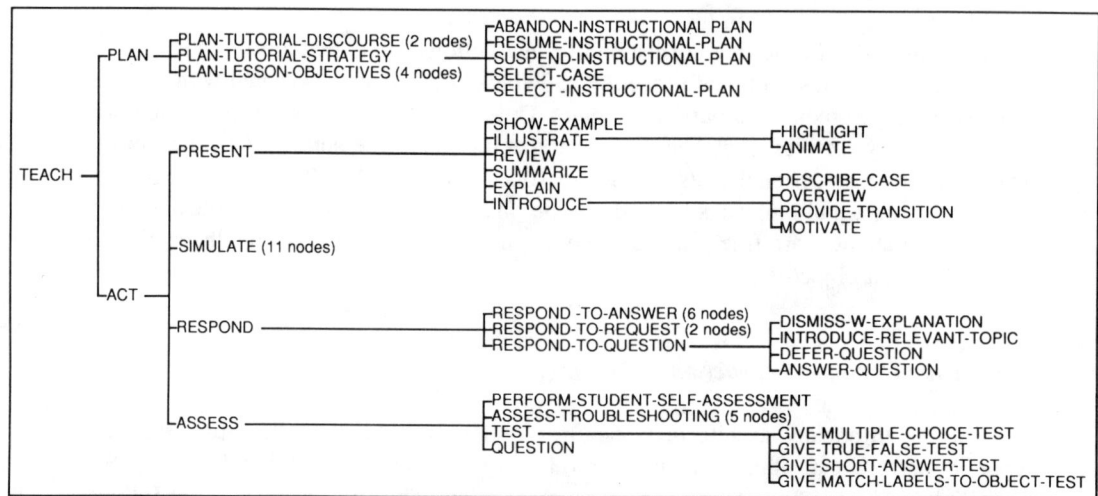

Figure 7 Categories of instructional actions for BB-IP.

- *Plan justifications.* Justifications for plan steps can be recorded and used to explain to the student, or to an instructional designer debugging a tutorial strategy, the reasons for the current instructional actions. These plan step justifications can also be used in replanning. Although plan justifications can be used for explanation in systems that only select but do not assemble, patch, or revise plans, such as GUIDON, those systems cannot use the justifications to replan.

The second key component of the blackboard architecture is the use of independent knowledge sources. This architectural feature supports dynamic planning by separating diverse kinds of knowledge while allowing their cooperation during the problem-solving process, in this case the construction of an instructional plan. The following kinds of knowledge are easily separated with this scheme.

- *Possible instructional actions.* New instructional actions, such as giving a new kind of test or explanation, can be added (or actions can be deleted) without requiring modification of existing tutorial strategies.
- *Tutorial strategies.* New tutorial strategies can be added without modification of the representation of instructional actions.
- *Planning strategies.* Different approaches to planning can be added without affecting the library of tutorial strategies or instructional actions.

Although systems based on discourse management networks separate out the first two kinds of knowledge, and thus improve on CAI where both kinds of knowledge are procedurally encoded, they do not separate out the third kind of knowledge. The second kind of knowledge can in turn be separated into different methods for curriculum planning, lesson planning (topic/skill selection and sequencing), strategic discourse planning (for preplanned presentation and assessment of topics), and reactive discourse planning (for handling student questions and requests).

Another disadvantage of DMNs is the lack of independence of knowledge sources when used in that architecture. With the extensions that Clancey has made to the basic DMN architecture, the t-rules and discourse procedures act as knowledge sources that select discourse actions, select future DMN states, and save values in registers. Such knowledge sources are not independent since they must communicate through registers, and there must be protocols for which knowledge sources set which registers and which read them. For example, some of GUIDON's t-rules set registers indicating the value of discussing particular MYCIN domain rules. Other t-rules responsible for presenting the rules must access these values. Clancey [5, p. 80] noted that

> Interactions of this kind are rare in GUIDON, but they illustrate the general problem of control when the decision about what to do is separated from its execution. The program must maintain a record of possible and/or intended actions, so that after all possibilities have been weighed, the program can return to the best plan and follow it. . . . For the moment, we must handcraft interactions between t-rule packets.

The blackboard architecture facilitates solutions to these problems: the agenda maintains a record of possible actions, the global data base records an explicit plan representation of intended actions, and knowledge sources are independent and communicate entirely through the blackboard.

The third key component of the blackboard architecture, agenda control, supports dynamic instructional planning by allowing the best instructional action (either an interactive action involving the student or further planning by the tutor) to be selected at any time with respect to possibly conflicting multiple goals. Agenda control supports global planning by allowing multiple goals to be considered simultaneously. It also allows actions to be favored that achieve more than one goal at a time. Although similar results can be obtained in theory by procedurally encoding heuristics that take into account all current goals and available actions, in practice this flexibility is difficult to obtain for complex domains. This difficulty arises from the unmanageable number of different possible sets of goals and actions that must be anticipated.

Scenario: Controlling a Programming Language Tutor

To illustrate the additional support that the blackboard architecture provides for dynamic instructional planning relative to discourse management networks, we will consider a hypothetical scenario. This scenario provides examples of how dynamic instructional planning in a blackboard architecture provides more flexible and effective instruction than a system based on discourse management networks that cannot so plan. We consider the operation of the tutor first when controlled by a dynamic instructional planner implemented in a blackboard architecture and then by a discourse management network. To consider a familiar domain, the scenario assumes that the tutor is teaching Lisp in a self-paced course. The student is called S and the tutor T below.

Scenario 1: Dynamic Instructional Planning in the Blackboard Architecture

T greets S and explains that the first order of business is to draw up a curriculum plan for the term based on the student's interests and the requirements set by the computer science department. T provides S with a brief questionnaire about his background and interests. S knows Pascal and is interested in learning Lisp for use in an AI course he is taking concurrently. He is planning on doing a term project in that class in Lisp, and is less interested in the theoretical underpinnings of Lisp (e.g., the lambda calculus) than in using Lisp for AI programming applications. T draws up a lesson plan given the inferred cognitive stereotype of the student (novice with only one Pascal course), the number of lessons expected in the course (20 lessons, each lasting about an hour to an hour and a half), and the student's interests. A different curriculum plan is drawn up for each student. Another student in a programming languages course might receive a much smaller number of lessons providing only a brief introduction to Lisp and concentrating on those unique features that distinguish it from other languages.

A typical lesson between T and S might appear as follows. T briefly reviews material covered in the last lesson and then explains today's lesson plan, which covers mapping functions and iteration. T proceeds to explain the concepts involved, relating them to analogous concepts in Pascal. T points out both the differences and similarities between iteration in Pascal and in Lisp. T proceeds to explain DO loops, but is interrupted by a request by S to explore. (The exploration mode of the tutor allows S to perform experiments of his own in the Lisp environment). T suspends its discourse plan to elaborate further on DO and lets S interact with the interpreter and editor. T monitors S during this time. S writes a simple DO to print out integers and their squares. Next S writes another DO, but this one loops forever since the exit test never becomes true. T's program analyzer, which has been checking S's DO loops for well-formedness, detects this error. T breaks the infinite loop and explains the problem. The student edits his DO, retries it, is satisfied, tries some other examples, and finally exits the exploratory mode. T has also monitored the time involved and would have reacquired the initiative if too much time had been spent exploring or too little progress appeared to result from the exploration. When T reacquires the initiative, it is satisfied that S has learned those aspects of DO that it was about to explain. Both the further explanation of DO and an assessment of the student's ability to use DO are now obviated by the student's demonstrated capability during the exploratory session. T replans to omit these parts of the lesson and then proceeds to discuss mapping constructs.

We pick up the scenario a few lessons later. Before moving onto more advanced topics, such as object-oriented programming in Lisp, T switches to a case-method style of instruction for this particular lesson. Each programming exercise tests some of the skills that S should have acquired and retained. S has only minor problems with the first two exercises but his solution of the third exercise has multiple bugs. (T uses a program analyzer such as TALUS [8] that can detect multiple bugs in student programs; a PROUST-like [9] analyzer is used when T teaches Pascal.) T plans which bugs to address and in what order. It switches to a Socratic question-and-answer discourse strategy to track down and remediate likely misconceptions about iteration indicated by the most important program bugs. T focuses on those bugs related to iteration, and quickly explains and repairs the more simple remaining bugs for the student. (T could let S repair the bugs but decides that the time is better spent addressing underlying misconceptions.) T switches to expository instruction to review those aspects of iteration the student appeared to have trouble with and then resumes case-method instruction to assess the student's capabilities in other areas covered.

At the end of this assessment and review of basic skills, T explains the student's progress so far and what remains to be learned. The remaining curriculum is reviewed, and the student is reminded that it can be revised (within limits) to accommodate his specific interests. The student expresses an interest in covering one of the optional topics (pattern matching in Lisp), because he has decided to build a theorem prover for his term project in the AI class he is taking concurrently. Since T cannot delete or compress any of the remaining lessons because they cover material mandated by the computer science department, it asks S if it is OK to add a short lesson on this. S agrees, so T inserts into the curriculum plan a lesson on using Lisp to write simple pattern matchers. S is now satisfied within the overall curriculum plan, requesting only that more time be spent on working through examples in the future.

Still later in the lesson, T is explaining object-oriented programming by focusing on an application that is used as a running example throughout the discussion. Much less expository instruction is given; more time is spent motivating concepts through examples and giving the student exercises to solve with the tutor's assistance. At times, S switches to exploratory mode, examines the case library, and requests that T solve one of the cases.[6] When this happens, T tutors opportunistically—it explains new material that arises in the case or material that it believes the student is weak on. It also replans to omit any similar cases that it had intended to cover, unless it appears necessary

---

[6] The available library of cases that the student can select from changes to include only those that the student should also be able to solve at that time.

based on its assessment of the student's understanding of the tutor's problem solving.

Scenario 2: Opportunistic Tutoring with a Discourse Management Network

Now we reconsider the same scenario assuming that *only* a discourse management network controls the tutor. In this case no individualized lesson planning can be done without recourse to mechanisms outside the discourse management network such as an agenda or auxiliary planning mechanism. Alternatively, the DMN could be extended so that some states correspond to lesson planning decisions that result in the selection of a curriculum plan from a plan library. But then states now represent both lesson planning actions and discourse management actions, undercutting the perspicuity of the DMN. Even with this extension, the curriculum planning is not nearly so customized as the blackboard tutor provides. S cannot interact with the DMN tutor T' to formulate a customized curriculum that achieves his objectives while satisfying department requirements. T' is also less capable of customizing individual lesson plans than T. Although the student can change the current topic with T', he cannot interact with T' to draw up an intended sequence of topics or cases for a particular lesson.

Let us assume that the DMN tutor is used either as a problem-solving monitor that tutors opportunistically (like GUIDON) or as a Socratic question-and-answer tutor (like MENO-TUTOR). In either case the tutor is not the primary source of instruction [10]; instead it assumes that the student is somewhat familiar with the material already, but his knowledge can be refined and is likely to have misconceptions. Thus, the tutor must be used with some auxiliary form of instruction such as a workbook (like Anderson's Lisp tutor [11]) or classroom instruction.

The DMN tutor is best suited to handle one particular class of tutorial strategy, either Socratic question-and-answer or another opportunistic approach.[7] Its ability to handle expository instruction is limited, and its inability to plan dialogue hampers it once it leaves the stereotypical discourse situations for which DMNs are best suited. When S gives the DMN tutor a program with multiple bugs, it cannot plan which bugs to address, which to ignore, and what order (and how) to discuss the bugs, because these are not local decisions (e.g., one bug may mask other more serious bugs or two similar bugs could be discussed together as resulting from the same misconception). The DMN tutor also cannot plan a series of cases along with the topics to discuss for each case to cover a particular set of topics. The ability to defer discussing one topic now because a subsequent case will better address the same topic requires foresight that the DMN does not have.

The DMN tutor does not allow the student as much freedom to express initiative as the blackboard-based tutor. The student cannot interrupt the tutor's presentation with requests that may change assumptions underlying the instructional plan currently being followed. Thus, in the earlier examples in which the student interrupted the tutor and entered an exploratory mode, or requested a change in tutorial strategy, these degrees of flexibility are not readily supported in the DMN tutor. As before, any specific example in the scenario could be handled by a special-purpose DMN, but in general the combinatorics prevent a single DMN from handling the possibility of these kinds of interruptions at any time, for any tutorial situation. The blackboard-based tutor is better equipped for

---

[7] Even though GUIDON can exhibit many different *discourse* strategies for presenting MYCIN rules or assessing the student's knowledge of these, its overall tutorial strategy is opportunistic.

these interruptions, because it has an explicit modifiable plan representation and can apply knowledge about replanning at the time the interruption occurs.

This discussion is not meant to imply that S cannot benefit from the DMN-controlled tutor T'—only that T' is quite limited compared to the blackboard-based tutor T. Most previous work in intelligent tutoring systems adopts an opportunistic tutoring framework similar to that of T', which can be well supported by discourse management networks. Many successful tutors have been built in this framework (e.g., GREATERP [11], WEST [12], WUSOR [13], and GUIDON [4],[8] while tutors such as T are still only hypothetical. The computational overhead in the blackboard architecture is also undeniably greater than that of the simpler discourse management network. However, the point of these scenarios is to illustrate that we can still build much more powerful tutors, encompassing both opportunistic and highly customized expository instruction, and that research in planners and architectures to support them, such as blackboards, is justifiable.

Summary

This paper compares the blackboard architecture to discourse management networks and concludes that the blackboard architecture better supports dynamic instructional planning. This paper also argues that dynamic instructional planning allows more flexible and effective instruction than tutors that have no plan or only a pre-stored plan. To support these arguments, the key architectural features of each architecture were compared and discussed in terms of the support or limitations they provide for dynamic instructional planning. A hypothetical scenario illustrating control by the two architectures also points out differences in capabilities.

Because discourse management networks are not sufficient in themselves to support the full range of behavior desirable of intelligent tutoring systems, especially the ability to customize lesson plans, to generate globally coherent instruction, to select the best action based on multiple conflicting goals, and to allow student initiative that may alter a tutor's current plan, research in blackboard-based dynamic instructional planners is justified. It is also clear that such blackboard-based planners are not strictly necessary to build opportunistic tutors that do not perform lesson or dialog planning and which limit student initiative. In these cases, DMNs are sufficient, and sophisticated tutors, such as GUIDON, can and have been built with that architecture.

This paper has not addressed the planning knowledge or the kind of planning required to realize a dynamic instructional planner such as the one described in the scenario above. [The planning required is an instance of a larger class of dynamic planning problems characterized by environments that are uncertain, dynamic, and multiagent. In this case the environment is

- *Incomplete*, because the tutor has limited access to the student's cognitive state
- *Uncertain*, because this access is indirect and the tutor's inferences may be wrong
- *Dynamic*, because the student's knowledge changes during and between tutorial sessions
- *Multiagent*, because the tutor and student cooperate to facilitate the student's learning,

---

[8] These tutors all adopt an opportunistic tutoring strategy, even though WEST and WUSOR act as coaches for a student who is playing a game while GUIDON and GREATERP assist students in problem solving for particular cases.

and where

- *Results of actions are uncertain,* because the tutor cannot predict with certainty the results of its actions.

To specify a blackboard architecture as preferable to one based on discourse management networks to handle these kind of planning difficulties does not provide an ITS designer with a solution to the control problem, although it is a step in this direction. *Any kind of planner can be implemented in the blackboard architecture.* Exactly what planning techniques are best in this domain, how to implement a practical dynamic instructional planner in the blackboard architecture, and the knowledge required for such planners are some of the difficult issues that remain to be addressed in future research.

Acknowledgments

I would like to thank Marshall Farr, Joe Psotka, Perry Thorndyke, and Lee Brownston for reviewing drafts of this paper. This work was sponsored jointly by the Manpower, Personnel and Training R&D Committee of the Office of the Chief of Naval Research (under contract N00014-86-C-0487); the Air Force Human Resources Laboratory; the Naval Training Systems Center; and the Naval Personnel Research and Development Center.

References

1. S. Vere, "Planning," in *Encyclopedia of Artificial Intelligence,* New York: John Wiley and Sons, Inc., 1987.
2. B. P. Woolf and D. D. McDonald, "Building a computer tutor: Design issues," *IEEE Computer,* 17:9, 61–73, 1984.
3. B. P. Woolf, "Representing complex knowledge in an intelligent machine tutor," *Computational Intelligence,* 3:1, 45–55, 1987.
4. W. Clancey, "Tutoring rules for guiding a case method dialogue," *International Journal of Man-Machine Studies,* 11, 25–49, 1979.
5. W. Clancey, *Knowledge-based tutoring,* Cambridge, MA: The MIT Press, 1987.
6. W. R. Murray, "Dynamic instructional planning in the BB1 blackboard architecture," Technical Report R-6168, FMC Corporation, August 1988.
7. D. M. Russell, "IDE: The Interpreter," in J. Psotka, D. Massey, and S. Mutter (eds.), *Intelligent tutoring systems: Lessons learned,* Hillsdale, NJ: Lawrence Erlbaum Associates, Inc., 1988.
8. W. R. Murray, "Automatic program debugging for intelligent tutoring systems," *Computational Intelligence,* 3:1, 1–16, 1987.
9. W. L. Johnson and E. Soloway, "Intention-based diagnosis of programming errors," in *Proceedings of the National Conference on Artificial Intelligence,* 162–68, 1984.
10. E. Wenger, *Artificial intelligence and tutoring systems,* Los Altos, CA: Morgan Kaufmann, 1987.
11. B. Reiser, J. Anderson, and R. Farrell, "Dynamic student modelling in a intelligent tutor for Lisp programming," in *Proceedings of the Ninth International Joint Conference on Artificial Intelligence,* 8–14, 1985.
12. R. R. Burton, and J. S. Brown, "An investigation of computer coaching for informal learning activities, *International Journal of Man-Machine Studies,* 11, 5–24, 1979.
13. J. C. Stansfield, B. Carr, and I. P. Goldstein, "Wumpus Advisor I: A first implementation of a program that tutors logical and probabilistic reasoning skills," Technical Report AI Lab Memor 381, Massachusetts Institute of Technology, 1976.

Chapter 5

Two Approaches to Simulation Composition for Training

DOUGLAS M. TOWNE
ALLEN MUNRO

Behavioral Technology Laboratories
University of Southern California
250 North Harbor Drive, Suite 309
Redondo Beach, CA 90277

> **Abstract** *Interactive graphical simulation provides the basis for an attractive approach to training operation and maintenance of complex systems, but it has demanded considerable specialized skills to produce realistic simulations and rich instructional environments. The Intelligent Maintenance Training System (IMTS) was developed to allow subject matter experts to produce interactive simulations, supported by intelligent and adaptive instructional processes, without involving programming skills. The instructional delivery functions in IMTS take care of interacting with the student, selecting problems, guiding the student when diagnostic progress stalls, and recording performance data. Intelligent instruction can be produced by a nonprogramming subject matter expert, without the labor associated with implementing of an instructional strategy, student-computer interactions, and adaptation to individual student proficiency. This chapter provides a detailed description of these functions and some directions for the future. A rich set of examples are drawn from actual work by novices and experts, constructed by attendees at two four-day seminars for potential IMTS authors.*

Interactive graphical simulation represents an attractive approach to training operation and maintenance of complex systems, but it has demanded considerable specialized skills to produce realistic simulations and rich instructional environments. The Intelligent Maintenance Training System (IMTS) was developed to allow subject matter experts to produce interactive simulations, supported by intelligent and adaptive instructional processes, without involving programming skills. An explicit objective of IMTS was to automatically generate instruction for training troubleshooting skills, based on a deep simulation model of the device [1–5].

The instructional delivery functions in IMTS take care of interacting with the student, selecting problems, guiding the student when diagnostic progress stalls, and recording performance data. Intelligent instruction can be produced by a nonprogramming subject matter expert, without the labor associated with implementing an instructional strategy, student-computer interactions, or adaptation to individual student proficiency.

The authoring environment of IMTS permits the construction of simulations based on either of two quite different approaches. In one, a deep model of the device is composed by manipulating graphic objects corresponding to the physical objects in the real device. In the other, simulations based on the behavior of the device as a whole are built up by creating tables that specify that behavior. The former approach is called *deep simulation*; the latter, *surface simulation*.

106 *Tools and Environments*

An advantage to IMTS is that it provides editors for composing both deep and surface interactive graphical simulations without using computer programming. The resulting simulations are presented in an environment that permits students to directly manipulate graphical controls and test equipments and to observe the effects of these manipulations on simulated indicators. Also, IMTS includes a generic expert that generates instruction in fault diagnosis. The instructional functions operate identically for surface and deep simulations because the data they operate on is unaffected by the manner in which the simulation is produced.

Researchers designed IMTS with a concern for productivity of authoring interactive simulations, responsiveness and accuracy of the graphic representations, and automatic generation of instructional support for the individual learner.

The Student Interface

The IMTS simulations can depict the simulated device or system in a variety of different ways, including schematically and in a front-panel format. Large simulations are divided into multiple graphic simulation scenes, each of which depicts some portion of the whole device in a graphic window. Students can bring any scene into the simulation window by selecting its name from a hierarchical list, or by selecting its scene icons from a graphical representation of the whole system.

Figure 1 shows the IMTS screen during student use. The largest window displays any one scene in the simulation. The scene shown in the figure depicts a portion of a jet aircraft engine-starting system. The window at the left can contain instructional messages to the student. The shaded window above the scene window is used by the student to store important objects from many different scenes. Because these copies are manipu-

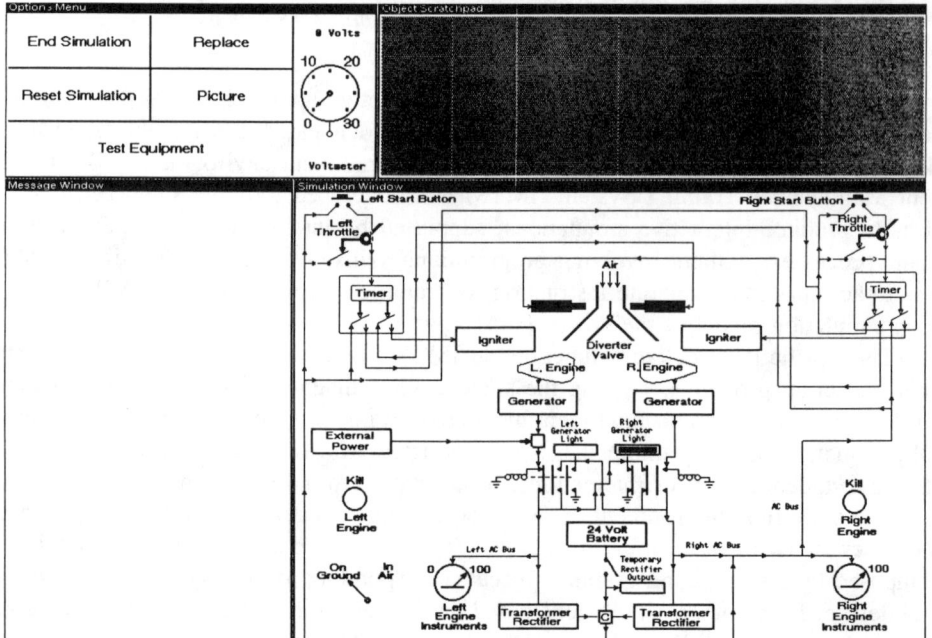

Figure 1.

Figure 2.

lable and graphically dynamic, just as are the originals, the student can minimize the need to change scenes by collecting critical parts here. Students may wish to see the physical appearances of the graphic objects in simulation scenes. The *Picture* command is used to produce such a view for a selected object. A Schematic representation of a rotor brake object (found on an IMTS simulation scene) is shown to the left of its digitized representation (Fig. 2). In surface simulations, the *Picture* button brings up a videodisk image of the object, if the simulation author has associated the object (in its current state) with a frame number on the videodisk.

The student manipulates some of simulated objects by use of the mouse. When the object, such as a switch, changes state in response to the student, it correspondingly changes its appearance, and it usually causes other objects to change their states. In addition to manipulating controls and observing simulated front-panel indicators and internal actions of objects, students can examine values at object ports using simulated test equipment. When they work with large simulations, students are sometimes discovering things about the behavior of the simulated device of which even the authors of the system were not aware. This is possible because authoring can be done at the level of component objects, with IMTS handling the emergent behavior of the entire device.

During troubleshooting training, an artificial expert on diagnostic testing called *Profile* [6–8] generates evaluative commentary and advice. The nature of this commentary and advice and how it is generated is described by Towne and Munro [2–4].

Deep Simulations

Deep simulations are based on detailed descriptions of the behavior of device component types, such as switches, relays, and lights. The deep simulation approach has two significant advantages over a table-lookup style of simulation. First, it permits more robust simulations, which respond correctly to virtually any action by a user or the simulated environment. These actions may include placing the device in any mode of operation, including erroneous modes that could harm the real device, and in transitional modes that result as a byproduct of achieving other modes. The actions may also include the injection of one or more faults, either by the user or by instructional routines.

Second, object-oriented models can be developed relatively rapidly, since the developer is not required to describe the behaviors of the system. To employ the model-based approach, the author produces a graphic diagram representing the device. This is done by selecting predefined objects from a library and positioning their graphic forms to represent the device. Behind the scenes, the functional forms of the objects are assembled into a functional model of the device. When an object is acted on, either by the user or by an adjacent object, its graphic representation changes and it propagates new values to its neighbors in the device. These objects in turn react and influence other objects.

In cases in which a new device model can be produced from objects already in the object library, the authoring function is quite straightforward. Where new objects must be produced, however, the author must understand the functions of the objects. Thus the deep simulation approach is preferred to the surface approach whenever the author has a thorough understanding of the behavior of the components of the simulated device.

IMTS Deep Model Approach

Behavior Modeled at the Element Level

Development of IMTS was heavily influenced by earlier research on a simulation-based system called STEAMER [9–11]. STEAMER demonstrated the value of giving the user access to graphical indicators—such as gauges and indicator lights—that can be used like real objects to monitor the status of the simulated device. But the underlying simulation in STEAMER is a device-dependent simulation, programmed to reflect the particular behaviors of that steam-generating system. IMTS was designed so that the entire simulation can be produced using predefined, generic objects. This approach has the advantage of permitting faster and easier simulation development, for the class of systems that can be simulated in this manner. The STEAMER approach has the advantage that it can simulate virtually any system, but at considerably higher development cost.

One other advantage to constructing the simulation of predefined objects is that models of failed components also can be created and inserted into the simulation. This failure insertion can be done by the student, to observe and learn about effects; it can be done by the instructional routines in IMTS, to set up an instructive diagnostic problem; or it can be done by the simulator if a current mode and/or failure condition causes a new failure.

Local Propagation of Effect

Every generic object has a set of behavior rules for each of its states. One rule determines when the object will transition to the state. Other rules, called *performance effects*, determine the values of the *ports* of the object in that state. Ports are points on an object that are associated with the passing of values to and from other objects. In terms of the represented world, ports are electrical, hydraulic, or mechanical connections.

When a student changes the state of a simulated control object, the object's performance effect rules determine new values for some or all of its output ports. These values are passed on to the neighboring objects, some of which may change state as a result of their new input values. These objects will, in turn, pass values on to the objects to which they are connected. In a complex simulation, hundreds of objects may be affected by a single manipulation, and thousands of port values may be recomputed.

Complex system-level behaviors are derived from simpler component-level behaviors. This permits accurate free-play simulations without requiring authoring an immense number of combinatorial effects (as did some earlier simulation training systems developed by this research group, described by Towne and colleagues [7, 12, 13]).

A major advantage of generating system behaviors from the detailed functional model is that the author is not concerned with where or when abnormal symptoms will appear in the simulated system; the effects are computed according to both connectivity and object behavior. Thus, abnormal symptoms produced by a failed part might not show up until the signal reaches a particularly sensitive indicator, or an abnormality might create other abnormal symptoms in a direction opposite to the direction of signal flow. These types of symptom propagations occur in the real world all the time and must be seen by students for them to become truly proficient.

The Bladefold Simulation

The largest deep simulation constructed thus far with IMTS is the Bladefold simulation. This is a 13-scene simulation with hundreds of specific objects. This simulation represents the bladefold/spread/brake mechanism of a military helicopter. Figure 3 displays one of the 13 Bladefold scenes.

The Bladefold system uses a complex combination of electrical, mechanical, and hydraulic parts, and it is not functionally modular. A change in one component may have

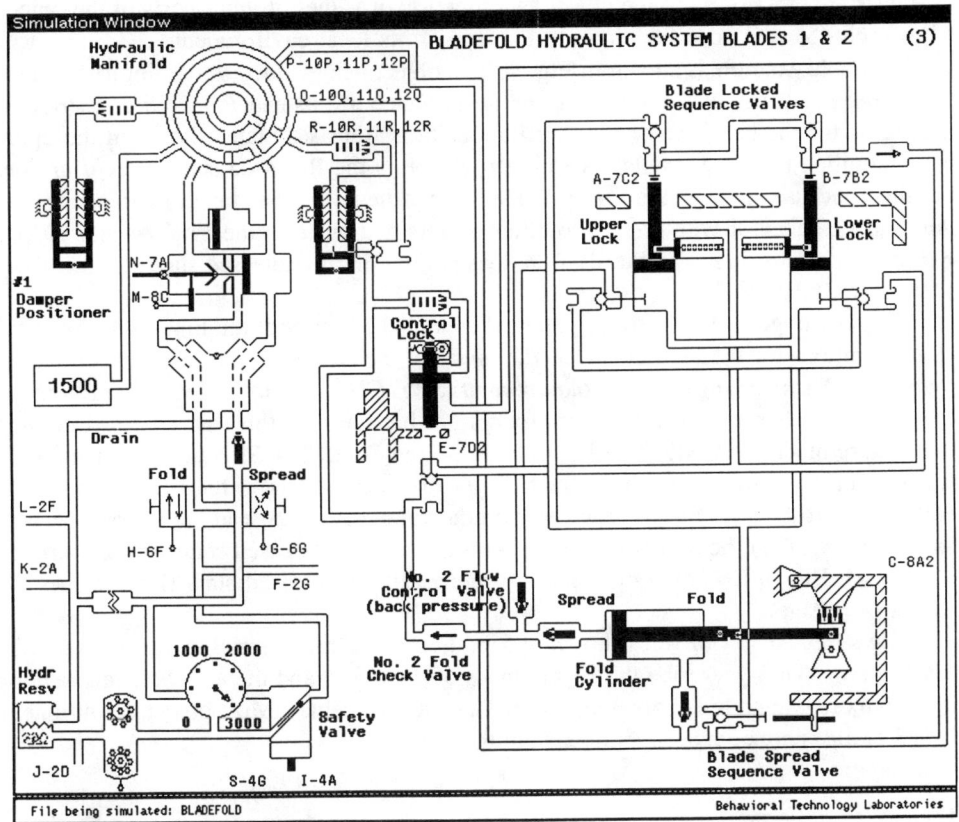

Simulation: SH-3H Helicopter blade folding system
Number of Scenes: 13
Authors: Quentin Pizzini and David Surmon (University of Southern California) with Bill Johnson (Search Technology)

Figure 3.

an effect on many other active components distributed throughout the system. Thus, in the deep simulation, a student's manipulation of a switch may result in the activations of hundreds of other objects (many of which will be on off-screen scenes, but must be processed nevertheless). Despite this complexity, performance of the simulation is quite acceptable. A worst-case time to completely simulate a response to a student action is approximately 14 seconds, which is about half as long as the real-world device response. Many graphical effects are displayed during this process, so the student is actively engaged. For a more thorough description of the Bladeford simulation, see Appendix D and F of *IMTS User's Manual* [5].

Generic Object Authoring

New simulations may require some object types not available in any library of generic objects. In these cases, authors can add new generic objects to their libraries. There are two major steps to building a new generic object. First, each of the different ways that the object can appear is drawn on the screen. Second, the behaviors for the generic object are specified in terms of the conditions under which each state exists and the resulting effects produced by the object under each condition. The effects are expressed as values at output ports of the object as a function of values at input ports of the object. Thus each object is specified entirely in terms of its local environment.

A simple example is shown in Fig. 4. The object being added to the library is called a *StartButton*. It has two graphic appearances, which are named *Released* and *Pressed*. The rules for the behavior of the StartButton in its Released state, shown in the upper rule definition box, are simple. The object will be in the Released state if the user does nothing (because the object is spring-loaded), or if the switch has failed in an open state. An object's behavior with respect to other objects is defined in the *Performance Effects* part of a behavior rule. The StartButton does nothing in the Released state (Performance Effects: NONE).

The object goes into a Pressed state if it is clicked on with the mouse. Look at the rule definitions for the Pressed state—the lower of the two behavior definitions boxes in the figure. The meaning of *State Requirement: SWITCH* is that the object will take on the Pressed state when it is clicked. This box defines behavior for this state of StartButton in two failure modes: NONE—that is, no failure—and FAILED-OPEN. If there is no malfunction in the StartButton, there are two performance effects in the Pressed state: (1) electrical port A of the StartButton takes on the value of port B, and vice versa, denoted as (A ↑ B), and (2) the StartButton then transitions to its Released state. If the StartButton is FAILED-OPEN, the object merely returns to its Released state (i.e., it does not pass any voltage).

The simple behavior rules shown in Fig. 4 describe both normal and malfunctioning StartButtons. They prescribe the appearances of the object and the effects it can have on its neighbors (which are expressed in port value assignments). Most object definitions in IMTS are as simple or even simpler than this one.

Scene Composition

IMTS simulations derive from libraries of generic objects. As the object instances are positioned in the diagram, any input/output ports close to ports on adjacent objects are automatically noted as being connected and can therefore exchange values.

In Fig. 5, an instance of the generic object called *Pipe-Horizontal* has just been

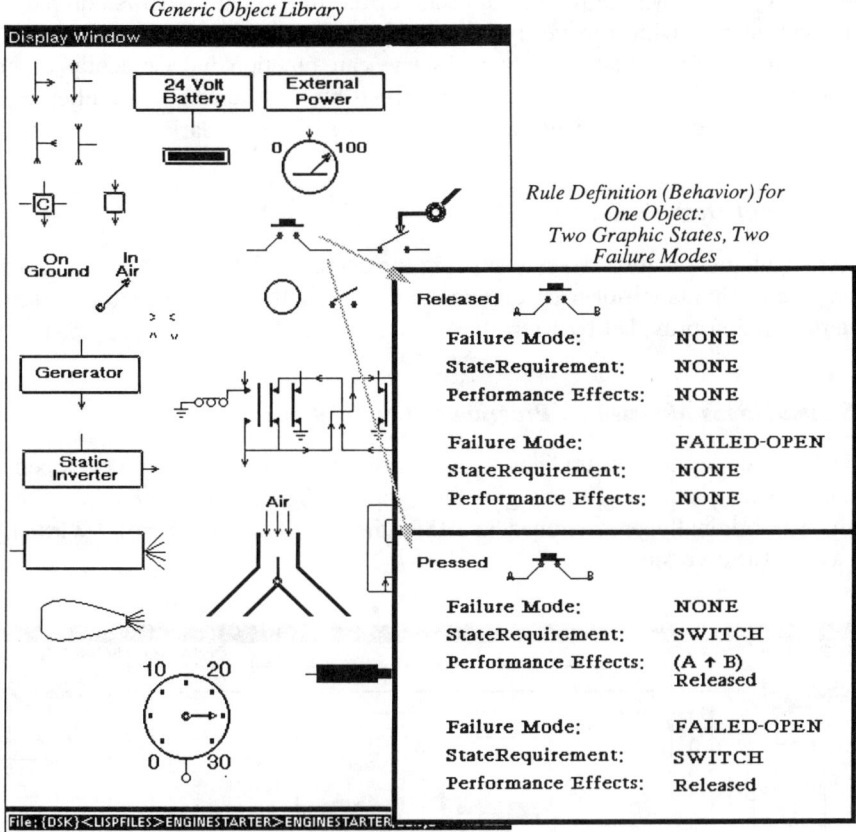

Figure 4.

created and positioned between two other objects (a function is provided that allows stretching and cutting pipes and wires to fit). The small black marks at the ends of the pipe are visual cues that its ports were automatically connected to ports on the adjacent objects. These marks do not appear in the student environment, and they can optionally be toggled on and off in the authoring environment.

A textual dialog about the currently selected object appears in a box to the left of the scene window, naming the object and its current state, along with the names of the

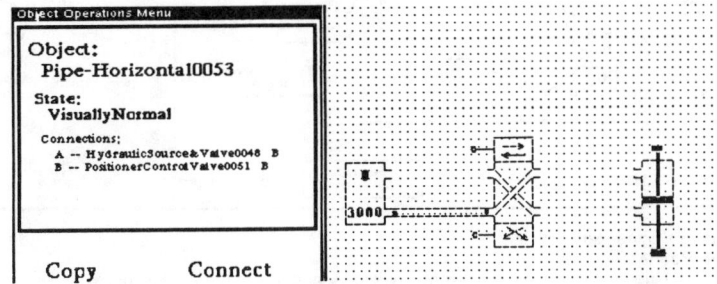

Figure 5.

connected objects. When an individual scene of the functional model is completed, the author may interact with it to verify its behaviors. The scene editor provides tools for setting input values manually, so that a scene can function independently of others. When all the individual scenes have been verified by such testing, the author connects them by identifying the port connections that cross scene boundaries.

An Example of Deep Simulation

Another training simulation developed using the deep simulation tools in our laboratory is a one-scene simulation of a jet engine starting system. This simulation, based on a device description provided by Kieras [14], is shown in Fig. 6.

IMTS Simulations Modeled on Programmed Simulations

One IMTS deep simulation imitates a simulation that was previously developed using conventional computer-programming methodologies. Figure 7, the jet engine oil cooling system, was originally programmed in LISP; the IMTS version behaves virtually the same as the LISP version.

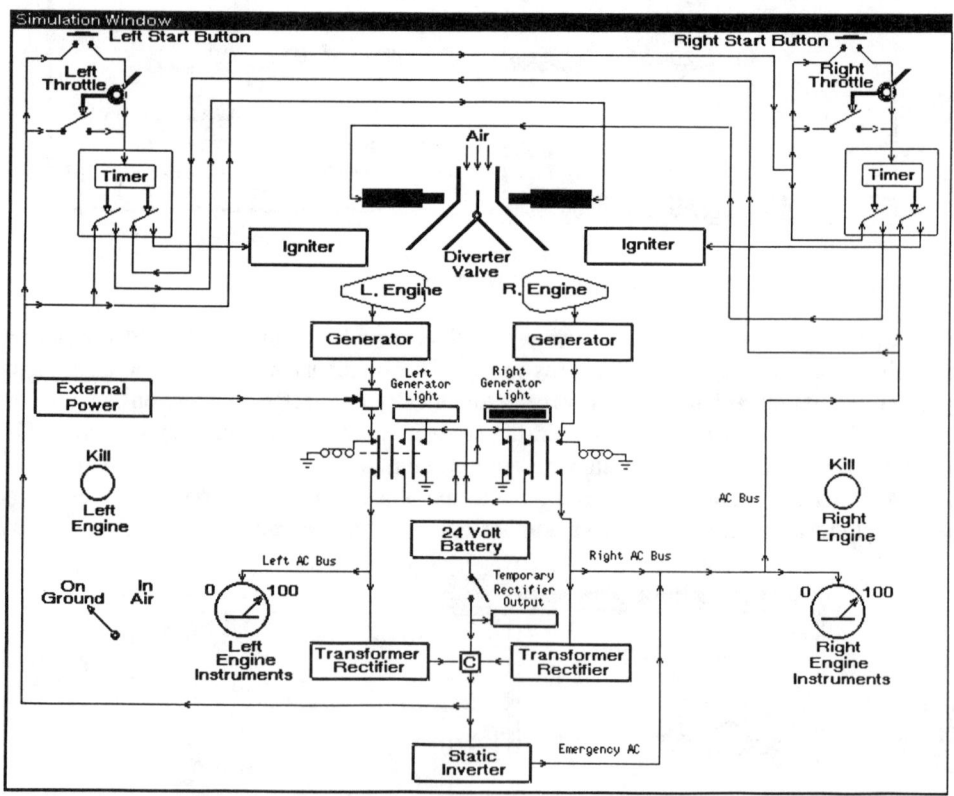

Simulation: Engine starter system
Number of Scenes: 1
Authors: Quentin Pizzini and Douglas M. Towne. Adapted from Kieras (1988)

Figure 6.

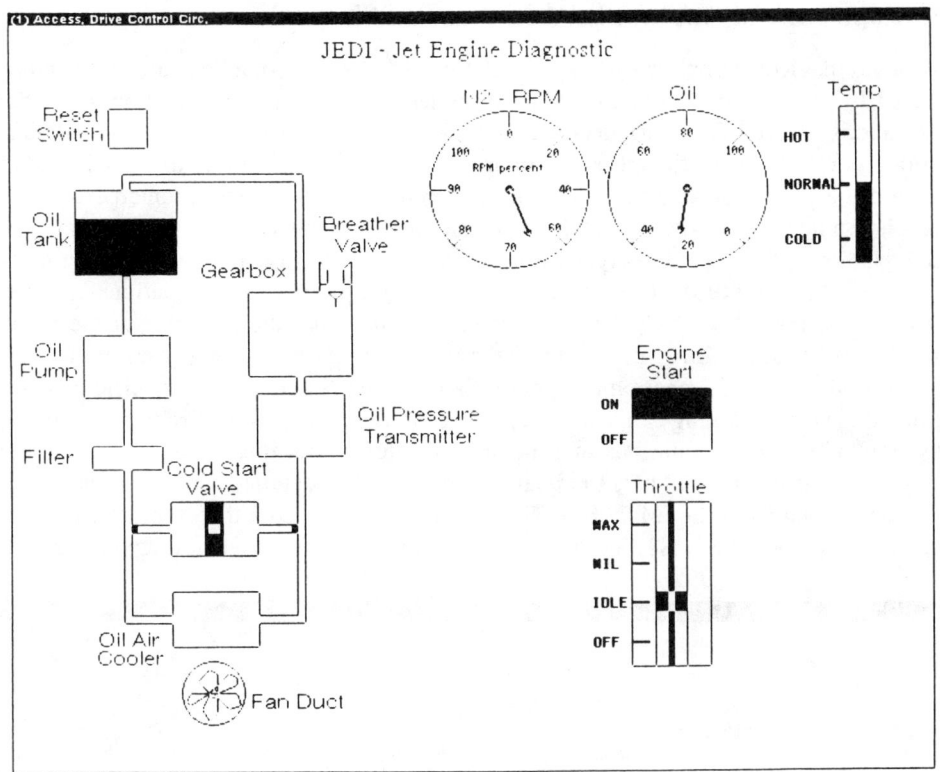

Simulation: Jet engine oil cooling system
Number of Scenes: 1
Authors: Quentin Pizzini, David Surmon, and Allen Munro.
Adapted from:
Keskey, L. C. & Sykes, D. J. Expert systems in aircraft maintenance training. In *Proceedings of the IEEE Westex Conference,* Anaheim, CA: 1987.

Figure 7.

IMTS Simulations Developed by Authors at Other Institutions

Using the IMTS authoring tools, a number of IMTS deep simulations have been developed by researchers at other institutions. Three such simulations are shown in Figs. 8–10.

IMTS Simulations Developed During Authoring Seminars

Several interesting deep simulations have been constructed by attendees at two four-day seminars for potential IMTS authors. One of these was held in December 1987; the other took place in January 1989. Three simulations built by participants during the last two days of seminars are shown in Figs. 11–13. All three simulations are functioning deep simulations. Some of these simulations were built quickly because they made use of generic objects from previously developed libraries. However, most of them also contain some new objects that were defined for that simulation. Experience suggests that both the object-definition and the scene composition tools of IMTS can be utilized productively to create new simulations after approximately two days of experience.

Development and Applications of Derivative Simulations

For many devices a complete IMTS simulation will be too complicated to be comprehended by introductory students. IMTS provides a mechanism for creating simplified simulations derived from the detailed functional model. These simplified simulations operate correctly, even though some or many of their critical parts are not included, because the parts that are shown obtain their behaviors from their counterparts in the complete functional model. This allows the simplified models to appear to operate correctly, although they could not really operate in the real world without the missing parts.

The IMTS authoring feature that supports these derivative simulations is called *yoking*. One object can be yoked to another, meaning that the behavior of the yoked object will be determined, not by the behavior rules of its own generic type and the inputs produced by adjacent objects, but rather by the behavior of the specific object to which it is yoked. Yoking can also be used to rapidly create physical representations of systems. These representations may be appropriate for training device operation and fault diagnosis using front panel indicators. In Fig. 14, a detailed functional model has been developed from the IMTS Generic Object Library. Then a three-object scene was produced by yoking S1, S2, and the indicator light to their schematic counterparts in

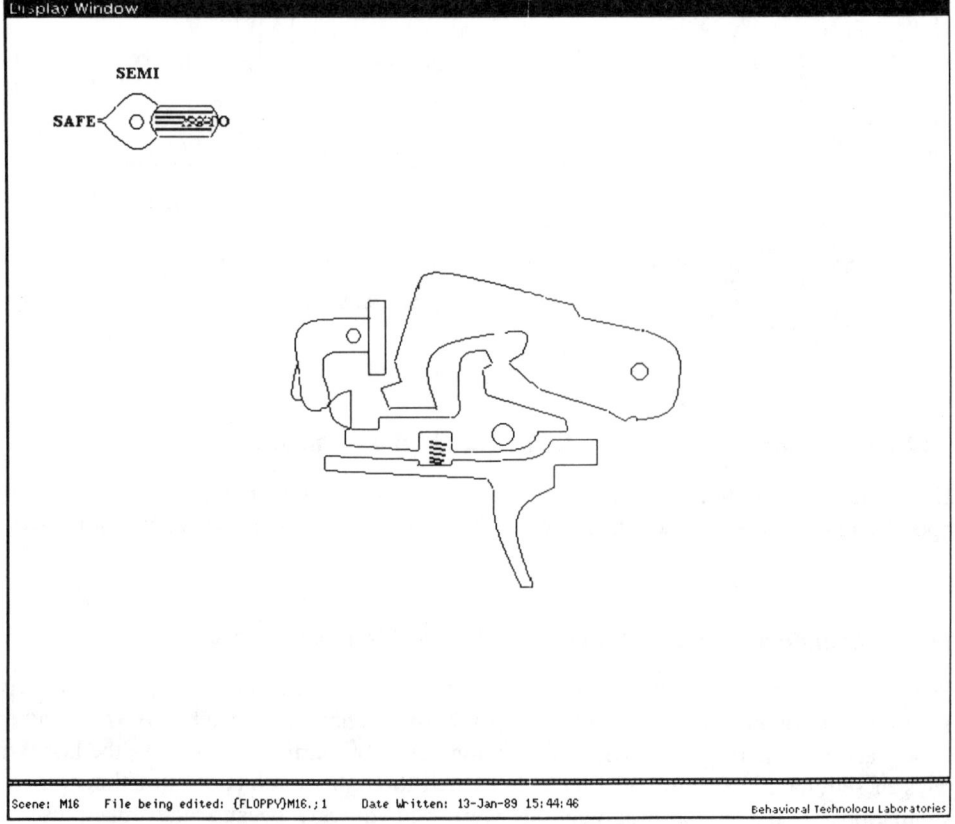

Simulation: M16 Trigger Assembly
Number of Scenes: 1
Author: Randy Morlen (Air Force Human Resources Laboratory)

Figure 8.

Simulation Composition for Training 115

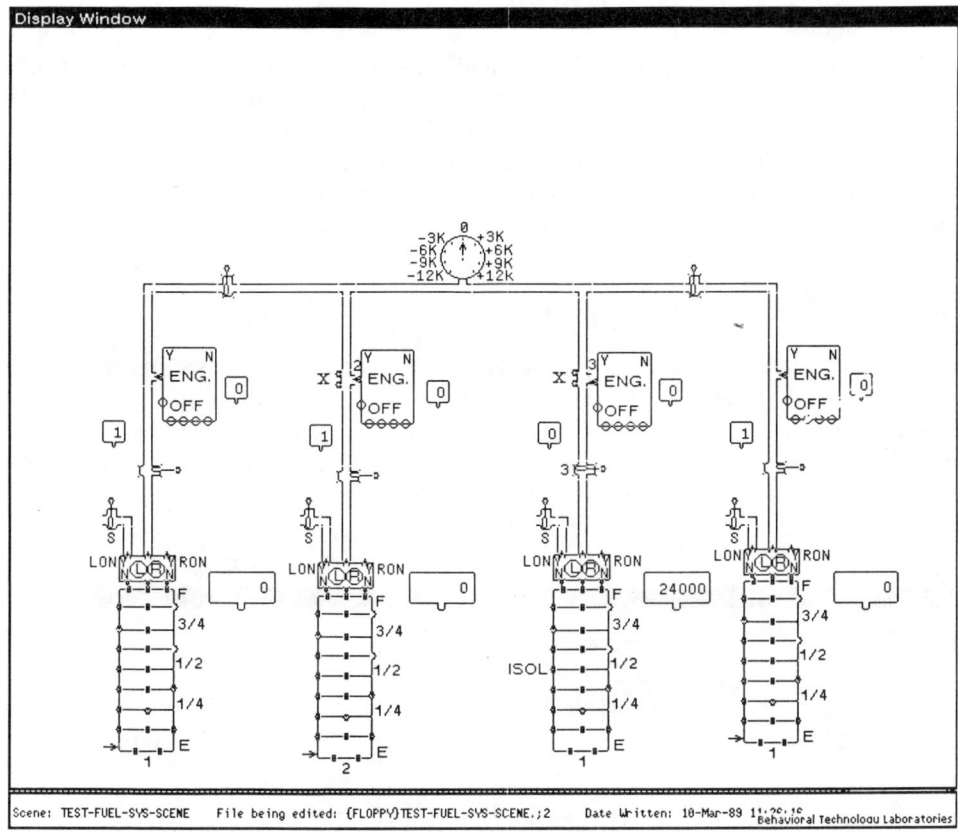

Simulation: Fuel System
Number of Scenes: 1
Authors: Mike Gick (Texas Instruments)

Figure 9.

the Detailed Functional Model (the detailed IMTS simulation model). If a student sets S1 on the front panel view, the indicator light on the front panel view will come on because S1 was automatically set in the functional model, and the indicator light there was determined to be in the ON state. When IMTS simulations are used to support training in equipment operation, this more physical form is usually desired, although the underlying schematic form may be a powerful explanatory tool.

The Bladefold helicopter simulation previously described was augmented by two researchers from the Navy Personnel Research and Development Center using this technique. They developed the scene shown in Fig. 15, which displays a front panel for the helicopter, in about two hours. Each of the lights and switches was yoked to a corresponding object in the 13-scene schematic representation. This scene is very useful to instructors and students, because it provides in one viewing place all the indicators and controls used in normal checkout procedures. Because the scene is visually and functionally similar to the control panel in the real helicopter, it can aid students in bridging the conceptual gap between the actual device and the schematic simulation scenes.

Surface Simulations

Surface simulations are developed by explicitly specifying behaviors of objects in relation to other objects in the particular device. For example, an alarm light might be specified as coming on if the main power switch is on and the high-power control setting is greater than 100. Authors use the surface simulation methodology when devices are too complex to describe in terms of the independent behaviors of their components or when the behaviors of low-level components are not central to the application.

Previous Surface Simulation Systems

The surface simulation methodology in IMTS is closely akin to that used in some earlier training systems developed at Behavioral Technology Laboratories [6, 12, 15]. The technique provides a way to specify complex system behaviors in terms of specific conditions rather than by enumerating the countless combinations of switch settings and failure states. The conditions express the settings of switches and the existence or absence of failures.

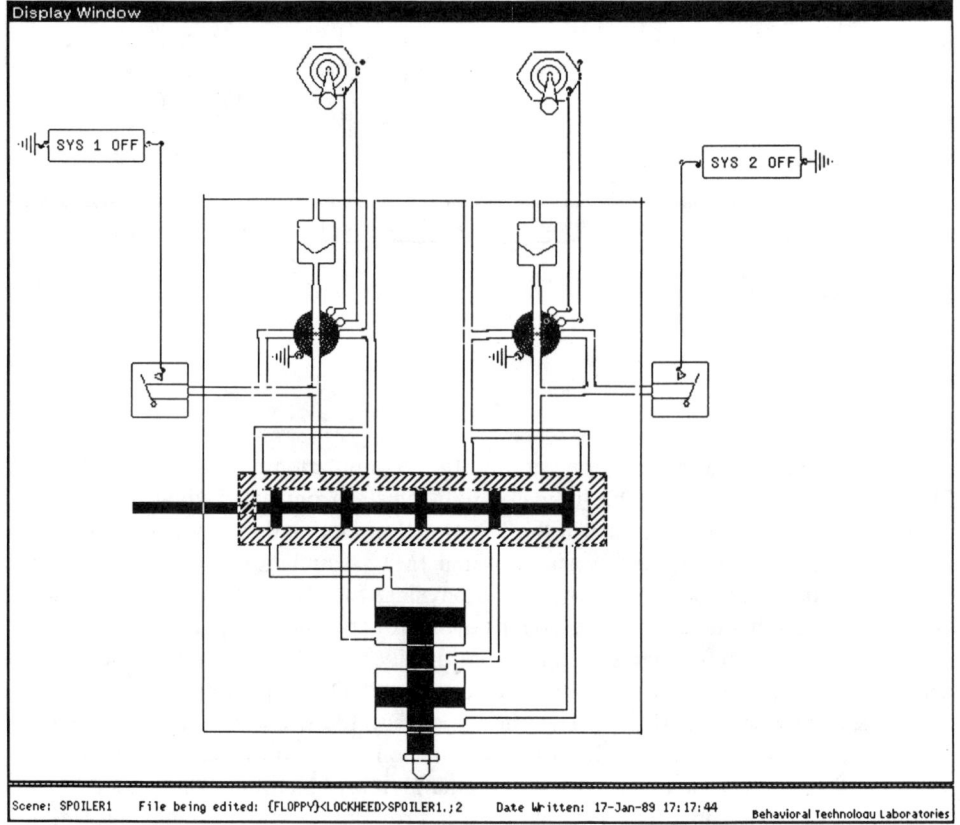

Simulation: Spoiler
Number of Scenes: 1
Authors: Tom Holzman and Bob Elm (Lockheed Georgia)

Figure 10.

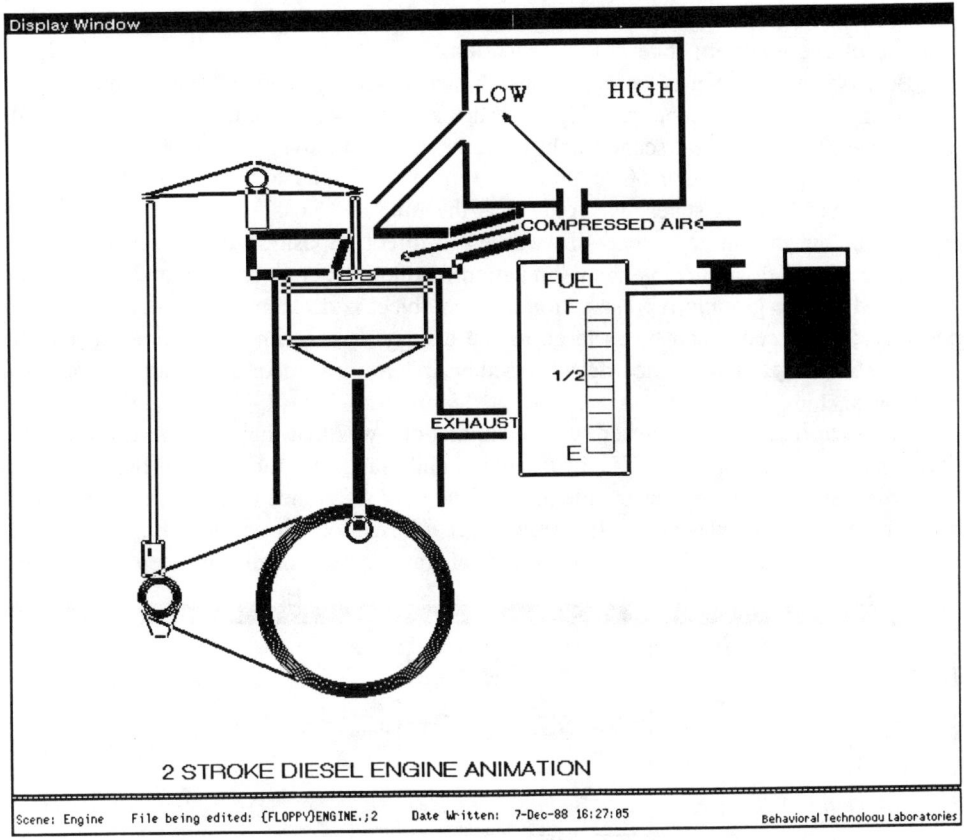

Simulation: Internal combustion engine
Number of Scenes: 1
Authors: Russ Hunt and William Johnson (Search Technology)

Figure 11.

Although the surface simulation approach is considerably less robust than the deep simulation approach, it does provide a way to simulate systems when the workings of components are either not fully understood by the simulation developer or are too complicated to justify the effort entailed. Thus, a technician having extensive field experience with a system could produce a fully accurate surface simulation by working out the various effects of front panel configurations and malfunctions on indicators and test points, even though the individual might not fully understand the functions of the individual units. Generally, the applicator prepares this specification for a modest number of different operating and maintenance modes, thereby avoiding the need to specify astronomical numbers of situations. The key limitation, therefore, is that a learner is restricted to modes of operation that have been predefined.

A second limitation of surface simulation has been overcome in IMTS. The most clumsy aspect of the Generalized Maintenance Training System (GMTS) and Electronic Equipment Maintenance Trainer (EEMT), its commercially produced version, was the medium employed to simulate the device. For GMTS this medium was randomly retrievable color microfiche images; for EEMT the medium was videodisc. Although the videodisc version retrieved and displayed images quickly and reliably, it suffered from the same limitation as the microfiche version, namely that fixed photographic images were required of every system state in the simulation.

Because the number of combinations of switch settings and indicator readings was astronomical, the only recourse in these earlier systems was to minimize the contents of each scene. Thus a typical scene might contain four to six switches and a few indicators and would require between 60 to 200 different photographs to reflect all the different combinations of object states. Not only was the time and cost to produce these images substantial, but the limited scene size worked in direct opposition to the desired goal of providing a realistic and convenient simulation for the learner to observe and manipulate.

In IMTS, the graphic representation of each object is determined and displayed independently; thus scenes can be as large as the display screen can accommodate, and the viewer of a simulation cannot detect whether it was produced using surface or deep techniques.

The final limitation of earlier surface simulations was that they embraced no instructional or diagnostic expertise. The instructional and diagnostic functions in IMTS operate for surface simulations and deep simulations. The only difference between the two, from a developmental viewpoint, is that the fault-effect information for surface simulations must be supplied by a human expert, whereas it is generated automatically for deep simulations.

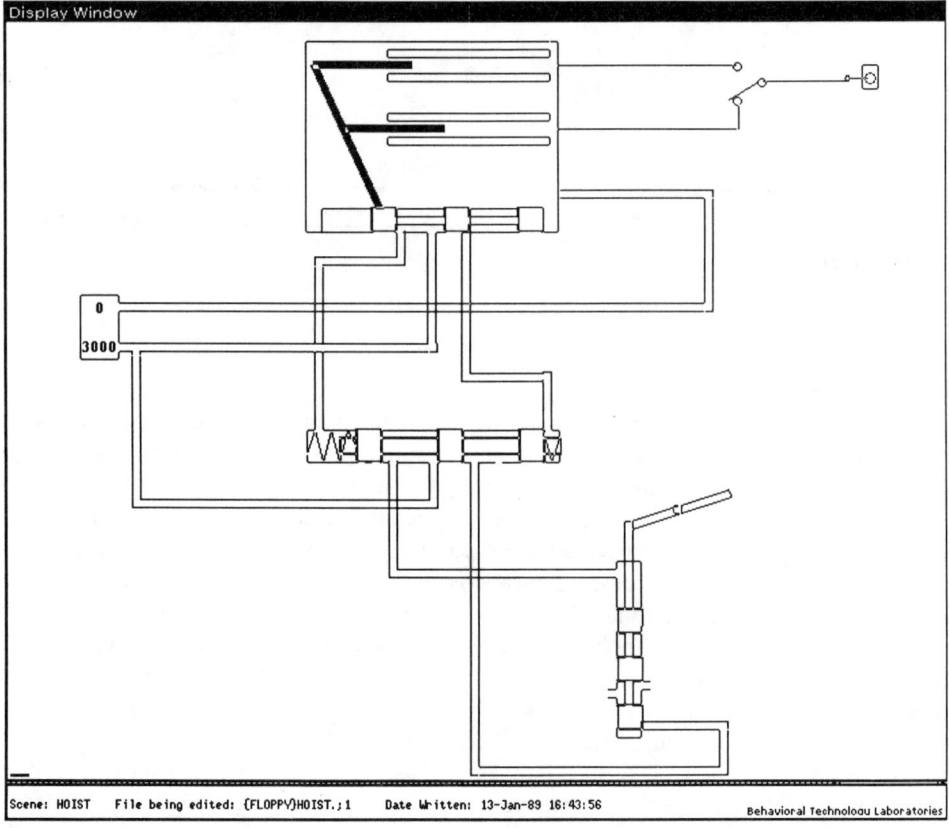

Simulation: Munition Hoist
Number of Scenes: 1
Author: William Murray (FMC)

Figure 12.

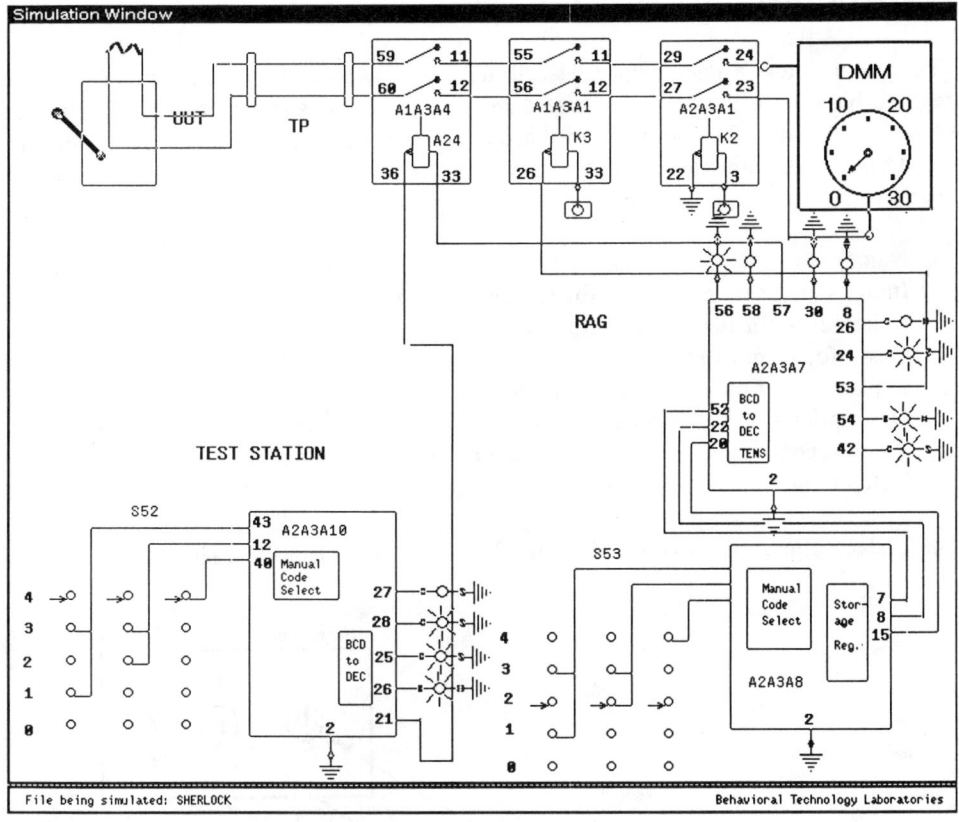

Simulation: F-15 Manual Avionics Test Station
Number of Scenes: 1
Author: Marilyn Bunzo (Learning Research and Development Center, University of Pittsburgh)

Figure 13.

Overview of Surface Simulation Authoring

The concept of yoking, used as a convenience in deep model building, is central to constructing surface models. The graphical objects in surface simulations are positioned into scenes just like deep model scenes. Their changes in appearance, however, are determined from *surface objects,* which are simply data records to which they are yoked. The only function of these surface objects is to control the appearance of the objects in the surface scene (i.e., they cannot be used in scenes themselves, as they have no graphic component).

Four editors are used to define surface simulation behaviors (Fig. 16). The *surface object editor* is used to create the data objects just described. The *mode editor* is used to define equipment modes of interest in terms of the states of specific objects, such as switches. The *test editor* is used to define tests, which are indicators and/or test points viewed in particular modes. The *surface matrix editor* is used to declare what values a test should exhibit in various failure conditions.

Surface Object Data

The author enters a data record describing every switch, indicator, jumper, or other replaceable object in a device. These data records are the surface objects. The graphic objects derive their behaviors from these data records and from others that specify the important modes and tests for the device. The data record includes the following attributes:

 Name
 Initial state—set at run time for test points and indicators
 Reliability—for diagnostic guidance
 Cost—for diagnostic guidance
 Replacement time—for diagnostic guidance
 Description—used for discussing with the student
 Video scene—name of the video scene on which the object appears
 Default value

The surface object editor (Fig. 17) is used to enter such data for a simulation.

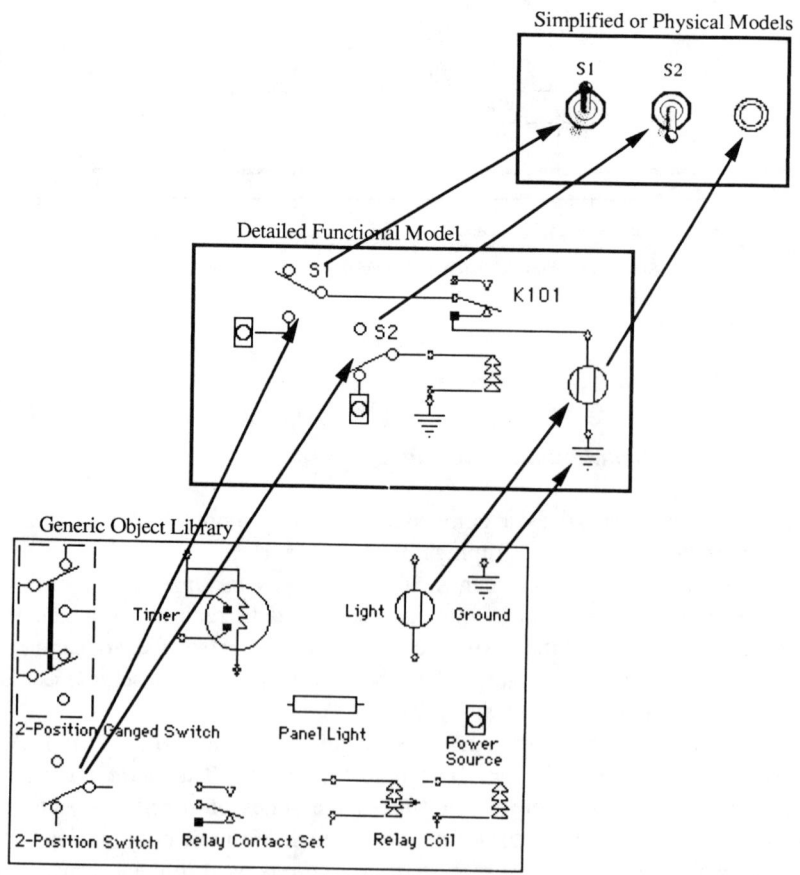

Figure 14.

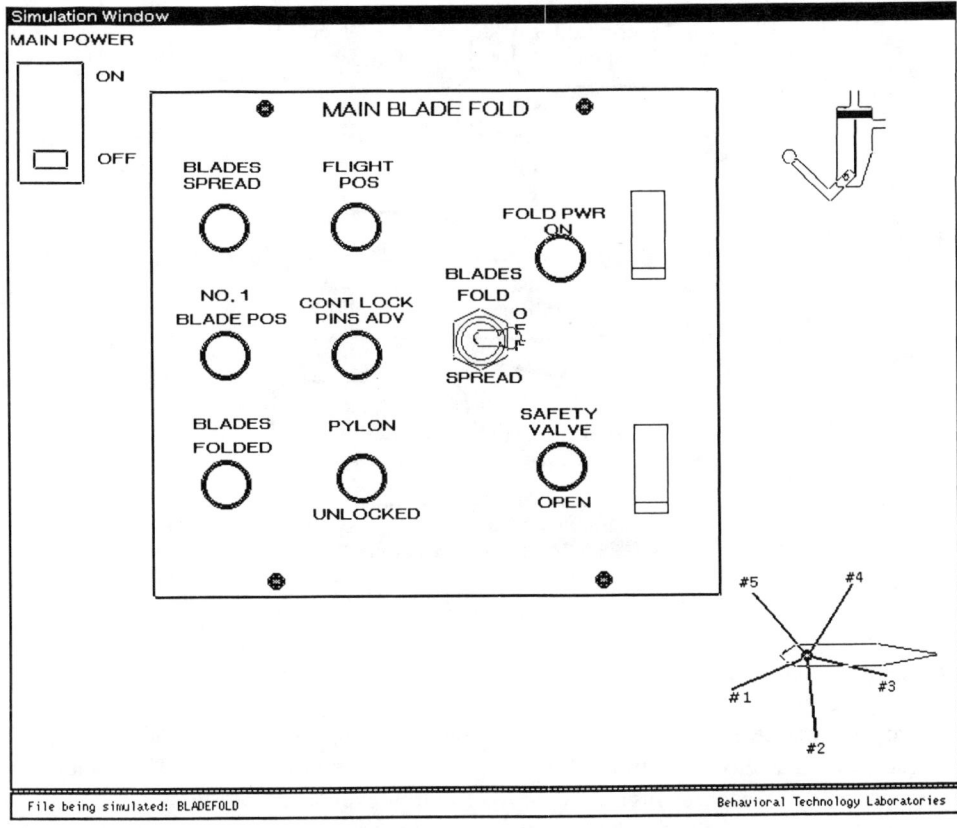

Simulation: SH-3H Helicopter blade folding system — 14th scene
Number of Scenes: 1
Authors: Vern Malec and Mike Cowan (Navy Personnel Research and Development Center).

Figure 15.

Modes and Tests

Authors define the *modes* in which a surface simulation will operate (Fig. 18). A mode is a combination of object states. For example, the power switch being set to *on* is a simple mode of interest. A more complex mode is power *on* and standby *off* and number 1 engine switch set to *start*. Mode information is edited with the mode editor.

A surface matrix editor is used to specify how indicators and test points behave based on the defined modes in a number of malfunction states. For each mode that is relevant for an indicator, the value or state of the indicator is specified for each malfunction. The fault-effect matrix in Fig. 19 is identical to the one generated automatically for deep simulations. The only difference is that in surface simulations the symptom data must be supplied by a human expert.

During a simulation, when the student changes a switch, all the modes that refer to that switch check to see whether their truth has changed. Certain modes might become true as a result of the switch manipulation. For each mode that changes, all of the indicators that refer to that mode are updated. If an indicator's state changes, then the graphic object that represents that indicator is redisplayed in the new state.

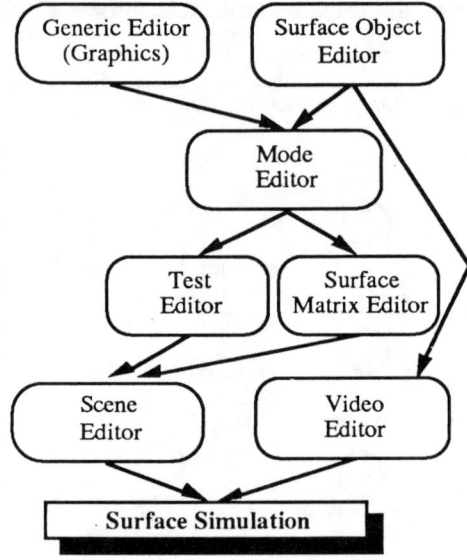

Figure 16.

The WSC-3 Simulation

The largest surface simulation constructed thus far with IMTS is of the AN/WSC-3 satellite communication system used for voice and data communication. This simulation is comprised of 28 scenes. Figure 20 shows the top-level scene of the system. Most of the graphic objects in the top-level scene are icons that lead to more detailed views. When the student clicks on the Antenna Control Unit in the scene from Fig. 20, for example, IMTS displays the close-up view shown in Fig. 21, reflecting the current settings of the switches on this unit.

This scene illustrates the combinatorial limitation of using media such as videodiscs for interactive simulation. This scene contains 11 objects that exhibit 2 graphic states

Figure 17.

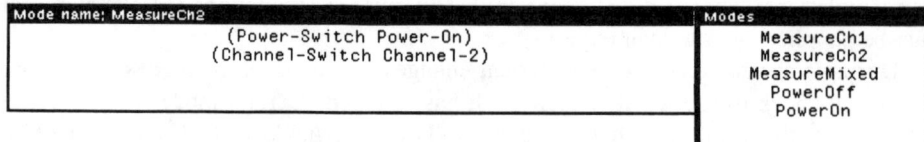

Figure 18.

Simulation Composition for Training   123

Tests \ **RUs**	PowerLightOn	Level	Select state		evelMixe
Power-Lig Burned-Ou	Off	NI	Normal Value LessZero Zero EightyFive SeventyFive Ten GreaterTh100		NIL
Power-Swi Failed-Ope	Off	NI			NIL
Normal	On	SeventyFive		Ten	EightyFive

Figure 19.

each, an Elevation Control with 13 discrete states, and an Azimuth control with 36 states. Theoretically, this scene could exist in 958,464 different states ($2^{11} \times 13 \times 36$). Of course, many of these states can only exist if there is a malfunction. Fully one half of the states, for example, involve lights being on with the Power switch off. This can happen if the Power switch is shorted to pass current in the off position. The deep simulation approach described earlier can accommodate this immense combinatorial situation easily. Any of the nearly one million states that can occur will be produced

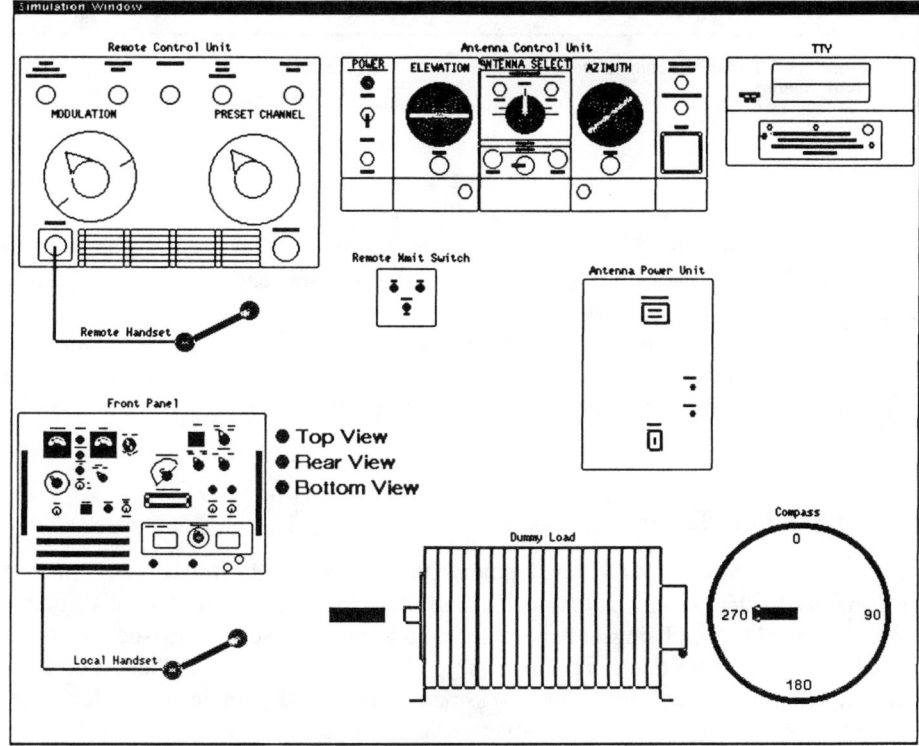

Figure 20.

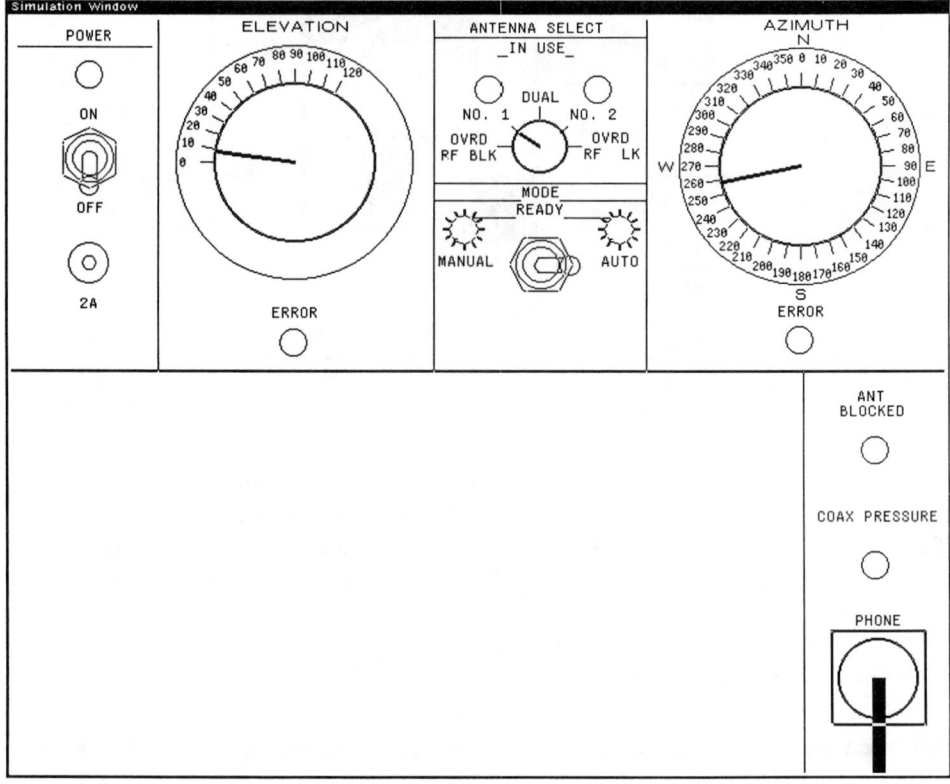

Simulation: WSC-3 Satellite communication system
Number of Scenes: 28
Author: Lee Coller (University of Southern California); data provided by Ron Renfro (Mantech Mathetics).

Figure 21.

when the conditions dictate. The surface simulation developed for this unit happens to describe only 256 states, but even this limited number of different appearances presents a clerical chore when using videodisc display.

Although videodisc is not a convenient medium with which to represent panels of equipments in each of the possible modes, it is quite attractive as a means to display test equipment readings. When the student performs an oscilloscope reading on the WSC-3, IMTS displays the waveform image from videodisc.

Expanded Training Range

The objectives for future IMTS development are to expand the range of diagnostic instruction and to add new instructional capabilities for nondiagnostic skills. In the diagnostic arena IMTS is being enhanced to include part-task exercises such as fault-effect drills (Where could this fault show up?); possible-cause drills (What could have caused this symptom?); and symptom-assessment drills (Is this indicator or test-point reading normal?). New functions in nondiagnostic environments will provide instruction in areas such as equipment familiarization, theory of operation, and performance of procedures.

These expansions of instructional range will be implemented by allowing the author to manipulate the IMTS simulation in a demonstration mode that can be played back by individual students at their own pace. In this authoring mode, the expert can explain why various actions are performed, how they are performed, and why the device responds as it does. In addition, the author can pose questions for the learner to address. Our objectives continue to stress the automatic handling of low-level instructional interactions, so that the authoring process is similar to that of personally demonstrating and explaining technical matter to an individual student, rather than to computer programming.

Acknowledgments

The work described here was sponsored by the Cognitive Science Program and the Perceptual Science Division of the Office of Naval Research, by the Navy Personnel Research and Development Center, and by the Air Force Human Resources Laboratory under ONR contracts N00014-85-C-0040, N00014-86-K-0793, and N00014-87-C-0489.

References

1. D. M. Towne, A. Munro, Q. A. Pizzini, D. S. Surmon, and W. B. Johnson, *Development of intelligent maintenance training technology* (Technical Report No. 106), Los Angeles: Behavioral Technology Laboratories, University of Southern California, May 1985.
2. D. M. Towne, A. Munro, Q. A. Pizzini, and D. S. Surmon, *Representing system behaviors and expert behaviors for intelligent tutoring* (Technical Report No. 108), Los Angeles: Behavioral Technology Laboratories, University of Southern California, February 1987.
3. D. M. Towne, A. Munro, Q. A. Pizzini, D. S. Surmon, and J. Wogulis, *ONR final report: Intelligent maintenance training technology* (Technical Report No. 110), Los Angeles: Behavioral Technology Laboratories, University of Southern California, March 1988.
4. D. M. Towne and A. Munro, "The intelligent maintenance training system," in J. Psotka, L. D. Massey, and S. A. Mutter (eds.), *Intelligent tutoring systems: Lessons learned*, Hillsdale, NJ: Lawrence Erlbaum, 1988.
5. A. Munro and D. M. Towne, *IMTS user's manual*, Los Angeles: Behavioral Technology Laboratories, University of Southern California, February 1989.
6. D. M. Towne, "A generalized model of fault-isolation performance," *Artificial Intelligence in Maintenance: Proceedings of the Joint Services Workshop*, Boulder, CO, 1984.
7. D. M. Towne, "A generic expert diagnostician," in *Proceedings of the Air Force Workshop on Artificial Intelligence Applications for Integrated Diagnostics*, Boulder, CO, 1986.
8. D. M. Towne and M. C. Johnson, *Research on computer-aided design for maintainability* (Technical Report No. 109), Los Angeles: Behavioral Technology Laboratories, University of Southern California, February 1987.
9. M. D. Williams, J. D. Hollan, and A. L. Stevens, "An overview of STEAMER: An advanced computer-assisted instruction system for propulsion engineering," *Behavior Research Methods and Instrumentation*, 13:85–90, 1981.
10. J. D. Hollan, "STEAMER: An overview with implications for AI applications in other domains," presented at the Joint Services Workshop on Artificial Intelligence in Maintenance, Institute of Cognitive Science, Boulder, Colorado, October 4–6, 1983.
11. J. D. Hollan, E. L. Hutchins, and L. Weitzman, "STEAMER: An interactive inspectable simulation-based training system," *The AI Magazine*, 2:15–27, 1984.
12. D. M. Towne and A. Munro, *Generalized maintenance trainer simulator: Development of hardware and software* (NPRDC Technical Report No. 81-9), San Diego: Navy Personnel Research and Development Center, 1981.
13. D. M. Towne, A. Munro, M. C. Johnson, and G. F. Lahey, *Generalized maintenance trainer*

simulator: Test and evaluation in the laboratory environment (NPRDC Technical Report 83-28), San Diego: Navy Personnel Research and Development Center, August 1983.
14. D. E. Kieras, "What mental model should be taught: Choosing instructional content for complex engineered systems," in J. Psotka, L. D. Massey, and S. Mutter (eds.), *Intelligent tutoring systems: Lessons learned,* Hillsdale, NJ: Lawrence Erlbaum, 1988.
15. D. M. Towne and A. Munro, *Preliminary design of the advanced ESAS system* (Technical Report No. 105), Los Angeles: Behavioral Technology Laboratories, University of Southern California, December 1984.

Chapter 6

Integrating Intelligent Tutoring, Computer-Based Training, and Interactive Video in a Prototype Maintenance Trainer

MARK L. MILLER

Apple Computer, Inc.
20525 Mariani Avenue, MS: 76-3A
Cupertino, CA 95014

SCOTT R. LUCADO

Lucado and Associates
7362 East Beckwood Drive
Fort Worth, TX 76112

> **Abstract** *An investigation is reported into the feasibility of integrating intelligent tutoring systems, conventional computer-based training, and interactive video into a hybrid instructional technology operable on personal computers. A 6-month effort resulted in a demonstrable prototype, combining a rule-based expert system with an interactive video courseware module, for teaching Army recruits to troubleshoot M16 rifles. The integration of intelligent tutoring with computer-based interactive video instruction, deliverable on inexpensive platforms, could ultimately revolutionize maintenance training. Further research—to explore more complete design examples and to investigate alternative architectures for authoring environments—seems warranted.*

Introduction

The Problem

Western economies face an alarming yet ever-widening gap between the complexity of job tasks and the skill levels of the individuals expected to perform those tasks. This chapter reports on an investigation into new technologies, such as intelligent tutoring systems (ITSs), for addressing this skills gap in the specific context of the United States Army. Although the Army was once highly personnel-intensive, it has become far more asset-intensive during recent decades. Equipment and weapons systems have become more elaborate and more expensive, requiring greater skills to operate, maintain, troubleshoot, and repair. By contrast, the armed services face difficulty in attracting and retaining qualified technical personnel. Many new recruits need remediation in basic skills such as reading and algebra.

The most disturbing effect of this gap between technological complexity and personnel capability is reduced operational readiness. Properly functioning modules are returned to depots for repair. Depots send improperly functioning units back to the field for deployment. More and more, the services depend on civilian experts and instructors, who command higher pay scales yet tend to be less familiar with military doctrine and procedures. Conversely, the availability of excellent technical training appears to be a key factor in recruiting high caliber individuals.

Although this investigation was performed in an Army context, the skills gap plagues our entire society. These and other considerations argue for creating and exploiting the most advanced instructional technologies that can be delivered cost-effectively and high volume. Unfortunately, communication between the (ITS) community and the traditional instructional design (ISD) community responsible for developing most current computer-based training (CBT) has been limited. Only a handful of courseware authors are conversant with ITS or related technologies, even though these practitioners are the most likely source of talent for future ITS development. It is hoped that the work reported here will help to stimulate dialogue between the two communities.

Potential Impact of Intelligent Tutoring Systems

One approach that has captured the imagination of researchers and educators is through ITS (also termed *intelligent computer-aided instruction*). The goal of ITS research is to provide a new plateau of instructional capability by integrating artificial intelligence (AI) techniques, such as expert systems (ES), into advanced learning aids. This goal includes incorporating task knowledge, culled from human subject matter experts, as well as insight into student reasoning, skill acquisition, and tutorial strategies, gleaned from skillful human instructors. Figure 1 shows the major modules in typical ITS architectures [1].

Unfortunately, because of the complexity of authoring, most systems have been weak in all but one or two of these modules, leading to a widespread misunderstanding that ITSs ignore fundamental cognitive principles (such as putting the learner in control and encouraging active exploration). Although a series of efforts, from SCHOLAR [2]

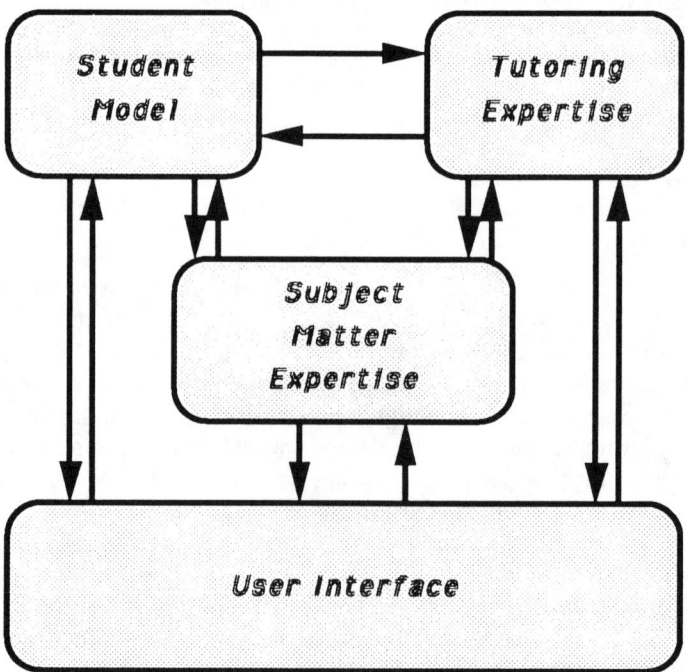

Figure 1. The major modules in an intelligent tutoring system.

through SOPHIE [3] to current projects—such as Bolt, Beranek, and Newman's HAWK Radar Tutor [4]—have moved us ever closer to the original goals, ITS research to date has lacked the impetus that ES technology gained from a few landmark success stories, such as DEC's famous XCON system [5]. Moreover, the most impressive prototype systems constructed to date have operated primarily on large computers (such as the DECSYSTEM 2060), or expensive, dedicated LISP workstations (such as the SYMBOLICS 3670). For the technology to achieve its potential, a more practical approach leading to an incontrovertibly successful application is needed.

Potential Impact of Computer-Based Training Using Interactive Video

Even on the most expensive equipment, ITS researchers have tended to settle for minimal graphics with less-than-convincing realism. Many ITS systems are entirely text-based; but, especially in applications where the target population lacks skill in reading comprehension, a purely text-based user interface is unacceptable. By contrast, sophisticated CBT authoring systems have been developing around interactive videodisc (IVD) technology, providing a range of multimedia options including full motion video, high resolution still-frames, audio output including speech, and graphics-over-video capability.

Unfortunately, interactive video—despite its random access to massive quantities of visual information—has too often been used as if it were merely a substitute for videotape. Typical CBT courseware incorporates lengthy sequences of linear video, interspersed with a lock-step, multiple-choice quiz format, leaving students bored and frustrated. All but the most sophisticated authoring tools force developers into rigid, frame-based, exhaustive branching structures, the very antithesis of the ITS goal. What is needed, then, is to integrate the best of intelligent tutoring, CBT, and interactive video, starting from sound cognitive principles (such as active involvement of the learner), and to deliver this powerful combination cost-effectively.

Relation to Prior Work

Although a thorough literature review is beyond the scope of this chapter, readers with varying backgrounds may seek entry points for further research. Two primary threads influenced the project. The early ITS background is best summarized in a collection by Sleeman and Brown [6]; more recent work is reported in a collection by Psotka, Massey, and Mutter [7]. The systems approach to training (SAT), prevalent in the CBT community, grew out of early work on such systems as PLATO and TICCIT [8].[1] Fletcher [9] has been a consistent proponent of ITS, CBT, and IVD technologies for enhancing federal and corporate training.

Technical Approach

Target Population

The target population for the training system was selected to be broad-based and typical of the Army's overall technical training requirements. Every new recruit undergoes M16

---

[1] The terminology, instructional systems design (ISD), may be more familiar to some readers. The U.S. Army Research Institute recommends the alternative phrase, systems approach to training (SAT).

Riflery Qualification, regardless of later specialization. Hence, educational backgrounds span the spectrum from those lacking a high school diploma to those with advanced degrees. Basic skills such as English literacy and vocabulary can be crucial issues in the design of training systems for this audience. Without such multimedia capabilities as interactive video, CBT would be completely inaccessible for large segments of this target population.

Selection of Content Example

There were several reasons for selecting M16A1 rifle troubleshooting as the content example for our preliminary investigation. Training soldiers on this widely used small arm had obvious relevance in the larger context of the funding organization. However, current generation ES technology seemed to lend itself better to troubleshooting than marksmanship. Also, this rifle, viewed as an example device for maintenance training, is both simple enough to analyze thoroughly, yet complex enough to require nontrivial knowledge and diagnostic reasoning. This was estimated by considering such factors as the number and relationships of its parts, the length of time devoted to each topic in basic training, and the style and format of available maintenance manuals. Finally, the Army was able to provide existing videodiscs illustrating a variety of information about the M16 rifle family, thereby enabling us to quickly incorporate both motion video and realistic still-frames.

Target Hardware

One key objective of the investigation was to determine the feasibility of delivering ITS-based maintenance training, integrated with interactive video, on inexpensive, widely available hardware. To this end, a personal computer compatible with the IBM PC/AT was specified. This computer family, based on the Intel 80286 micro-processor and Microsoft's MS-DOS operating system, normally supports a maximum of 640 kilobytes (K) of random access memory (RAM). Prototypes were configured with hard disks, enhanced graphics adapters, and color monitors. Some stations were augmented (for development convenience) by a two-megabyte memory extension card simulating a fast disk (RAM-disk). Additionally required were the ONLINE graphics-overlay board, a Sony multi-synch monitor, a Sony LDP 2000 videodisc player, and a Microsoft 2-button mouse. Figure 2 illustrates this equipment setup. For ultimate delivery, the Army prob-

Figure 2. Equipment setup for prototype system.

ably would have required replacing this prototype configuration by an EIDS system,[2] necessitating: (a) modification of the video-graphics device driver; and (b) reduction of RAM utilization from 640K to 512K.

One alternative approach might have been to use two networked personal computers, with one dedicated to running the video-graphics courseware, and one dedicated to running the expert module. This would have been easier to implement, but twice as expensive to deliver. A simpler alternative would have been to use a personal computer supporting additional memory address space. By specifying EIDS, the Army could attempt to deliver an advanced training solution in high volume on a standardized, low-cost platform.

Selection of Expert System Tool

Often ITS developers begin by developing a new set of tools for knowledge representation and inference before tackling the ITS project itself. Although there is always a need for better tools, many projects run out of resources before addressing the interesting issues in the design of intelligent learning environments. Part of the design philosophy for this project was to use off-the-shelf components wherever possible, including an existing ES development tool.

Because of the strict memory limitations, a C-based tool, rather than a LISP-based tool, was crucial. OPS5+, a product of Computer * Thought Corporation, was chosen.[3] This tool uses a byte-code internal representation for production rules, offering a careful balance between execution speed and memory usage. (Compilation to assembly code tends to require more memory per rule.) A key consideration in the selection of this tool was flexibility, including its ability for rules to call on user-written C code (such as for interfacing with the video module). Other factors included source code availability, our familiarity with the internal operations, and the possibility to modify the product if necessary. Although our hope was to interface unmodified, off-the-shelf modules, these characteristics provided an important "hedge" against unforeseen difficulties. Finally, it was thought that a straightforward rule-based paradigm—as opposed to a more elaborate hybrid package incorporating rules, frames, and other representations—would suffice for capturing the particular subject matter, and more readily lend itself to existing student modeling paradigms.

Selection of CBT/IVD Authoring Tool

During the tool evaluation stage, several PC-based tools for authoring CBT systems using interactive video were also examined. In this category, the TENCORE+ author-

---

[2] U.S. Army training and dissemination programs typically specify the electronic information delivery system (EIDS), a proposed standard platform for computer-based training, manufactured by Matrox Electronic Systems LTD of Quebec. EIDS is an MS-DOS-based personal computer supporting video-graphics overlay with a standard memory configuration of 512 kilobytes. The existing prototype M16 trainer operates on IBM equipment, Zenith equipment, and other "compatibles" running MS-DOS 3.2, occupying 640 kilobytes. The existing version can also be configured with Sony's LDP 1000 laserdisc and Mouse Systems' 3-button mouse.

[3] At the time that this work was performed, the first author and several other contributors were employed by Computer * Thought Corporation, which has subsequently ceased operations. Tools comparable to OPS5+ are currently available from other vendors. OPS5+ was derived from the OPS5 expert systems language, which was originally developed in LISP at Carnegie-Mellon University [10].

ing system, which resembles the TUTOR language of PLATO, was chosen. In the course of its adaptation for PC delivery, TENCORE+ had undergone dramatic enhancement in video-graphics device support and overall functionality. There were several additional reasons for its selection. One of these reasons was the considerable experience of the personnel from Rediffusion Simulation, Inc.[4] (RSI) in using it. This experience included prior development of a user-friendly, menu-oriented interface, which RSI provided for use on the project. Once again, a major factor was flexibility. Unlike many authoring tools, TENCORE+ does not lock the courseware author into a specific lesson format or organization. In particular, the capabilities to write data files readable by the expert system, PAUSE to the operating system, and execute arbitrary programs were crucial.

Design Methodology

Knowledge Acquisition and Task Analysis

Initial project efforts focused on acquiring M16 riflery knowledge through a variety of sources. Reading materials included both informal field manuals aimed at infantry recruits [11] and formal technical manuals intended for depot personnel [12, 13]. This was followed up by formal classroom instruction, including hands-on disassembly of the rifle and marksmanship practice on the range. One of the classroom instructors also served as our consultant and primary subject matter expert, leading small group sessions in which a variety of questions were answered and hypothetical cases analyzed. Much of this training time was captured on VHS-format videotape for later review and analysis. Some project personnel had prior small arms experience (such as with the AR15 rifle); all others participated in a formal M16A1 Qualification Course.

Project personnel with predominantly ITS backgrounds studied the materials from a knowledge engineering perspective, asking themselves questions such as: What is the knowledge required to troubleshoot rifle malfunctions? How should this type of knowledge be represented? Project personnel with predominantly CBT/SAT backgrounds examined the same materials from a task analysis perspective, asking themselves questions such as: What are the instructional objectives for this module? Which partial or simplified tasks should be taught first?

Knowledge Representation

The strength and the weakness of any expert system tool is that it constrains one's epistemology; without such constraint, one might consider many alternative representations.[5] OPS5 notation resembles LISP code in appearance; whereas, tools in the EMYCIN genre tend toward a more English-like rule notation. The OPS5 language divides knowledge into working memory elements and production rules. Its inference engine operates in a strictly forward chaining fashion, without certainty factors or

---

[4] At the time that this work was performed, the second author and several other contributors were employed by Rediffusion Simulation, Inc., which was subsequently acquired and renamed Hughes Simulation Systems, Inc. TENCORE+ is currently sold by Computer Teaching Corporation.

[5] A thorough review of commercially available expert system tools would exceed the scope of this chapter. For an introduction to essential concepts and better known tools such as EMYCIN, please refer to [14].

built-in explanatory capabilities. However, such a lean philosophy can provide flexibility and performance advantages relative to backward-chaining tools. In OPS5+, in particular, a problem space can be searched using a variety of strategies, with meta-rules controlling the relative priorities of other rules. Our experience has been that a predominantly forward-chaining engine supporting meta-rules can provide greater flexibility in matching search strategies to the reasoning processes of subject matter experts. This flexibility can also facilitate modeling the reasoning processes of learners. However, diagnostic applications often tend to involve backward chaining. Fortunately, by using rules that establish goals and subgoals, backward chaining can be conveniently emulated within the framework of a forward-chaining engine. The M16A1 expert module uses a hybrid search strategy, derived from both technical manuals and SME protocol analyses.

Table 1 illustrates a typical format for troubleshooting information in technical documentation. Much knowledge of this sort can be encoded in the form of tables of working memory elements (or "facts"), rather than rules, taking advantage of what would otherwise be highly redundant structure across groups of rules for symptom clusters. The subject matter expertise being modeled is only one of several types of knowledge required for a complete ITS. Four other categories of rules were devised for the initial prototype, although implementation of the last two categories was left for future work:

(1) *Bookkeeping rules,* for updating rifle state or displaying trace information on a second screen for the benefit of the knowledge engineer;
(2) *Communication rules,* for controlling interactions with the video-based user interface, such as setting up a hint to be given by TENCORE+ on student request;
(3) *Student modeling rules*, for keeping student-specific parameters, session history and records, and, eventually, modeling the student's knowledge state and learning process; and
(4) *Tutorial strategy rules,* for encoding such knowledge as when to interrupt the ongoing interaction because of a serious student error and, eventually, how to adapt the content and style of feedback based on the learner's instructional history and cognitive style.

The expert system architecture per se does not impose a clear division of this knowledge into isolated modules. Interactions between different groups of rules are

Table 1
Typical Troubleshooting Information in Technical Documentation

Malfunction	Test or Inspection	Corrective Action
Failure to fire	Broken hammer	Replace hammer
Failure to fire	Broken firing pin	Replace firing pin
Failure to cock	Worn trigger nose	Replace trigger
Failure to cock	Missing disconnector spring	Replace spring
Failure to extract	Defective extractor pin	Replace extractor pin
Failure to extract	Badly pitted chamber	Replace barrel assembly
Fires on safety	Worn firing mechanism	Evacuate to direct support

controlled by such programming techniques as adding additional slots to the left-hand-side patterns for each rule class.

Systems Approach to Training

The process of developing computer-based instructional systems, using a traditional courseware authoring approach, typically involves four steps: (1) task analysis; (2) courseware design; (3) development and implementation; and (4) validation and evaluation. Step 4 is not considered here since it would have been premature.

Task Analysis. The task analysis stage parallels the knowledge acquisition stage in the development of expert systems. Knowledge engineers are generally less concerned with detailed task analysis of expert and student behavior; but their synthetic task requires at least as much rigor. Setting aside the technical jargon of these two seemingly diverse activities, however, the underlying processes are strikingly similar.

Shared by the practitioners of both cultures is an example-driven or case-study methodology. To understand a portion of a task or a chunk of knowledge, it seems necessary to observe its operation in a specific context. This intuition corresponds to the Piagetian observation that, even when formal operations are used on familiar topics, learners often rely on concrete reasoning processes on unfamiliar topics. Hence, a specific set of troubleshooting scenarios was chosen to guide further work. The expert system would be considered acceptable if it could solve each scenario using any approach that a student might plausibly find. The video-based courseware would be acceptable if it could set the stage for each scenario using realistic motion sequences, accept all plausible student responses to work through each possibility, and illustrate the modified rifle state after each step. Table 2 summaries the four scenarios that are supported by the current prototype.

Courseware Design. Because of the experimental nature of this project and the requirement to interface with an expert system, the courseware design process was somewhat unconventional. It was recognized that there was a need for an initial questionnaire to obtain information about the trainee's background. It was also recognized that there would eventually be a need for traditional tutorial presentations, to provide basic factual information about the rifle and its operation. However, it was decided to develop only minimal (or "stub") versions of these modules, to focus the available resources on ES integration for problem solving scenarios, where the benefits of incorporating ITS technology would be most apparent.

One key idea was to represent all student problem-solving steps as verb-object pairs.

Table 2
M16A1 Troubleshooting Scenarios

Symptom	Context	Solution
Failure to fire	Practice range	Safety on
Failure to feed	Combat	Mis-seated magazine
Fires with safety on	Cleaning	Return to depot
Failure to extract	Practice range	Carbon build-up

Every action taken by the student during the solution of a troubleshooting scenario is modeled as the selection of a rifle part and the selection of an action or test operation on that part. The student's complete problem-solving protocol can be concisely expressed in this format. Once this idea crystallized, the expert system was rewritten so that it used precisely the same set of rifle parts and actions in its own solutions. Thus, the student's solution and the expert's solution to any scenario can be directly compared.

Another key idea addressed the potential problem of excessive keyboard input and difficult terminology. An M16 riflery trainer that required recruits to type phrases such as, "check the extractor for carbon build-up," would be unworkable. Instead, a pointing device interface was specified for all interaction except the initial entry of essential user information such as name, rank, and serial number. The mouse was used rather than other pointing devices supported by TENCORE+ (such as touch panels), since accurate part-action selections would be required. During each scenario, all user input occurs through pointing and clicking on mouse-sensitive screen regions, such as to select a rifle subassembly, and verb menus, such as to select a particular action to perform on that part.

Development and Implementation. Since one key objective of the research was to demonstrate feasibility using entirely off-the-shelf components, existing video footage was used. To demonstrate a broad range of capabilities, different scenarios used the video display screen for motion video, for still frames, for graphics-only, and for graphics-over-video. Additional efforts were made to ensure that the audio channel was used for sound effects, for music, and for speech.

The objectives of this preliminary investigation were to demonstrate a range of capabilities, rather than to produce instructionally sound courseware. This is a crucial point in evaluating the resulting implementation. One example of this involves the use of video sequences to illustrate outcomes arising from student choices. It is plausible that interesting video sequences, especially those with audio accompaniment, function as positive reinforcers; but the current prototype may be as apt to provide such multimedia reinforcement for incorrect responses as it is for correct ones. This reflects neither an inherent limitation of the technology nor a lack of appreciation for these issues by the development team.

Design Principles for Interfacing Between ITS and CBT/IVD Modules

The following principles guided the design of the interface between the ITS and CBT/IVD modules:

Off-the-shelf components should be used whenever possible;
Decision-making and answer-analysis should be handled by the expert system;
User interfaces should be handled by the courseware system;
Each of the two major modules should be restricted to its own screen[6];
Conceptually, the intermodule interface should behave like cooperating coroutines, even though operating system limitations might force the implementation to observe a parent-child protocol;
Each module should be able to restart after interruption by examining a state vector

---

[6] The expert system uses the main screen to provide a trace of its reasoning for demonstrations and debugging; it would not be seen by trainees. The video-graphics overlay equipment requires a separate, multisynch monitor, which serves as the learners' screen.

specifying the current scenario, the state of (dis)assembly of the rifle, the steps taken thus far, and similar information;

The expert system should be able to solve each scenario from any legal state that could arise as a result of plausible student exploration; it should be capable of being reset after each student step to the state encountered by the student;

The intermodule interface should be simple and reasonably hardware-independent; and

The intermodule interface should be reasonably generic with respect to the content example; although some data structures might be specific to say, rifle troubleshooting, the overall design should be generalized.

Memory Architecture of the Prototype System

The mechanics of the system architecture were largely imposed by the 640K memory limitation of the MS-DOS operating system. The organization of random access memory is shown in Fig. 3.

OPS5+ acts as the driver. At each step, it creates a new copy of the TENCORE+ delivery module. Thus, whenever the courseware side starts up, it suffers from a malady that we dubbed "nonresident amnesia." After virtually every click of the mouse, TENCORE+ must reawaken, read a state vector from a disk file, ascertain what must be currently on the screen, and reconstruct the set of options currently available to the user. It surprised us to discover that speed, such as for screen refresh during intermodule switching, was not a major problem. The original design called for keeping key files on the RAM-disk, for faster access; however, for effective demonstrations it actually became necessary to insert delay commands, so that viewers could observe the intermodule switching operation.

System Integration

The long-range goal of the research is to deliver fully functional intelligent tutoring systems—incorporating interactive multimedia—on cost-effective personal computers. This implies a commitment to provide student modeling and sophisticated tutorial strate-

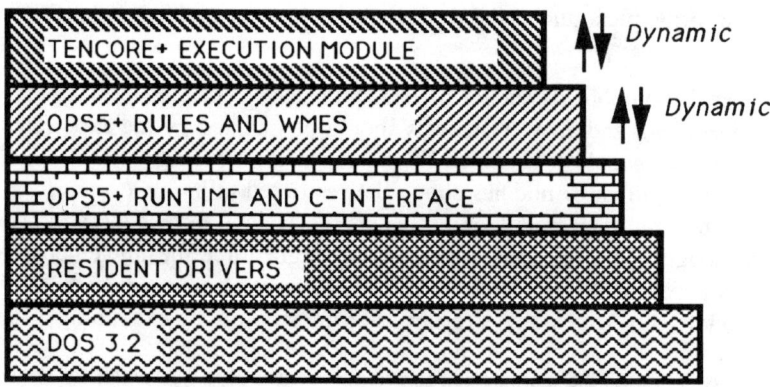

Figure 3. Memory organization of prototype maintenance trainer.

gies, as well as a representation of subject matter expertise. "Hooks" for adding planned capabilities have been designed into the prototype system.

The availability of an operational expert system for the task offered an opportunity to provide a level of interaction not possible using a traditional approach. Key issues included exploiting the expert system's availability despite the initial lack of meaningful student modeling. The expert module could be considered "semiarticulate," since it contained some explanatory capability regarding its own actions. This took the form of machine-readable commentary associated with important production rules. Currently, the system supports learning by imitating (giving a specific hint as to what its next step would be in solving the scenario), and learning by guided doing (giving a general hint, as to what overall question or hypothesis its next step would be trying to address). It also provides a variety of feedback after every student step, primarily in the form of encouraging comments such as, "You are getting warmer. Your last step was a good idea." This feedback is based neither on canned hints stored with the problem nor on rigid branching structures. It is generated at runtime and is based on the actual rule firings that occur in the expert system.

Since the state of the expert system is reset after each student step, it is forced to track the student's solution trajectory. It is far more helpful, from the student's perspective, to see how to finish your current solution—or to see why it is a blind alley—than it is to be told about some other approach prescribed by an expert. This sort of interaction—helping students solve problems their own way—is possible only in a system containing operational expert knowledge. Likewise, the system can tolerate legal but questionable moves by the student, such as side excursions that add steps to a solution but do not prevent completion of the exercise. This represents an area for a more complete tutorial module to determine when to intervene. Currently, wrong moves are classified as nonfatal or fatal; only on fatal moves is the student prevented from proceeding. Since this training is intended to address real-world situations, possibly including actual combat, in some scenarios it is possible for the student to get "killed." Vivid enactment of the implications of serious errors during situated training scenarios can be highly effective for learners.

Implementation

Approximately 200 production rules were encoded, capturing sufficient knowledge of M16A1 troubleshooting to solve a rich variety of scenarios, including a range of presenting symptoms (e.g., failure to eject a spent cartridge), allowing for multiple contexts (e.g., combat versus firing range), and handling some cases of multiple, interacting faults (e.g., bent extractor and carbon build-up in chamber). To conserve memory, a subset ES sufficient to handle the four scenarios was embedded in the prototype trainer.

The ES was interfaced to the instructional delivery module of the TENCORE+ CBT authoring system. Key design principles included allocating most decision-making and answer analysis processing to the expert system, whereas all user interface operations (such as screen input-output) were left to the authoring system. A set of mouse-sensitive screens illustrates the M16A1 in various states of disassembly, allowing students to interact with the system without typing or recalling the precise terminology for each rifle part. A set of pull-down menus allows the student to select various operations such as clearing the rifle, removing a subassembly, or cleaning a part. The set of rifle parts and operations on those parts available to students are identical to those available to the ES.

Overview of Sample Sessions

Figure 4 shows a top-level flow diagram for student sessions with the prototype. The system is started by calling a customized OPS5+ runtime. The expert system runtime loads in the rifle knowledge base and then initiates a call to TENCORE+, which starts at the *Main Menu* screen. Items in these menus are selectable either by moving the mouse cursor to the corresponding screen location and then clicking on the left mouse button or by keyboard entry. The screens involving *Background Information* are the only ones where keyboard entry is necessary; all other input is more conveniently handled with the mouse.

Selecting the *Introduction* menu item results in a brief textual and graphical presentation of introductory material about the system's overall objectives. Selecting the *Background Information* menu item begins a questionnaire about the trainee's name, rank, relevant background, level of education, length of service, and prior experience.

Selecting the *Tutorial Menu* results in another menu screen offering the following four choices: *Rifle Components, Rifle Operation, Rifle Troubleshooting,* and *EXIT to Main Menu*. Selecting one of these tutorial topics leads to a screen indicating that an introductory sequence about that topic would be made available in a fielded system. The last two items in the Main Menu, QUIT and Videodisc Checkout, are used to exit from the system and to calibrate the equipment, respectively. Selecting the *Scenario Menu*,

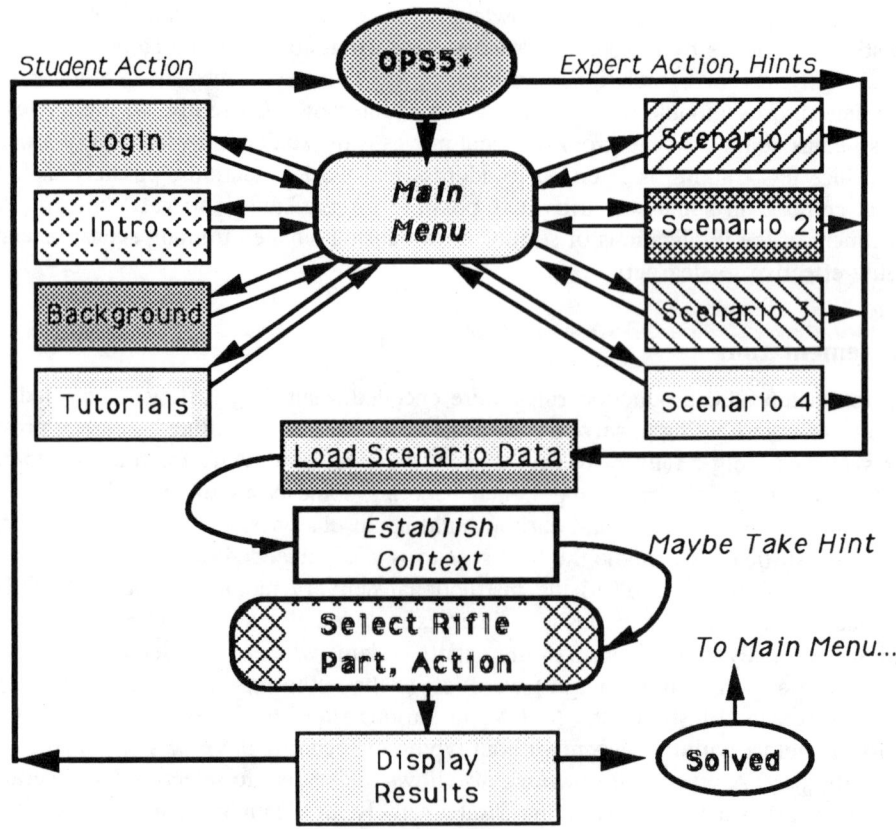

Figure 4. Flow of control and information during typical session.

leads to another menu screen, allowing a choice from among the four scenarios currently available: *Firing Range,* rifle does not fire; *Combat,* rifle ceases firing; *Functional Check,* rifle fires on "safe"; and *Firing Range,* rifle ceases firing.

The trainee can select any scenario to pursue, in any order. Trainees may also repeat any scenario, exploring the effects of alternative decisions. In a more complete system, it might be desirable to offer somewhat less freedom, if the system knew enough to provide reliable guidance as to the best sequence of scenarios to explore. Given the current state of ITS technology, it is better to err on the side of leaving the learner in control whenever there is any doubt.

Example From Scenario Three

The third scenario illustrates a very dangerous situation, even though the context is a routine cleaning. It is also one in which doctrine comes into play. As a part of the cleaning procedure, trainees perform a functional check of the rifle. During this process, the trainee sets the selector lever to *Safe* and then squeezes the trigger. In this example, the trainee is asked to handle a serious malfunction: even though the safety is on, upon squeezing the trigger, the hammer falls.

The trace output, reproduced here, illustrates the expert system's reasoning in response to this symptom. The trace shows the name of each rule that "fires," a set of numerical indices indicating which elements in working memory were matched by the antecedent (condition pattern) of that rule, and other information about the system's solution trajectory. Notice that to inspect the lower receiver, the system backward chains to obtain the subgoal of separating the receivers, which in turn leads to the subgoal of first clearing the rifle. This corresponds, therefore, to the first action expected from the student.

>   87. scenario_3_load 146 158 148
>   88. check_firing_mechanism_1w_q 152 154
>   89. takedown_start 157 154
>   Need to separate the receivers before lower_receiver can be inspected.
>   90. clear_rifle_rule 158 154
>   91. takedown_instruct 158 159 154
>   The student should clear the assembled_rifle.
>   General hint will be: The rifle should first be put into a safe state.

Although the system could "cheat" (because, of course, the database for each scenario includes a description of the actual fault condition), it does not do so. Instead, it reasons through the problem as if it were being operated as a job aid; except that whenever it needs to make a specific observation or test, it probes the scenario database for that piece of information rather than querying the user. Having determined a preferred [part, action] pair, or "move," the expert system returns control to TENCORE+ to obtain the student's response to the identical situation.

Figure 5 illustrates the *Repairs* menu on the main rifle interaction screen, which provides the primary user interface within scenarios.[7] The rifle is shown in one or more views, illustrating its current state of assembly, with major parts and subassemblies

---

[7] Figures illustrating screen output have been edited to improve legibility. The content is nonetheless faithful to the implementation.

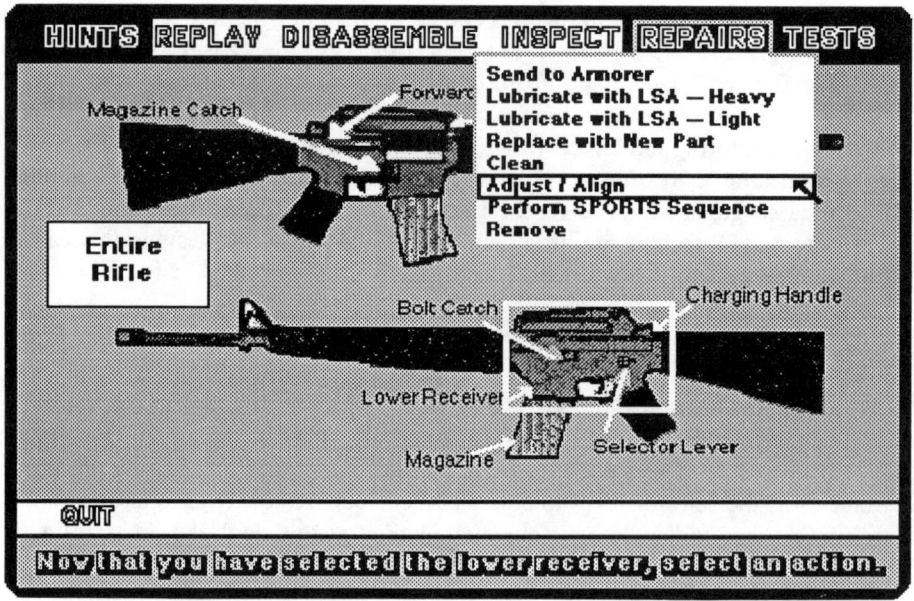

Figure 5. Repairs menu—main rifle interaction screen.

labeled. The entire screen is mouse sensitive (i.e., any rifle part, or the entire rifle, can be selected by pointing and clicking with the mouse).

Across the top of the screen, there is a stationary menu with the following items: HINTS, REPLAY, DISASSEMBLE, INSPECT, REPAIRS, and TESTS, representing the various actions that can be performed at each step. A similar menu bar appears toward the bottom of the screen, offering the options QUIT and RE-ASSEMBLE (when appropriate). Prompts, hint output, and other feedback appear in the region below the bottom menu bar. Most of the menu items, such as REPLAY (which reviews the scenario definition sequence), result in an immediate response by the system. However, HINTS, REPAIRS, and TESTS—the plural options, which are shown in a different color—offer pull-down menus for selecting more detailed actions, as illustrated.

In general, almost any [rifle part, action] pair can be selected; the student is free to explore the effects of various moves. However, certain pairs that would be meaningless, beyond the scope of the current implementation, or impossible because of the current rifle state are disallowed. For example, it is not possible to disassemble the entire rifle unless it has first been cleared and the magazine removed. If the student attempts such an illegal [part, action] pair, the system displays an explanation near the bottom of the screen and the requested operation is aborted. All other operations are carried out on request and are immediately reported back to the expert system for further analysis.

Assuming that the student in fact chooses to clear the rifle, as expected, the expert system would continue with the next step in its own troubleshooting plan.

     97. detect_ok_step 172 160 171 166
     98. sep_recv_rule 174 158 161
     99. takedown_instruct 158 176 174
     The student should disassemble the cleared_rifle.
     General hint will be: Need to take a closer look.

```
100. remove_takedown_goal 158 157 178
101. continue_ok 175 178 169 172 151 177 171 168 173 149
```
Scott Lucado made a correct choice in Scenario 3.
Clear the assembled_rifle was chosen.

A motion video sequence then illustrates the effects of the student's correct choice. The rifle is now shown with the receivers separated and the bolt carrier group removed. The expert module continues reasoning as follows.

```
114. detect_ok_step 284 177 296 196
115. query_database_positive 156 288 155 153
```
The expert system is querying the database to determine the following condition:
Are there worn parts in the lower_receiver?
The result of the query is YES.
The student should inspect the lower_receiver.
General hint will be: How old are the parts in the lower receiver?

When the trainee requests a general hint using the HINTS menu, the response that appears on the student's screen is precisely this text, as generated by the expert system during its own reasoning. Had the trainee asked for a specific hint in this situation, the response would have been, "Check the lower receiver for worn or broken parts."

Suppose that, despite these hints, the trainee incorrectly thinks that the fault lies with the magazine. This hypothesis can be freely explored. A motion video sequence, accompanied by a spoken explanation, reminds the student how to take apart the magazine. The final frame of the video sequence shows a magazine with no visible flaws.

The trainee can safely proceed to examine any number of other rifle parts, cleaning them, lubricating them, and inspecting them for possible explanations for the malfunction, *fires_on_safe*. For example, the bolt carrier group could be disassembled, the firing pin and retaining pins checked for wear or corrosion, and so on. Eventually, inspection of the lower receiver will reveal worn and possibly broken parts. A training system for depot personnel might continue with further maintenance actions from here. However, since this system is intended for field personnel, the correct solution is to evacuate the lower receiver to the armorer. (U.S. Army doctrine does not permit lower receiver disassembly by field personnel.) Through the HINT feature and other feedback, the student should eventually reach this conclusion, even if many side excursions are taken along the way.

Example From Scenario Two

Whereas the aforementioned scenario illustrates that students are free to browse and explore moves that are not directly on a solution path, the second scenario illustrates that the appropriateness of this style of interaction can depend on the context. Figure 6 depicts the opening sequence of the second scenario, wherein motion video provides a realistic firefight simulation.

The rifle is set on AUTO after inserting a new magazine. However, after firing one round, the rifle ceases firing. Diagnostic actions that might be recommended for the practice range are irrelevant here. In situations like this one there is only one right answer. The correct response, nicknamed SPORTS, should be practiced until it is completely automatic.

SPORTS is an acronym for the following sequence of steps: (1) Slap upward on the magazine (to seat it properly); (2) pull the charging handle back (look for ejection); (3) observe the chamber (check for obstructions); (4) release the charging handle (to feed new round); (5) tap the forward assist; (6) shoot the rifle (by squeezing the trigger).

The expert system solves this scenario in a single iteration using the SPORTS action. A trace of its reasoning follows.

    40. scenario_2_load 69 73 71
    41. combat_sport 75 77 76
    42. select_final_step 79 78 77
Rifle Troubleshooter found the answer!
SPORTS the assembled_rifle will solve the scenario.
General hint will be: You have only one chance. If you blow it, you're dead.
    43. retest_rifle_y 81
    44. scenario_begin 74 73 71 88 68

Unlike the open-ended exploration encouraged on practice-range scenarios, in the case of a wrong move here, a tombstone is displayed, bearing the student's name, rank, serial number, and "life dates" above the epitaph, "Died of Folly in Combat." Fortunately, after this stern lesson, the student is "revived," and can try this scenario again.

Figure 7 illustrates the video feedback provided when the trainee correctly selects the SPORTS action as his first step on this scenario. This is followed by congratulations and the opportunity to go to the other scenarios or quit the session.

Figure 6. Scenario 2—combat simulation using motion video.

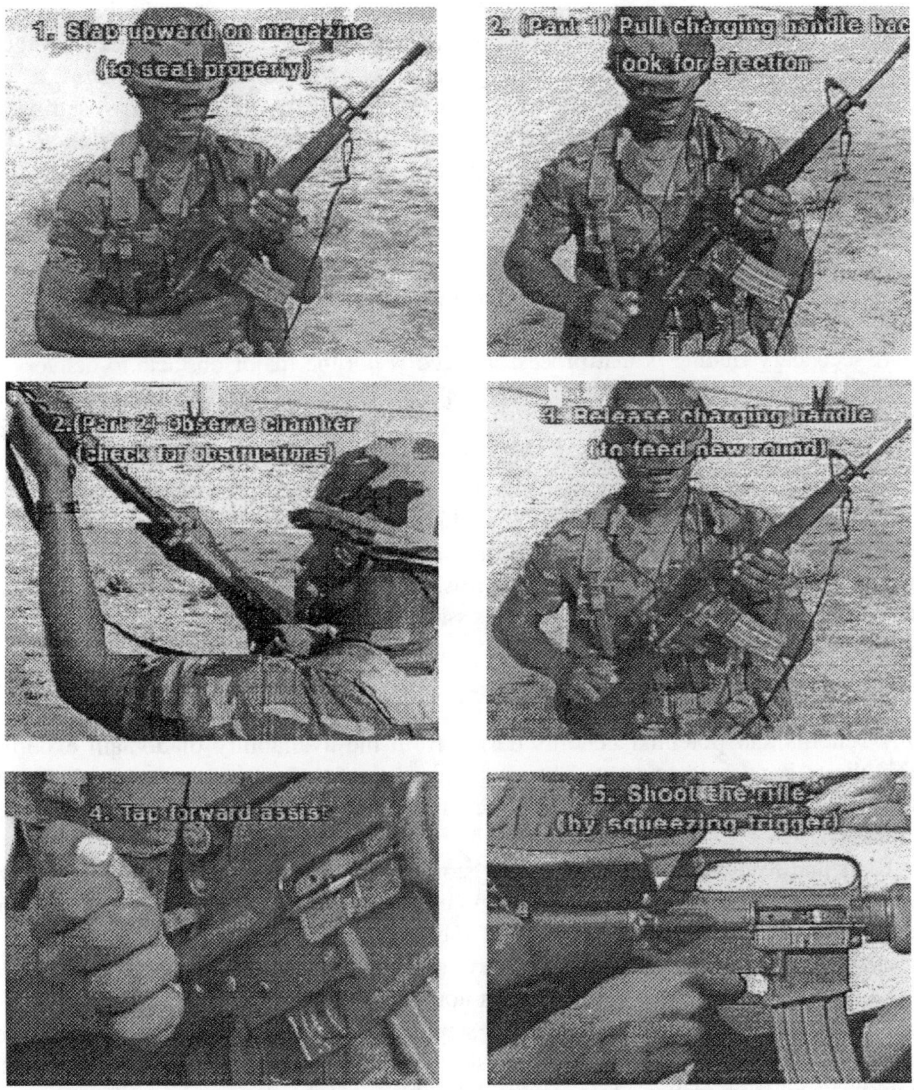

Figure 7. The SPORTS immediate action sequence.

Results

The prototype system described in this chapter is at a very early stage of development, representing slightly more than 6 months of work by a small team. Although the system is capable of more elaborate interactions than can be shown here, much additional work would be required to field a practical system. Hence, it would have been premature to measure its pedagogical effectiveness or contrast it with alternative forms of training. Nevertheless, the system is remarkably robust for an experimental prototype. It has been demonstrated nationwide at installations ranging from the Air Force Human Resources Laboratory to the Defense Language Institute. Viewer reactions to the capabilities exhibited at these live demonstrations have been overwhelmingly enthusiastic.

Our primary finding is that a carefully engineered ITS *can* be integrated with CBT courseware incorporating interactive video and delivered on cost-effective personal computers using off-the-shelf components. Moreover, much has been learned about the tools and techniques for constructing this class of systems. A key feature of the prototype is the ability of the expert system to single-step its solution, resetting its current state at each step to match the actual step taken by the student. This allows the expert system to track student solutions closely, even when multiple, acceptable solution trajectories exist or when the student has pursued a relatively harmless blind alley. At each step, the system is able to provide feedback and hints, based not on canned branching within the courseware but on the actual rule firings within the expert system. Provisions for student modeling and more sophisticated tutorial strategies were designed but not implemented.

A secondary finding is that processor speed is *not* the major obstacle to delivery on personal computers. However, the 640K memory limitation of MS-DOS caused many difficulties; the 512K limit of the standard EIDS configuration would represent a major obstacle. Although the prototype is PC-based, it is not a toy system. The task being performed has more real-world elements than do some larger prototypes running on far more expensive platforms. However, its subject matter is quite modest in complexity, compared to say, a radar system. As additional domain knowledge becomes required, the negative impact of limited memory on the feasibility of this type of training system would eventually become insurmountable and necessitate conversion to another platform.

Benefits Attributable to Intelligent Tutoring

Many benefits and potential benefits derive from the availability of domain expertise within the training environment. The most apparent of these are summarized here. We believe that additional benefits will surface as we learn more about how to apply this technology.

Hints and feedback are not canned or stored with the specific scenario, but instead represent advice that might be given by an actual expert placed in the student's current predicament. Depending on the degree of thought and care invested in structuring the knowledge base and associated explanatory information, trainees should therefore perceive a level of capability or insight that is not apparent in conventional courseware. By using the specific hint capability at each step, the student can essentially operate the system as a stand-alone job aid. Thus, the specific hint feature could ultimately support a usage spectrum from training to cooperative problem-solving to job aiding. The division of labor between domain expertise and instruction can enable instructional designers to concentrate on presentation and pedagogical considerations, including the appropriate use of multimedia capabilities. Answer judging, remedial sequencing, and similar decision-making, which can distract from other aspects of the design, can be turned over to the knowledge engineer and ES.

It is far easier to add new scenarios than would be the case with a traditional courseware module. This extensibility results from eliminating the requirement that the lesson author consider every possible student response and prepare suitable outcomes and branching. It suffices to ensure that the expert module correctly solves the new scenarios. (It is also necessary to ensure that the expert module can recover from student-chosen blind alleys on the new scenarios.) If a new scenario involves the same context and presenting symptom as an existing scenario, it may not even be necessary to prepare new video; modifying the database to include the new underlying fault may be all that is required. There is a possibility that the system could create new scenarios dynamically, either at its own initiative or at the request of the student, by applying standard variations to existing scenarios. For example, the student could conduct a "what if" analysis to better understand the expert's approach by inserting alternate or additional faults that might be manifested by the same presenting symptoms. Moreover, the training system could be made to generate variations on existing scenarios to exercise aspects of the knowledge (e.g., certain rules) for which the student has completed the standard materials without demonstrating mastery; this assumes a well-elaborated student model. The key to success for these types of capabilities is developing techniques to ensure that scenario variations cannot fall outside the coverage scope of either the expert system or the available video footage.

The potential exists for truly individualized tutoring, based on a fine-grained model of the student's knowledge and skills. The current prototype was developed to support such a student modeling system, using variations at the level of clauses in domain rules. Although some design work has been done, this feature has not been implemented in the current prototype.

Benefits Attributable to CBT/SAT and Interactive Video

What benefits or potential benefits, then, derive from the availability of a sophisticated CBT authoring environment supporting the systems approach to training, interactive video, and other instructional capabilities? A number of such benefits are apparent.

First and foremost, the instructional impact of the expert system is dramatically enhanced by the multimedia user interface in general, and the intense realism of video in particular. Until recently, many ITS systems have relied on little more than "glass teletype" modes of interaction. Their focus on knowledge engineering has been at the expense of human engineering. Such systems, although sometimes exhibiting impressive reasoning capabilities in laboratory demonstraitons, may turn out to be completely ineffective in field testing when confronted with real-world trainees, some of whom lack basic reading skills or prefer viewing television to learning.

The lowest levels of input analysis, such as collection and timing of key presses and pointing device input, are handled by modular device drivers localized to the instructional delivery package. Thus, the ES need not be concerned with individual keystroke analysis, such as whether the student indicated an affirmative response by typing *Y* versus *y* versus *YES*, or by touching the *YES* box on a touch-sensitive screen, or by clicking the left mouse button while the cursor was over the *YES* region. This level of detail can be suppressed by the lesson author, and the abstract response signaled to the expert system, for a more appropriate level of granularity.

Impossible or meaningless operations can be caught at the level of the authoring system and suppressed. This allows the expert system and the modeling modules to focus on the more interesting types of student errors, such as legal and meaningful but inappropriate steps for a given problem-solving context. The decision as to what level of granularity is appropriate for student modeling is of course a cooperative design decision to be made by the ITS specialist and the lesson author, and it depends on the context.

A lack of attention to SAT and related developments within the traditional instructional technology literature may have hampered the success of earlier but potentially successful ITS systems. This results from an unfortunate lack of communication between the communities. Depending on the degree of thought and care invested in instructional design, the pedagogical value of proposed intelligent tutors might be significantly improved. Realizing this potential benefit depends on the cooperation of development personnel in bringing disparate but complementary backgrounds into harmony. A division of labor between knowledge engineering and instructional design could enable ITS developers to concentrate on representing complex domain knowledge. Overall course organization, screen design, video production, and similar matters could be turned over to instructional designers, who tend to be more attuned to relevant considerations. Perhaps most importantly, this type of hybrid approach seems to offer the potential to create innovative learning environments that combine realistic, open-exploration simulations with sufficient guidance to challenge learners without frustrating them.

Applicability

Several distinct paradigms have been proposed in the literature for designing intelligent tutoring systems. Some approaches, for example, do not assume the availability of an operational expert system for the content to be taught. These approaches are especially relevant for topics such as computer programming, where expert systems technology has not yet succeeded in capturing the requisite performance knowledge.

The approach taken here has relatively broad applicability, but it also suffers from certain inherent limitations. In some learning contexts, conventional computer-based instruction may be most appropriate. In others, discovery learning may be essential. In other cases, an ITS might be appropriate, but based on an approach other than that taken here. The ability to rapidly construct an expert system for the rifle troubleshooting task was crucial to this paradigm. Yet, there are many topics where the development of a useful ITS system may be feasible, even though developing an operational expert system for the same task remains beyond the state-of-the-art.

When developing an expert system to perform the task is practical, the approach may be suitable, if the training situation can be characterized as a game, simulation, or problem-solving scenario. The approach is less relevant to topics that are primarily declarative or that emphasize rote memorization or motor skills. Thus, M16A1 rifle troubleshooting was studied, whereas M16A1 marksmanship seemed less appropriate as a topic for the demonstration system.

The single most frustrating and time-consuming aspect of the project was the 640K byte memory addressing limitation imposed by the MS-DOS operating system. Each of TENCORE+ and OPS5+ could easily fill the entire 640K; both benefit from memory extension options when available. To offer advanced maintenance training for devices much more complex than rifles, it would be essential to use a platform supporting a larger address space. Substantial additional memory would be required for expert-based maintenance training for devices as complex as, say, the avionics bay of a modern helicopter. (Although increased main memory [i.e., RAM] could be essential for adequate performance, the really critical parameter is actually address space, rather than RAM per se. Many MS-DOS platforms have a provision for adding RAM in the form of a RAM-disk, which can improve performance but does not increase the usable address space above 640 K.) Additional address space can be achieved on Intel 80×86-series processors either by "protected mode" extensions to MS-DOS (such as those developed by AI Architects, Inc.), or by conversion to OS/2. Alternatively, this requirement could be met by switching to Macintosh.

In spite of the unfortunately severe 512K memory limit imposed by EIDS, for troubleshooting domains involving low to moderate complexity, work-arounds within this restriction might be feasible. These would involve: developing highly compact, stripped-down versions of all tools; partitioning knowledge bases into small, modular clusters; and overlaying large segments of the system to the hard disk. The need to rely on awkward techniques such as overlays does not preclude applying the approach to somewhat more complex devices; but it does imply performance degradation, increased development cost, and an upper bound on complexity. The impact of these memory restrictions would become more pronounced as additional instructional features (such as student modeling) are added.

Discussion

The next major capability that should be studied using this class of prototype trainer involves student modeling. Currently, the data to initialize the student model is collected, based on the questions posed to actual students by instructors at the start of the rifle qualification course. Detailed evidence pertaining to student mastery is generated by the expert module at each step; this evidence could be used to model the student as a perturbation from the expert rules. Goldstein [15] refers to this as an "overlay model." The approach of resetting the ES's state to correspond to the trainee's situation after each step is related to Anderson's model-tracing technique [16].

Two levels of granularity are possible. At one level, the chunking consists of a checklist of rules mastered or not mastered. When the student makes a move that agrees with the expert move for that situation, this is positive evidence that the student has mastered the corresponding expert rule(s). When the student makes an incorrect move, this is negative evidence with respect to the corresponding expert rule(s). Although there may be multiple rules leading to the same choice, this potentially combinatorial problem is minimized by resetting the expert after each move. Also, potential ambiguities can be resolved through differential diagnosis, using situations where one rule applies but another does not.

Complexities arise in the presence of hints. If the trainee takes the specific hint and then makes a correct choice, this choice should not be treated as positive evidence. If the trainee takes the general hint and then makes a correct choice, it should be treated as positive evidence, but it should receive less weight. Modeling problems can also arise when the student selects a correct choice after first exploring a garden path; this situation can probably be handled by assuming correct rules with incorrect priorities or conflict resolution strategies.

A finer level of granularity involves localizing flaws in student rules to particular clauses in the antecedent or consequent part. A trainee model might be missing a clause from the left-hand side of a given rule, for example, or have an unnecessary condition on a rule's applicability. The right-hand side (conclusion) of a rule might be overly specific, and so on. One key to the feasibility of this type of modeling involves restricting the structure of domain rules. For example, disjunctive conditions can be factored out into separate rules, a discipline that was in fact observed in building the rifle troubleshooter ES.

A two-stage modeling scheme could prevent the more fine-grained approach from searching too large a space of possible student models. Working first on a coarse granularity model (at the level of entire rules) and then refining the model to specific flaws within the buggy rules is analogous to the use of an abstract problem space in AI planning research. A sensible strategy for further work, therefore, would be to experiment with a modeling system using this two-tier design. Eventually it will be necessary to understand student modeling in the context of more elaborate knowledge representations including such elements as frames and model-directed reasoning.

In future work, we expect to explore the idea of generating alternative scenarios (using the same video setting and symptom, for example, but different underlying faults) for differential diagnosis during student modeling. The results of modeling could then be used to guide all of the following: scenario selection (or generation) for remediation; presentation of hints within a given scenario; and modification of feedback after a given step.

Another area for further investigation involves tutorial output. One intriguing idea is to provide additional categories of hints and advice, leaving the selection of hint category and the timing of hints under control of the student. Just as the student can currently request either a general hint or a specific hint, other classes of hints could be offered in the menu. As examples, one hint type could rule out certain possibilities; another could suggest a class of similar faults; a third could refer the student to previously solved, related scenarios. The current plan is to break down the *specific* hint category into a *part* hint and a separate *action* hint. With this easily-implemented enhancement, the student could get advice on which rifle part to consider without being told what needs to be done to it. This idea could be generalized to allow hints that indicate which subassembly, or which group of related actions, and so on. The hint categories need not be rifle-specific; the same idea would be applicable to many equipment maintenance tasks. Eventually, by keeping records of student preferences and learning patterns, we may learn enough to begin constructing systems that really use the session history and student model to select the best hint category automatically, spontaneously presenting hints when needed.

Eventually we will need integrated authoring environments that support intelligent tutoring, more conventional CBT arising from the SAT methodology, and interactive multimedia. Ideally, such systems would be accessible to subject matter experts who are neither professional programmers nor instructional designers. Before attempting to define the specifications for such sophisticated, integrated tools, however, it seems prudent for the community to acquire much more experience with successful implementations using ad hoc, hybrid techniques and off-the-shelf tools, along the lines pursued here.

Conclusion

This project has demonstrated the feasibility of integrating an intelligent tutoring system, CBT/SAT courseware, and interactive video for delivery on inexpensive personal computers. The primary obstacle to scaling up the approach is the requirement for additional memory address space. The memory limitations imposed by MS-DOS and its derivatives such as EIDS can be partially circumvented by awkward but well-understood programming techniques for instructional topics of bounded complexity.

Several prevalent themes from the ITS research literature are exhibited in the prototype system. The role of content expertise in driving the instructional dialogue is intriguing and unique. Other key concepts, such as student modeling, although incomplete or absent in the prototype, are nearly within reach. A suitable hardware platform could be delivered in high volume in a practical and cost-effective form. The hybrid authoring environment, based on an interface between shelf software products, could be made completely generic with a small amount of additional work, providing a test-bed for a near-term, hybrid authoring approach. The instructional paradigm, although not universal, is applicable to a myriad class of federal and corporate training needs. These include fault diagnosis for complex equipment and other tasks that can be characterized as problem-solving scenarios, gaming environments, or simulations.

Future investigations are likely to include generalizing the approach, extending it to more complex tasks, implementing student modeling and tutorial components, and initiating formative evaluations. Conversion to EIDS had originally been contemplated; conversion to the Macintosh II now seems preferable. [A preliminary feasibility study concluded that a Macintosh II version could provide a compelling benchmark. This could use Orange Micro's *Alive!* (or similar) board to support video in a window and graphics overlay from a Pioneer 4200 (or similar) videodisc player. It could use the native mouse and window system, including standard Macintosh pull-down menus (or the floating tool palettes of its *HyperCard*® software) for TESTS and REPAIRS. NASA/COSMIC's CLIPS and Authorware Inc.'s *Authorware Professional* could replace OPS5+ and TENCORE+, respectively. The Macintosh's larger memory, MultiFinder operating system, and Inter-Application Communications support would significantly facilitate extensions to more complex equipment.] The first author is currently investigating maintenance training for commercial aircraft, in collaboration with a leading manufacturer. Ultimately, after the community has gained some additional experience with sophisticated hybrid systems, using currently available tools, we hope to explicate the requirements for a new generation of integrated authoring environments.

Acknowledgments

The authors are indebted to numerous people and several organizations for ideas and efforts culminating in this chapter. Significant contributors included Tom D. Conkright, Ph.D., A. Ramsay Ellis, Jr., Arthur James, Alice Marie Miller, Janet L. Miller, Karen Pogoloff, Bob Stevenson, and Michael J. Wild. Dr. Dennis Irons of Texas Instruments contributed important insights regarding tool selection and instructional systems design.

We are grateful to our subject matter expert, Master Sergeant Tom Coppock, for sharing his insights and anecdotes on M16 riflery training. We also wish to thank Staff Sergeant Eric Weaver, Technical Sergeant Allen West, the 136th Combat Arms Training and Maintenance Unit, the Texas Air National Guard, and Carswell Air Force Base, for allowing us to attend and videotape their training sessions.

We also thank Doug Lising and Tom Wooten, at Fort Eustis, for helping us to obtain three key videodiscs; the U.S. Army provided all motion video sequences and video still frames, including those used for illustration herein. Special thanks to Marshall Farr, Stan Kostyla, Joseph Psotka, Merryanna L. Swartz, and the Army Research Institute for the Behavioral and Social Sciences, for helpful editorial suggestions, for valiant efforts to keep us on track, and for having the vision to encourage and fund these sorts of investigations.

The authors are also grateful to the Federal Systems Group of Apple Computer, Inc. for supporting the Macintosh® II Computer Conversion Feasibility Study. Laureen Larson at Apple helped by diligent proofreading and in countless other ways.

"Apple," "Macintosh," "HyperCard," and "Mac" are registered trademarks of Apple Computer, Inc. The origination of other trademarks is clarified by the context.

References

1. B. P. Woolf, *Context-dependent planning in a machine tutor* (doctoral dissertation), Amherst, MA: University of Massachusetts, Department of Computer and Information Science, 1984.

2. J. R. Carbonell, "AI in CAI: An artificial intelligence approach to computer assisted instruction," *IEEE Transactions on Man-Machine Systems*, 11:4, 1970.
3. J. S. Brown, R. R. Burton, and J. de Kleer, "Pedagogical, natural language, and knowledge engineering techniques in SOPHIE I, II, and III," in D. Sleeman and J. S. Brown (eds.), *Intelligent tutoring systems*, New York: Academic Press, 1982.
4. J. De Bruin, D. L. Massey, and R. B. Roberts, "A training system for system maintenance," in J. Psotka, D. L. Massey, and S. A. Mutter (eds.), *Intelligent tutoring systems: Lessons learned*, Hillsdale, NJ: Lawrence Earlbaum, 1988.
5. J. McDermott, "R1: A rule-based configurer of computer systems," *Artificial Intelligence*, 19:1, 1982.
6. D. Sleeman and J. S. Brown, *Intelligent tutoring systems*, New York: Academic Press, 1982.
7. J. Psotka, D. L. Massey, and S. A. Mutter (eds.), *Intelligent tutoring systems: Lessons learned*, Hillsdale, NJ: Lawrence Earlbaum, 1988.
8. S. M. Alessi and S. R. Trollip, *CBI: Methods and development*, Englewood Cliffs, NJ: Prentice-Hall, 1985.
9. J. D. Fletcher, "Intelligent instructional systems in training," in S. J. Andriole (ed.), *Applications in Artificial Intelligence*, Princeton, NJ: Petrocelli Books, 1985.
10. C. Forgy and J. McDermott, "OPS: A domain-independent production system language," *Proceedings of the Fifth International Joint Conference on Artificial Intelligence*, 1977, 933–939.
11. Department of the Army, *The M16A1 rifle: Operation and preventive maintenance*, D.A. Pamphlet 750-30, Washington, DC: Department of the Army, U.S. Government Printing Office, 1969.
12. Department of the Army, *Organizational, direct support, and general support maintenance manual (including repair parts and special tools list): Rifle, 5.56 MM, M16 (1005-00-856-6885), Rifle, 5.56-MM, M16A1 (1005-00-073-9421)*, Technical Manual TM 9-1005-249-24&P, Washington, DC: Department of the Army, 1983.
13. Colt Firearms, *M16A1 Rifle: Operation and maintenance instructions*, Colt Manual CM101, Hartford, CT: Colt Firearms, 1980.
14. D. A. Waterman, *A guide to expert systems*, Reading, MA: Addison-Wesley, 1986.
15. I. P. Goldstein, "The genetic graph: a representation for the evolution of procedural knowledge," in D. Sleeman and J. S. Brown (eds.), *Intelligent tutoring systems*, New York: Academic Press, 1982.
16. J. R. Anderson, C. F. Boyle, and Brian J. Reiser, "Intelligent tutoring systems," *Science*, 228 (4698):456–62, 1985.

III
Implementations and Evaluations

Chapter 7

Semantically Constrained Exploration and Heuristic Guidance

HENRY HAMBURGER
AKHTAR LODGHER

George Mason University
Fairfax, VA

Abstract *Exploration can be an effective learning experience, if suitably constrained and guided. Moreover, it can provide this benefit for specifically targeted formal skills in the arithmetic curriculum. This paper presents two complementary techniques for promoting success in computer-based exploration environments. Semantic constraints on eploration cut out meaningless options, and heuristic guidance facilitates search on the basis of a heuristic function with both cognitive and problem-solving components. We have implemented an environment that permits semantically constrained exploration for subtraction, as well as a related environment that facilitates the transition to paper-and-pencil subtraction. The authors tested the system in individual hour-long sessions with more than twenty children in grades 1-3.*

Introduction

Exploration that leads to discovery can be a motivating and effective learning experience, but exploration can also degenerate into groping, leading to frustration and misconception. These twin observations, or variations on them, have long been present in the teaching literature; see [1, pp. 73–75] for discussion and references in the realm of arithmetic. This paper presents two complementary techniques for promoting success in computer-based exploration environments: semantically constrained exploration (SCE) and heuristic guidance (HG). The constraints in SCE are intended to cut out meaningless options that would only be distractions from useful exploration. HG further facilitates the learner's search by offering warnings and hints on the basis of a heuristic function with both cognitive and problem-solving components. The strategy is to provide useful individual attention without speculating excessively on the student's exact thoughts. We have implemented an environment for SCE in the domain of subtraction and are extending it to provide HG. The first two sections of the paper introduce and discuss SCE and HG. Each presentation makes extended reference to subtraction, for which some domain analysis appears in the third section.

Semantically Constrained Exploration

We seek to create an environment in which search will most often be gratifyingly successful and educationally beneficial. Semantic constraints are our first strategic move in the pursuit of that objective. The imposition of constraints has the effect of prepruning the search space and abstractly underlies the learner's possible actions in the environment. The trick is to make the constraints tight enough so that the search space is of proportions manageable for the target population of learners, yet not so tight as to undermine the interest and educational value of the activity.

The constraints we use have the additional property of being semantic in that they have a reasonable interpretation in the story or description of the environment and therefore make immediate sense to a learner. This claim is related to the idea of conceptual fidelity [2], and is elaborated below. In sum, the rationale for SCE is to facilitate hypothesis formation (active thought) by making good hypotheses readily available, and yet to relieve the learner of the cognitive load of memorizing seemingly arbitrary strictures. In the first three subsections, we introduce a particular SCE, justify it pedagogically, and relate it to other techniques. The fourth subsection is on representation, and the final one treats the transfer of skills to traditional representation.

Coinland: An Example of an SCE

To fix ideas about what an environment for SCE can look like, we provide a sketch of a particular one, Coinland, which we introduced in [3]. Coinland provides an environment for exploring elementary arithmetic operations. By doing so, it supports active problem-solving in a domain, arithmetic, in which many children can succeed at problem-solving [4]. At the same time, it enforces and thereby facilitates the learning of specific arithmetic principles, namely, place value representation and the conservation of quantity. Coinland is accompanied by an alternative but related representation, Numberland, which embodies the same concepts but has a more numerical aspect. Numberland moves the learner closer to the traditional world of pencil-and-paper arithmetic performance.

To set the stage for Coinland, let us introduce an apprentice in a fictional Coinland. This project may then be seen as providing a simulated apprenticeship [5]. Our Coinland is real (implemented) but imaginary (made of screen images). Imagine instead a Coinland that is imaginary (fictional) but real (made of physical entities, not computed ones). This fictional Coinland is a society without writing, in which people exchange goods and three types of coins, called Is, Xs and Cs. The different value levels of the coins give rise to conventions about exchanging them: ten Is for an X and ten Xs for a C.

Now, in this society there live an artisan and her son, the apprentice. The artisan has several product types, each with a price, and she keeps on hand a supply of coins, for making change. She is training her son to handle transactions in her shop, so that one day she can go off and do materials research. Lacking writing, she records the various prices by leaving a set of coins with each product type. Lacking (abstract) arithmetic, she tells him to extract the price of an object from whatever coins are proffered by a customer, having first executed coin exchanges as needed to make such extraction possible. Is this instruction sufficient? If not, what more does the apprentice need? When these folks make change correctly, do we credit them with knowing subtraction? And if not, what more is there to it? We seek answers to these questions in our computational Coinland (henceforth simply "Coinland"), to which we now turn.

The student in a Coinland subtraction problem plays the role of a cashier in a store. A customer buys an item for some given price, selected by the system. The price, which is the subtrahend (the number to be subtracted), appears on the screen in numerical form. The money tendered corresponds to the minuend (the number from which the subtrahend is subtracted). This number is also picked by the system and takes the form of coins, shown as color-coded circles (yellow for pennies, etc.). The student's task is to give the correct change, keeping a set of coins identical to the price, thereby subtracting the price from the initially tendered amount. A change-making machine is available. For simplicity here, we shall use only pennies, dimes, and dollars, which suffice to introduce base-10 subtraction of three-digit numbers with borrowing across zero, as well as easier problems with one- or two-digit numbers. Varying the set of denominations gives an even broader range of problems, supporting the learning of additional concepts. Within the existing framework one could introduce a thousands column and/or a decimal column, or even move to base-5 arithmetic, with pennies, nickels and quarters.

The Coinland subtraction screen appears in Figure 1, which permits a more detailed description of the task. The particular situation shown is partly through the solution of the problem 861 − 307. Initially, the minuend, which is the money from the customer, was represented in the top row in the form of coins, in this case 8 dollars, 6 dimes and a penny. However, in the figure the student has already exchanged a dime for ten pennies to get a total of 11 pennies in the take-in tray, and then lowered 4 of them to the give-back tray. Notice that during solution, the student no longer has available a statement of the problem. Although that information is not necessary for the task, an earlier version of Coinland did keep it on the screen in numerical form, but it seemed to interfere more than help. In the second row appears the price of the item, expressed as a number, which serves as the subtrahend. The student's task is to get the correct amount of change into the "give-back" tray in the bottom row. This row starts out empty and should end up holding a representation in coin form of the answer to the subtraction problem (the tendered amount minus the price). The student can tell whether the problem has been solved correctly, since if it has, the take-in tray will hold a set of coins equal in value to the price.

Each of the two trays has three compartments for the three denominations. The student can manipulate the situation in two general ways, as exemplified above: (1) by shifting coins from one place to another, and (2) by converting between adjacent denominations. Shifts can be in either direction, from the take-in tray to the give-back tray, or, if necessary, from there to the take-in tray. A single shift can involve any number of coins, provided they all come from a single compartment of one tray (and are therefore of the same denomination) and go to the corresponding compartment of the other tray.

The learner navigates around the screen with the four arrow keys, picks up and drops coins with three specially labeled function keys, and uses the ENTER key to select options on the screen. (We also have an implementation in which a mouse controls movement and the mouse button triggers all actions). To shift coins, a user must (1) move the cursor, which plays the role of a hand, to the intended source compartment (using the arrow keys), (2) press the "PICK ONE COIN" key once per coin to be picked up, (3) move to the destination, and (4) press the "DROP ALL COINS" key. When a coin is picked up, it turns white at its source until it is dropped at its destination. One might say the white coin connotes a ghost or trace of the picked coin. The ghost

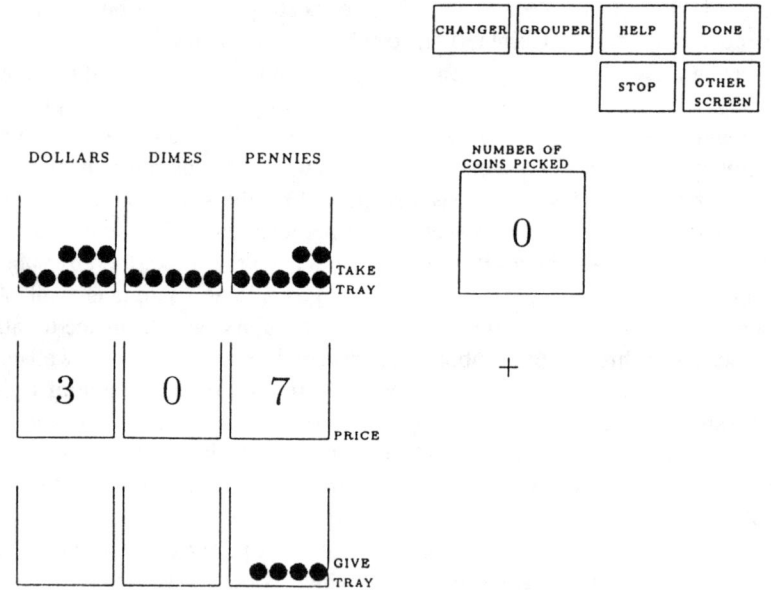

Figure 1 The Coinland screen during solution of the problem to give change from 861 when the price is 307. The student has exchanged one dime, leaving 5 of the original 6 dimes, and obtaining 10 extra pennies for a total of 11. Four of the 11 pennies have been lowered to the "give-back" tray and 7 remain in the "take-in" tray.

disappears when that coin is dropped at its destination. The number of coins picked up but not yet dropped appears on the screen as a numeral in a special box. There is a "DROP ONE COIN" key which can be handy for adjustment when you pick up too many coins. Dropping one coin back to its source restores color to a ghost coin.

Also on the screen are a changer and grouper. The changer converts a dime into ten pennies or a dollar into ten dimes. To use it, one simply clicks on it after having picked up a permissible input. The output coins go automatically into the appropriate compartment of the tray its input coin came from. The operations of the grouper are precisely inverse to those of the changer. Optimal plans for subtraction eschew the grouper, but it may be needed by some learners to undo unnecessary use of the changer. Also, the grouper is crucial earlier in the arithmetic curriculum for addition problems as well as for the teaching of place value, and we plan to use it later with multiplication.

Another important feature of our environment is the provision of an alternative representation. The student can move back and forth at will between Coinland and a Numberland with the same conceptual underpinnings, carrying out an equivalent search. Facility with the translation process carries the learner close to the ultimate transition, transfer of knowledge from the computer screen to the world of pencil and paper. This step is important, since "strong semantic understanding . . . in no way guarantees ability to use [a corresponding algorithm]" [6]. The grand strategy is that the student combines learning in Coinland, where things make sense, with the understanding that everything in Numberland has its equivalent in Coinland, to achieve not only correct performance in Numberland, but also the conviction that there too everything makes sense. A fuller discussion of Numberland, especially its role in promoting the transfer of skills, is the topic of the last subsection of this section.

Pedagogical Significance

This section highlights and justifies some design decisions, first concerning the interface and then the nature of the constraints on exploration. A screen world such as Coinland establishes conventions about the appearance of objects and the operations on them. For the student to readily understand, retain, and use such conventions, they should map accurately and consistently to simple familiar objects in the real world. What do children in the early grades know about the world that the conventions of Coinland draw upon? At the simplest level, children from before the age of one year exhibit a sense of object permanence, an expectation that objects, although they can be moved, do not just appear or disappear without reason. In Coinland, there is object permanence except when the student activates the grouper or the changer, and even then, the conversion process is emphasized by animation of the coins entering and leaving. For example, if a dime goes into the changer, ten pennies are shown emerging one after the other, accompanied by sound effects for emphasis. The explosion of common nouns in the vocabulary of two-year-olds is one form of evidence that youngsters realize that there are types or categories of objects, so that things that look the same tend to behave the same. Coinland coins of the same denomination are indistinguishable in appearance and behavior. Children also know, before school age, about the notions of ownership, keeping, giving, and buying that our scenario draws upon.

A specifically arithmetic notion for which a convention must be established is magnitude correspondence, the idea that larger numbers correspond to larger amounts of something. Dienes blocks gives a complete correspondence of volumes to numerical magnitudes, whereas Coinland provides a similar correspondence within denominations but not across them, since coins of different denominations are distinguished only by color, not by size. Across denominations we draw instead on conventions established by the money system, with which most of the children are familiar [7]. The changer and the grouper reify the exchange operations, giving a realistic procedural interpretation of the equivalence between one dime and ten pennies, and between one dollar and ten dimes. The changer and grouper conserve monetary quantity in Coinland, and they conserve numerical quantity in Numberland. Admittedly, conservation of quantity is a less transparent notion than object permanence, but children do typically encounter and master coin exchange, not to mention coin-operated machines, before they reach formal multi-digit subtraction.

A final word on the pedagogy of the interface is in order. A major and robust finding in psychology is the efficacy of active as opposed to passive learning. For this reason, we have elected to have the student actively operate upon the set of coins that end up constituting the answer. We rejected an alternative arrangement in which the answer to a problem would have emerged passively, comprising the remaining coins (the "remainder") after removal of an amount equal to the price.

Let us now move past (or through!) the interface to explore the benefits of constrained exploration. Conservation of quantity constrains the search space, making it manageable and instructive. Perhaps most importantly, the conservation property permits circumvention of documented types of misperformance in subtraction [8]. Take the error of borrowing directly from the hundreds to the ones column. This action simply cannot be expressed in our representation. In other words, the privilege of making this error is an option that the student unwittingly relinquishes by acclimating to the Coinland environment.

A few examples will further illustrate the potential benefits of the semantic constraints. The conservation of quantity has as its simplest role the gentle enforcement of one-digit subtraction facts. In the problem $8 - 3$, suppose the learner begins by moving, say, 6 pennies to the give-back tray instead of 5, in effect, trying to break 8 into 3 and 6. This is an attempted violation of the conservation of quantity, which the system does not support, in the sense that only 2 pennies (not 3) will be left in the take-in tray. The system provides the same type of enforcement of intra-column subtraction facts like $13 - 4 = 9$ whose need arises after borrowing.

Consider now the role of semantic constraints in the more challenging example of $100 - 7$. This problem requires subtraction from 0 and borrowing across 0, both of them operations known to cause difficulty. In the Coinland version of this problem, the take-in tray has a dollar at the outset and must in the goal state hold 7 pennies and nothing else. What can the student do to achieve such an outcome? Since the dollar is the only object available at the outset, there are only two possible moves: shift the dollar to the give-back gray or exchange it for ten dimes. Of course, one hopes the student will realize that getting dimes in this way satisfies a precondition for getting the needed pennies. The student will have learned this way of getting pennies earlier in the curriculum from two-digit subtraction problems and from preliminary exercises with the changer.

We hold that such problem solving has a much better change to emerge in the constrained search space of our SCE environment than in the relatively bushy search space of unconstrained paper-and-pencil subtraction that gave rise to the repair theory of bugs [9]. (A bushy space is one that tends to have many options at a point.) In the pencil world, there are many operations, and their various conditions are not easy to remember or determine. In Coinland there are very few operations and the tests for their legal applicability correspond closely to what would be physically possible with real coins. For example, if you haven't got something, say, a dime as in the preceding example, then you just can't do anything with it or to it.

Technique

To put the technique of SCE into perspective, we compare it to related techniques and consider some variations. Exploration of an environment is a learning mode that is characterized by the fairly high degree of control that the student is permitted to exercise, in comparison to other learning modes or educational techniques. On the spectrum from relatively free exploration to drill and practice, Coinland occupies a middle position, allowing the learner to control some aspects of the situation but not others. Specifically, our learner is free to choose moves within a problem, even in violation of the tutor's suggestions, but it is (currently) the tutor that chooses the problem, with a view to progressing in a specific syllabus. With many other techniques there is substantially greater reliance on the direct presentation of material. That presentation may take place via language, either in written form or (ultimately) spoken, and either undirectionally or interactively via questions, with the learner doing either the asking or the answering. As a complement or an alternative to language, material can be presented spatially, with diagrams, animation, pictures, or motion pictures.

Within the particular realm of exploration environments, one can distinguish at least three flavors, according to whether the concepts to be explored constitute a knowledge base, a programming language, or a problem in some domain such as subtraction. The latter two, programs and problems, provide an interesting comparison, as can be seen by looking at some important similarities and differences in the exploration of Coinland and of Papert's world of turtle geometry [10]. First, the turtle's primitive actions of walking, turning, and drawing certainly conform to the above demand for an accurate mapping from an artificial learning environment into familiar aspects of the real world. A further similarity is that with both the turtle and the coins, the learner's task is arguably to assemble primitive operations into an algorithm that solves a problem. The crucial difference is that programming the turtle requires an explicit statement of the looping and branching for a particular problem, whereas in Coinland, as in paper-and-pencil arithmetic, one needs only implicit knowledge equivalent to a correct algorithm. The Coinland learner need not be able to state a correct algorithm to be deemed correct, just be able to act as if she were executing one. (An analogous case arises in speaking a language, where one need not be able to state the grammatical rules, just produce sentences governed by them).

To consider more explicitly the issue of explicitness, suppose we shared Papert's worthy aim of promoting explicit exploration of algorithms. Then perhaps we should have invented a Coinland programming language. About the simplest form such a language could take is one that allows one-digit subtraction to include code something like that shown just below. Although a child might work successfully at this level of complexity, the move to multidigit numbers would require a more complex language, specifying columns. An explicit procedural solution encompassing borrowing is demanding indeed, since it requires nested loops. (On the difficulty for preschoolers of even implicit handling of nested loops, see [11].)

```
GO-TO       TOP;
DO          PICK-UP-ONE
UNTIL       TOP = BOTTOM;
GO-TO       ANSWER;
RELEASE;
```

Methods involving props like coins and Dienes blocks have been used without computers for a long time [12]. The computer adds flexibility in the creation of new environments and variations in them (as well as a potential for individual attention that may otherwise be in short supply from human teachers). An example of this flexibility is the choice of whether or not to permit the commingling of coins of different denominations. In the current version of Coinland, there simply is no mechanism for putting down coins in a compartment not of their own denomination. It would be easy enough to change the implementation to let the student choose an arbitrary target location, but we have chosen not to, since that would expand the search space. This design decision is easy to implement computationally, but in traditional setting with physical coins the prohibition of commingling could only be expressed as a rule that the student would have to remember, thereby potentially interfering with concentration on important conceptual matters.

Another design decision that the computer facilitates involve a technique introduced below, the use of partially covered compartments in the answer tray, to discourage removing coins from the answer once they are put there. The computer also deftly accommodates large numbers of coins, no small consideration in view of the observation that "teachers usually stop using objects to represent mathematical processes in second or third grade, when the numbers get bigger, [and] the objects become more unwieldy" [7].

To take a final viewpoint for comparison, regard Coinland as a one-person game. In conformity with Malone's [13] important observations about games, Coinland has no conceptually wrong moves, just inefficient ones. In this context, taking more moves than necessary to reach a correct answer means nothing worse than achieving a less than maximum score. Consequently, there need be no inhibitions about trying things out. A recent survey of some relevant game-like techniques is in [14].

Representation

Good representation can help make ideas clear or even obvious by facilitating their mapping to analogous subject matter that is well understood. As noted above, good mapping is one desideratum we claim for the Coinland representation; another is versatility. Versatility is desirable for a representation, because it means that once the conventions have been learned, they can be used for the learning of a broad range of subject matter. The applicability of Coinland transcends subtraction, clearly extending downward in the arithmetic curriculum to addition, counting forward and backward and finding the larger of two numbers. Moreover, we intend to use most of its representational conventions for more advanced subject matter, including multiplication, division by very small integers, the use of a radix other than 10, and the addition and subtraction of small mixed numbers. We have hopes that Coinland will make it possible to teach beginning division in a particularly meaningful way, independently of multiplication, relying directly on an experience that many children have been exposed to: dividing things to share them.

The use of objects in the representation, in place of numerals, is a move toward concreteness, versus abstractness. It also entails more entities and actions, be they abstract or concrete, and in this sense it is relatively detailed. For example, in place of the one symbol "5," there are five objects, and the solution process for "8 − 5" involves three operations of picking up, rather than accessing the single atomic fact that 8 − 5 = 3. By failing to afford a way to use the atomic fact, the representation makes the learner show the validity of that fact in the course of using it. Erroneous beliefs about such facts can be discovered in this way.

In this regard, our approach bears an interesting relationship to that of White and Frederiksen [15] in the domain of electrical circuits. In their system, the learner starts with a qualitative model that moves away from quantitative detail. One of their rationales is that the simplicity of this representation makes it suitable for beginners. Teaching a simplified version may even have a benefit beyond the learning period, because it has been found in studies of physics problem solving that experts make greater use than do novices of qualitative models in the initial approach to conceptualization. This result may in part be a pedagogical artifact, but presumably a part of expertise is knowing when and how to suppress detail and focus on what is important.

Why then do we choose for our subtraction representation one that involves extra detail for the beginner? The answer to this question reflects conventional wisdom in the matter, as voiced by Glennan and Callahan [1]. Briefly, the details are included in order to demystify what may otherwise appear to be arbitrary aspects of the algorithm. The simple arrangement allows the tray to serve as a visual version of a *list*. The elements of this list are the numbers of coins in each compartment. Since a number now appears as a list of smaller numbers, instead of just a digit-string, it is possible to show 42 as either

([• • • •][• •]) or ([• • •][• • • • • • • • • • •]).

(On the screen the 12 coins are stacked in rows of 5 for easier comprehension.) Thus, the results of regrouping can be accommodated with no extra representational machinery either visually or with respect to the way things are interpreted. We simply make the compartments big enough to hold more than 9 coins.

Contrast this situation to the case of pencil and paper. Just when the important idea of regrouping is to be conveyed, the pencil and paper student is confronted with new conventions demanding the crossing off of digits and the writing of digits in strange new places. To regroup 42, for example, one crosses out the 4 and writes a 3 and a 1 in slightly different positions within their relative columns and with different kinds of interpretations. At this point, moreover, the 3, 1, and 2 are all active, with no clear boundary between any two of them, and yet they do not collectively mean 312.

Transfer Via Numberland

Transferability concerns the question of how smoothly the student will be able to move from capabilities acquired in the computer-based instructional environment to target capabilities in more conventional venues and representations. As noted earlier, transfer need not be automatic; for some learners with some techniques, it must be explicitly learned. Even if a new environment is superior to a traditional method, its value may be undermined or partially offset by difficulty or failure when it comes to transfer.

To facilitate transfer, we have developed a second, alternative screen, called Numberland, that is intermediate in appearance between the Coinland screen and the look of pencil-and-paper subtraction, and that permits solution of the same subtraction problems. Once the student has succeeded at the Coinland version of a problem, the system presents that same problem in Numberland. The student can then use her own valid knowledge of how to proceed in Coinland to instruct herself in Numberland, by toggling back and forth between the two representations, solving the problem in both, more or less in parallel. This capability is implemented for subtraction; ideally, learning the translation between the two screens would begin in the simpler reals of counting, regrouping, and addition, so that its occurrence with subtraction would constitute a continuation.

Numberland looks much like Coinland, though there are some crucial differences. Figure 2 shows the Numberland screen for the same problem as Figure 1, at the same stage of solution (even though the "11" here has no counterpart in Figure 1, a point we will return to). The most noticeable difference is that in Numberland all the coins give way to numbers. For example, the numeral "8" appears, in Figure 2, corresponding to the eight dollar coins in a cup in Figure 1. The three keys for picking up and dropping coins are still active, as are the arrow and ENTER keys. Moreover, the same sequence of keystrokes that solved a problem in Coinland will also solve it in Numberland. This is because at an underlying conceptual level, the two "lands" are identical. Not only is Numberland internally unified with Coinland but, equally important, it is externally unified with the world of paper and pencil, as will be spelled out momentarily. It is this twofold linkage that suits it for the role of fostering transfer.

Further description of Numberland will reveal its external similarity to paper and pencil. In particular, the original problem statement is *not* destroyed by student actions here. The actions that in Coinland move five out of the eight dollars from the take-in tray down to the give-back (answer) tray now cause a "5" to appear down in the answer. For the duration of those actions, the "8" is still present, but it turns white, like the dis-

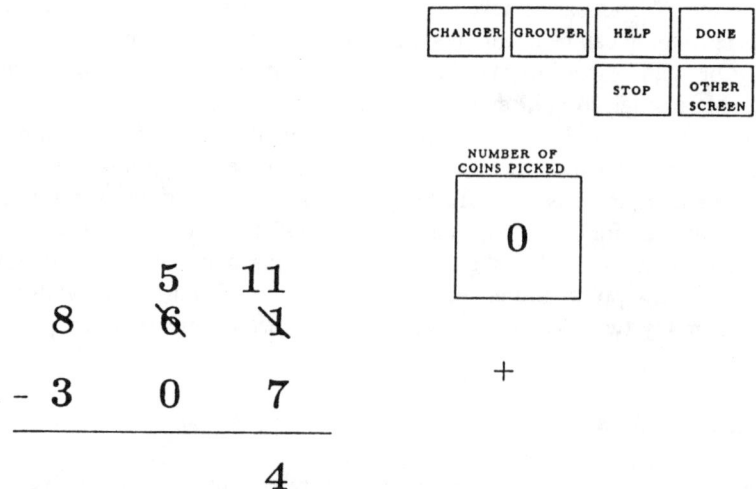

Figure 2 The Numberland screen during solution of the same problem, 861 − 307, as in Figure 1, at the same point in the process. The same keystrokes have occurred as in Figure 1. Conceptually, there are only 7 pennies left in the upper row, even though the number 11 is apparently active there. Therefore, a student who attempts to pick up more than 7 pennies at this point is thwarted, and there is an offer of help.

turbed coins in Coinland, and regains its color at the end. Thus the "8" ends up looking as if nothing had happened, which does not correspond to Coinland (where eight pennies would be reduced to three) but does correspond to the task on paper. A further partial correspondence to the world of paper and pencil is the appearance of the effects of using the changer. In Figure 2, the learner has executed the keystrokes that in Coinland exchanged a dime for pennies in the take-in tray. In Numberland the result is that the minuend, which is 861 or (8 6 1) in list form, becomes (8 5 11) in list form. Notice that the "6" and the "1" are still visible, to provide continuity, but each is now overstruck with a slash. Of course in the figure, the "11" does not mean that there are 11 pennies still in the take-in tray, because four have been lowered so that only seven remain. A learner who forgets this and tries to pick up more than seven pennies is notified why that is not a possible move. The "11" does reliably signify that there are 11 pennies altogether, because we do not allow coin exchange in the give-back gray, in further correspondence to the conventions of paper and pencil.

One last potential contribution to transfer is the slotted, or partially covered, cup. It is harder to extract coins from a slotted cup than from the open ones seen up to now. Slotted cups complicate the environment but will simplify the search. They are to be introduced after the learner has already become familiar with the environment. Recall the observation that a small search space should help reduce inefficient and possibly unenlightening search. The slotted cup would be the external representation of a tightening of the search space. It represents a disincentive, which may be regarded as a variable constraint, in contrast to an absolute one. Specifically, we attach a disincentive to the action of taking coins out of the compartments of the give-back (answer) tray. Even in the absence of such extractions, there would still be enough flexibility to allow an interesting search. The rationale for discouraging that kind of move is that it is invariably evidence of an inefficiency, caused either by the move itself or by an earlier move.

More to the point in this discussion of transfer, altering the answer (the pencil and paper analog of removing coins from the answer) is not permitted in the subtraction algorithm that is taught in most U.S. schools. With that algorithm as the target skill, this variable constraint is a bias toward successful transfer.

To communicate the disincentive visually, the give-back compartments, which otherwise appear as cups without tops, would be drawn with partial tops: each top appearing as a line with a gap, to indicate the side view of a slot. The story is now that it is easy to drop coins in through the slot but hard to get them out once they are in, since the hand that carries coins does not readily fit through the slot. The game score reflects this difficulty by penalizing extractions from slotted cups. Notice that whereas real physical impediments might frustrate a student, the graphical impediments are easily overcome, yet continually provide a reminder and challenge.

Heuristic Guidance

Heuristic guidance (HG) of a student's exploration in problem solving is complementary to semantically constrained exploration (SCE). HG shares SCE's general objectives and is envisaged for use with it, providing guidance within the constraints. HG differs from SCE by being responsive to the choices, especially the most recent choice, that the current student has made in attempting to solve the current problem. Also, the suggestions in HG need not be heeded, unlike constraints. The first subsection places HG in context and presents the basic idea. We then begin to assess its potential for success outside subtraction by looking at some domain characteristics. Finally, we provide the specifics of generating warnings and hints, in terms of a state-based analysis of a subtraction scenario.

The Context and the Basic Idea

In HG, an electronic tutor uses a heuristic function with both cognitive and problem-oriented components to guide exploration. Since such guidance might otherwise be based on a model of the student, formed through a process of diagnosis, HG can be seen as a strategy for reducing the need for diagnosis. It is also quite possible to use HG and student modeling techniques in tandem. In any case, reduced dependence on diagnosis could be highly beneficial, because diagnosis is impossible to do perfectly and difficult to do well, that is, with great detail and reliability. Even for subtraction, it is complex [16]. We shall try to show that, at least in subtraction, HG can generate the kind of advice that is needed for successful learning. After an introductory subsection, we explore limitations on the applicability of the approach and then provide a detailed example of how warnings and hints can be generated.

As envisioned so far, HG can take the form of warnings and hints. When the HG tutor issues cautionary advice to the student, its reasoning, though not necessarily its actual output, goes something like this: "I don't know why you made that move, but it doesn't achieve any obvious improvement and doesn't appear to make things look simpler for you." This sort of warning does not depend on hypothesizing a student model, say, by ascribing a partially deformed version of a standard algorithm. Instead, the method requires the ability to evaluate the new state that results from the student's current move, in comparison with the evaluation of the previous state. We use the word "state" here to refer to a complete specification of all the significant aspects of the current situation, such as the number of coins of each type in each tray. In a more positive vein, in contrast to issuing a warning, the evaluation of states can also provide the basis for a hint, suggesting a move to a specific alternative state.

To be selective about when to give warnings, an HG electronic tutor needs a heuristic function with both problem-oriented and cognitive components. From a purely problem-solving viewpoint, a heuristic evaluation function may take into account how similar a state is to a goal state (solution) with respect to various attributes. Such information concerns the problem, not the solver. It ignores cognitive limitations and possible strategies to deal with them. Two other classes of candidate traits for desirable states are more explicitly cognitive in nature, namely, the various aspects of simplicity and prominence. A prominent state is, more or less by definition, easy to keep in mind, so it may be looked to as a subgoal. Simple states, by allowing low cognitive load, may make reasoning easier.

How to combine these various characteristics into a single value for the heuristic evaluation function is another aspect of tutorial strategy. One reasonable approach is to add within category and then combine the results disjunctively, by taking their maximum. Using a function formed in this way, a tutor can leave a student alone as long as he or she either is getting closer to a solution state or appears to be successfully pursuing simplicity or prominence.

Domain Characteristics

Not just any evaluation function will be good for HG, and some domains other than subtraction may not have any good ones. Two characteristics of the subtraction domain seem to help make it yield a suitable heuristic function. One is relative freedom of ordering in the move sequence (to be discussed shortly), and the other is a high degree of informational independence. Both of these characteristics show up in the fact that the desirability of ever borrowing (anywhere in any sequence of steps) into a particular column can often be confirmed or denied entirely from information local to the column itself. Relatively free ordering means that there are many equivalent paths, so it severely undermines the possibility of diagnosing in the sense of determining the intended path. HG, however, deals with states, not paths—neither a path chosen by a resident expert nor one attributed to a student.

A somewhat unorthodox approach to teaching subtraction is implicit in this discussion and is in fact adopted in Coinland: no preferred sequence of operations is enforced. For example, in any problem without borrowing, the individual intra-column (digit-wise) subtractions can occur in any order, unlike algorithms commonly taught that prescribe a right-to-left sequence. This neutral stance, which is a natural consequence of the design, is compatible with Self's third slogan: "Empathise with the student's beliefs; don't label them as bugs" [17].

Since order is not imposed, it becomes possible for a student to discover a benefit of (the standard) right-to-left order in some circumstances. If she prematurely shifts all the dimes to the answer, say, but subsequently needs a dime to exchange for pennies, she will have to (at least partially) undo that shift. She may notice this, react appropriately, and ultimately reach the goal state, though not via a minimal length path. Our current (non-HG) system-resident expert can readily detect that its own optimal solution is shorter and can inform the student that the "expert" finished the problem in fewer moves. Somewhat better are the slotted cups referred to above, which warn of inefficiency before the end of a problem. However, the earliest warning of all can be provided by HG, which, as seen below, will discern when a move first commits the student to a suboptimal path.

From this discussion of alternative paths, it may already be clear that there are many correct algorithms for subtraction. Various ones are taught around the world, so it is not possible for everybody to have the same one. That being the case, we may relax about letting children invent their own if they can. To stress the existence of variety, we express, in terms of Coinland moves, a little known but simple and correct algorithm that we call "prepare-and-do." It has the unusual virtue that it covers all cases, including borrowing across zero, with an unvarying sequence of steps: (1) go through the take-in tray from left to right, using the changer at every position; (2) perform all intracolumn differences (in arbitrary order) and (3) go through the give-back tray, from right to left, grouping wherever possible. Coinland does not, however, support all possible algorithms. One that it does not support in any obvious direct way is one in which borrowing is accomplished by incrementing the subtrahend, because that method violates conservation of quantity; see [18].

This discussion has been based on what we have asserted is the relative looseness in the ordering of steps to solution in Coinland subtraction problems. To be a little more precise about it, moves in this problem space are partially commutative, in the following sense. For any two operations, A and B, such that it is possible to apply them successively in either order to some state, the order is immaterial. That is, if both A(B(s)) and B(A(s)) exist, they are equal. This makes for a space with many cycles and many options at a state. Most to the point, suppose that a student's method dictates the setting of a particular subgoal and that a particular state corresponds to that subgoal. Then there will typically be several equally good paths to that state from the current state and more than one possible first step. Thus, it is not unusual for the set of possible first steps for one approach to overlap those of another approach, making the steps in the overlap ambiguous clues for determining what algorithm, if any, the student is using. Although over the course of time the system might gather many clues in hopes of resolving ambiguity, the student might change (learn) in that same period.

To this flexible domain HG brings flexible tutoring by (1) granting validity to more than one algorithm and (2) allowing exploratory deviation from an algorithm. In contrast to HG, one can envision a tutor with flexibility only on point (2). Such a tutor might allow a student to take more than one step off the standard path but guide the too-deviant student back onto the track. Note, however, that determining how far someone has deviated is not just a simple matter of keeping track of how long ago he or she left the designated path, since despite a long excursion the student may now be almost back on track. This point is especially relevant in a domain with flexible move ordering.

Before leaving this discussion of HG, we put it in some context by mentioning two interesting alternative strategies, each of which renders diagnosis simpler, rather than avoiding it. Singley [19] makes the student post goals, which is like partial self-diagnosis in that it turns over to the student some of the responsibility for figuring out what he or she is up to. In Anderson's Lisp tutor [20], a complex intelligent tutoring system, the price of simple diagnosis is the use of tight constraints on exploration, a tack that, in its severest form, would reduce exploration to mere tracking, with neither of the two types of flexibility described above. On the potentially negative side for HG, it may be subject to the weaknesses of the "weak" method of search, notably inefficiency. However, if it is true that there are pedagogical advantages to allowing exploration, and if we agree that we want to get them, it seems that we have already bought into the idea of search. It is our endeavor first to constrain that search, as we have done by introducing the

Coinland environment, and then to guide it, as we are now suggesting here by introducing the use of a heuristic evaluation function.

To sum up, we have argued that a substantial degree of order-independence in a domain tends to make diagnosis hard, since a step doesn't reveal an algorithm, but that same property tends to make HG, with its state-based evaluation functions, a reasonable alternative for guiding the tutorial process. If this is so, the two techniques are complementary, and each should be used where it is suitable.

Hints and Warnings for Subtraction

To devise a system of HG for a domain, one must first decide what counts as a state and what are the significant attributes of states. For the Coinland version of the subtraction domain, we abstract a state space that consists only of situations that can exist immediately after the completion of a shift of coins or a conversion between denominations. We thereby exclude from statehood those situations that arise from just picking up or unpicking coins and/or moving the cursor. Although we have earlier justified having the learner pick up one coin at a time, those detailed actions have little or no significance for solving the problem at hand and indeed have no analog in the world of pencil and paper. From this viewpoint, picking up coins is a micro-operator used to form operands for the principal operators of shifting and exchange.

Obvious candidates for significant attributes of a state are the numbers of coins in various compartments. We shall use all six of these numbers. Less obvious are attributes that summarize the problem-solving history, like the net number of dimes converted to pennies (changer uses minus grouper uses) and net number of dollars converted to dimes. We shall use both. These last two numbers are not visible on the screen so they are less appropriate for use in explanation, but they can help decide whether to issue a warning, and they are useful in generating unexplained hints. (The number of coins picked up is not a useful attribute, since it is always zero in the situations allowed as states.) These eight state attributes are not all independent, since for example, the combined pennies in the two trays, minus 10 times the net number of dime-for-pennies exchanges, is invariant. Two other similar constraints hold, so that three of our attributes could be dropped without loss of information, though with some loss of simplicity in the discussion below.

From a purely problem-solving viewpoint, the most desirable value for a compartment of the take-in tray is its target value, the corresponding digit of the price, because this is its correct final value in any solution. Of course the correct final values for the give-back compartments are also desirable values for those attributes, but they are not given in the problem statement, as the price digits are. Therefore these values, like the net conversion numbers, are used only to formulate unexplained hints.

We now work through a scenario with a specific problem, 503 − 7, showing how warnings and hints come into play. In accord with the discussion above, the initial state in denoted 5,0,3;0,0,0;0,0. The first six numbers indicate how many coins are in the compartments of the two trays, and the last two are the net number of exchanges of, respectively, dollars for dimes and dimes for pennies. The correct final state is 0,0,7;4,9,6;1,1. Suppose the student begins by lowering all the dollars to get 0,0,3;5,0,0;0,0 and then exchanges a dollar within the given-back tray to get 0,0,3;4,10,0;1,0.

In the pencil-and-paper world, corresponding operations are incorrect with respect to the standard algorithm, and indeed the second one would require an extension of the usual representation. Nevertheless, they do appear to have some possible merit in terms of our attributes, so no warning is given. In particular, the first move meets a take-in tray objective of being equal to the implicit price digit of zero in the hundreds column. Introducing a zero is also regarded as a potential cognitive benefit of simplification. The second move in the scenario brings both the give-back dollars and the net number of dollar exchanges to their correct final values. Note that exchanges are counted without regard to the tray in which they are performed. Strange as these moves may seem, it may indeed turn out to be wise to have tolerated them, since they do flow from the not incorrect solution strategy, "give enough dollars back to the customer so that it is the customer who must make the change."

Converting another dollar at this point does deserve a warning. That move yields the state 0,0,3;3,20,0;2,0, thereby disturbing some goal values, with no compensating benefits. To generate a hint, we use an algorithm that loops through the columns from right to left. In each column it first tries to achieve what we will call G1, the visible objective of equating the take-in compartment to the price digit, by means of a shift in one direction or the other. If G1 is not already satisfied, it can be achieved provided the total number of coins of the current denomination is greater than the price digit, but by an amount no more than 20 (the capacity of a compartment). For example, if a (confused) student has 14 pennies in each tray while solving a problem in which the price is 5, the system should not suggest lowering 9 pennies to get the required 5 above but an impossible 23 below. In the present example, with only 3 pennies present altogether, there is no shift that can get the take-in tray to have 7.

If G1 is already satisfied or cannot be satisfied by a legal shift, the hint generator instead seeks a hint for achieving G2, getting closer to the correct net number of exchanges. Preference is given for an exchange within the take-in tray. In the present example, the target number is 1 and the current number is 0, so the hint would be to get ten pennies for a dime by using the changer. Since there is no dime in the take-in tray, the hint here would be to change a dime from the give-back tray. Suppose that the student accepts this hint, so the state become 0,0,3;3,19,10;2,1. If a hint were requested again, from this state, it would be to move 4 pennies up to the take-in tray, giving that tray 7 pennies, its target number, and making the new state 0,0,7;3,19,6;2,1.

If yet another hint is requested, the hint generator algorithm will find that G1 and G2 are satisfied for pennies, so it will move to a consideration of dimes. There, G1 is satisfied but G2 is not. Since there has been an excess of changer uses, 2 instead of the 1 called for, the hint is to use the grouper. Since 10 dimes are not to be found in the take-in tray, the hint is to group 10 dimes from the give-back tray, thereby solving the problem. Interestingly, this move happens to be the inverse of an earlier (poor) move. There is a rationale for this move that can be expressed in terms of the objectives of Coinland subtraction and can therefore be cited when communicating the hint to the student. This is the requirement that, in the final state for every problem, no compartment in the give-back tray may have more than 9 coins.

Before leaving this discussion of HG for subtraction, we comment on its use with two particular algorithms, first Ikeda's algorithm [21] and then the standard one. The advantage of Ikeda's method lies in its intriguing property of not requiring the student to learn the harder subtraction facts like $12 - 7$ and $17 - 9$, in which the minuend is greater than 9. To solve $62 - 37$, the student in effect regroups 62 not into 50 and 12 but rather into 52 and 10, then takes 7 from the 10, and to the resulting 3 adds the 2 taken out of 52 to get 5 in the ones column of the answer, leaving 50 from which to take 30 to get a 2 in the tens column of the answer.

HC accommodates Ikeda's method without having been explicitly designed for it, as will be seen momentarily. When subtracting 62 − 37, Ikeda's method corresponds to lowering 2 pennies, then regrouping, then lowering 3 more, then finishing off by lowering 2 dimes. The first move scores well because it creates an empty compartment, regarded as an aspect of simplicity or relatively low cognitive load. Note that it would be difficult if not impossible to infer, over several problems, with a possibly changing student, whether this kind of move is really being used systematically and therefore deserves to be credited with possibly being part of an algorithm, or is just a part of some haphazard searching.

Finally, suppose one wishes to push the student toward move sequences compatible with the standard algorithm. One can do this with an "advanced" version of the game, formed by introducing the extra constraint that coins may not be taken out of the give-back tray, even for a penalty in the score. In this version, it makes sense to warn against certain moves in our example scenario that did not receive warnings in the above discussion. In particular, the 9-coin limit on give-back compartments applies to all states, via an explainable warning that says, in effect "That's too many dimes (or pennies) in the give-back tray for a final answer, and you are not allowed to take them out, so you better not put them in." It is worth noting that this constraint is not too terribly confining, as can be seen from the fact that it permits both Ikeda's method and prepare-and-do (see above), in addition to the standard algorithm.

The Syllabus

Both SCE and HG have their effects within a single problem, as the learner seeks a solution to it. Up a level and outside the individual problem is the broader issue of how to choose a particular problem, among the huge number available, to meet the need to work on a particular topic in a syllabus. This matter in turn raises the issue of what material goes into a syllabus and how to structure it for orderly coverage. The first subsection deals with syllabus issues with extended attention to subtraction. Then we turn to navigating that syllabus, that is, to selecting a topic and problem. The final subsection concerns the nature of subtraction.

The syllabus not only specifies material but also makes commitments to the granularity of its organization. For example, given that one-digit addition with one-digit answers is a topic, should adding 1 be split off as an introductory subtopic? The level of granularity to which the topics are dissected should be coarse enough for the units being learned to be coherent, yet fine enough to permit the explicit expression of constraints and preferences with respect to the order of presentation of material. These constraints and preferences structure the material and thereby determine in what order(s) it is to be learned. One important kind of constraint is the prerequisite relation, imposed between two topics or concepts. One topic is an absolute prerequisite of another if failure to master the first completely undermines the learning of the second. A variable strength version is also possible. Note that a syllabus closely resembles what is commonly called *domain expertise,* but it emphasizes declarative principles over compiled procedures. The earlier emphasis on conservation of quantity exemplifies the distinction and is in this regard indebted to insights that drove the evolution of Guidon II [22] away from unexamined expert performance rules.

The Subtraction Syllabus

Turning now to the specifics of subtraction, note first that there are roughly half a million ways to choose a pair of three-digit numbers for a subtraction problem with a positive answer (and about 50 million with four-digit numbers), but many of these are essentially equivalent in terms of cognitive demands. To carve up the space of possible problems into structured topics for a syllabus, we will appeal to the notion of subgoal and to other constructs from the study of artificial intelligence (AI), thereby pursuing a strategy that has potential use in domains outside arithmetic.

First, consider the requisite values of parameters in the goal state. To complete a Coinland subtraction problem successfully, one must get each compartment in the take-in tray to have a number of coins equal to its corresponding target digit. This requirement gives rise to three subgoals, one for each compartment or digit. These subgoals categorize problems according to whether or not a given subgoal is already satisfied in the starting configuration. This is equivalent for a subtraction problem to whether the top and bottom digit in a column are the same. Problems can be further divided according to whether in the starting configuration some subgoal can be achieved in a single move, as when the top digit in a column exceeds the bottom one. Two additional subgoals arise from the requirement that the dime and penny compartments in the give-back tray must each have no more than nine coins to represent a permissible answer.

Turning from subgoals to moves, we note that some starting configurations are of special interest for a different kind of reason, namely that they make certain initial moves impossible. For example, in the problem 100 − 7, the two initially empty compartments in the take-in tray constrain the learner to just two possibilities for the first move, specifically, exchanging the dollar for ten dimes or moving the dollar to the answer tray. This paucity of options is attributable to the semantic constraints of the Coinland environment. In this way, we hope and expect that Coinland will facilitate solution in the foregoing problem with its renowned pitfall of borrowing across zero. If that turned out to be true and if the success in Coinland were to transfer, it would strongly suggest that the original difficulty was an education artifact.

The foregoing example has the potential for a different pitfall. To follow the usual algorithm, one must undo satisfaction of an initially satisfied subgoal, specifically that of having the number of dimes in the take-in tray equal to zero, the target digit in the middle column. To preserve satisfaction of that subgoal, one could move the dollar down, make change to get dimes and then pennies in the give-back tray, and raise seven pennies to the take-in tray. This approach answers the question of what is left when 7 is taken from 100, rather than how much must be taken from 100 in order to leave 7 behind. Volpel [23] discusses such ways of thinking about subtraction.

There are two kinds of moves in Coinland, the shifting and the exchange of coins. Shifting is simpler in that it involves only one denomination at a time and conserves physical (graphical) objects, as well as quantity. Moreover, shifting is used in all problems (except those whose answer is zero). These considerations suggest that borrowing, which is equivalent to exchange (to a lower denomination), is the more crucial type of move and so should be considered at the top level in our taxonomy of problems. Of course, one already knows that borrowing is crucial; what is of interest is that this conclusion has flowed out of a general consideration of exploratory problem-solving moves.

In summary to this point, we have considered two aspects of problem-solving: move types and sub-goals. Consideration of move types has suggested the importance of borrowing, and our earlier discussion of subgoals showed that the relative sizes of zero, T and B are important, where T and B are, respectively, the top and bottom digits in a column. There are six distinct "B-T-zero patterns," as follows:

$$0 = B = T$$
$$0 = B < T$$
$$0 = T < B$$
$$0 < B = T$$
$$0 < B < T$$
$$0 < T < B$$

These distinctions yield 92 categories of subtraction problems, each with a three-digit minuend and a non-negative result. Most of these, 72 ($= 6 \times 6 \times 2$), arise because any of the above six orderings can hold in the ones and tens columns, along with either of the two orderings for which $T > B$ in the hundreds column. Only another 20 possibilities arise with $0 < T = B$ in the hundreds column, since the other two columns must then be consistent with a non-negative result. Moving to four-digit problems, for which there is some pedagogical justification, increases the number of categories roughly six fold. This is better than going up by a factor of a hundred, as is approximately the case for possible individual problems, but still of finer grain than is convenient or necessary for tutorial purposes. A coarser categorization is available at the level of borrowing patterns, which arise above from considering move types. Distinguishing three-digit problems according to borrowing, seven patterns arise, three of them having at most a single borrow:

NB: No Borrow
$\qquad T \geq B$ in each column.

B-ONES: Borrow only into the ONES column
$\qquad T < B$ for 1s; $T > B$ for 10s; $T \geq B$ for 100s.

B-TENS: Borrow only into the TENS column
$\qquad T \geq B$ for 1s; $T < B$ for 10s; $T > B$ for 100s.

The remaining four patterns all involve borrowing in both columns, so $T < B$ for 1s, $T \leq B$ for 10s, and $T > B$ for 100s. Various special conditions are indicated for 10s, according to whether or not $0 = T$ and whether or not $T = B$.

B-TWICE; Borrow TWICE, but with none of the special conditions below, so $0 < T < B$ for 10s.

BAZ: Borrow Across Zero.
$\qquad 0 = T < B$ for 10s.

UNDO: UNDO the initially met subgoal.
$\qquad 0 < T = B$ for 10s.

BAZ-UNDO: The equalities in BAZ and UNDO hold.
$\qquad 0 = T = B$ for 10s.

The last few distinctions are motivated not only by subgoal considerations but also by another general AI problem-solving notion, that of constraints on the order of moves. In particular it is the borrowing (exchange) moves that are constrained. In BAZ, borrowing into the tens column necessarily precedes borrowing into the ones column. In UNDO, the reverse order is strongly suggested, since with $B = T$ in the tens column, there is initially no apparent need to borrow into that column. BAZ-UNDO provides a basis for both the preceding arguments with their associated opposite orderings for borrowing. Confusion could arise if BAZ and UNDO were each encoded by a learner as a slightly over-general rule with antecedents of, respectively, $T = 0$, and $T = B$. Then both rules would be applicable under the conditions of BAZ-UNDO, giving conflicting advice.

One important reason for creating this taxonomy is to impose a partial order on the material by the prerequisite relations, depicted in Figure 3. The easiest pattern is at the bottom of the figure, the hardest at the top. Although B-ONES and B-TENS are shown as unordered with respect to each other, there is a rationale other than the prerequisite structure for presenting B-ONES first. Since it can be taught with only two denominations present, it will be taught before B-TENS if the tutor has a principle of using as simple a problem as possible when introducing a new problem characteristic.

There are at least five discernible granularity levels in an over-all taxonomy of subtraction problems. Beneath the top level (at which no distinction are made) is the level of borrowing patterns, then that of the B-T-zero patterns. A fourth level concerns distinctions between relatively difficult facts like $17 - 9 = 8$ and easier ones like $12 - 6 = 6$ (discussed below). Fifth and last is the level of individual problems. One might insert yet another level, concerning number of digits, right beneath the top level. Viewing subtraction within arithmetic brings the number of levels to seven.

Selecting a Topic and a Problem

Having structured the material, we turn to the question of navigating within that structure. Paying attention to the individual learner and responding appropriately could begin with such high-level decisions as the choice of subject matter and instructional technique. Having chosen subtraction as the subject matter, the tutor may encounter a student

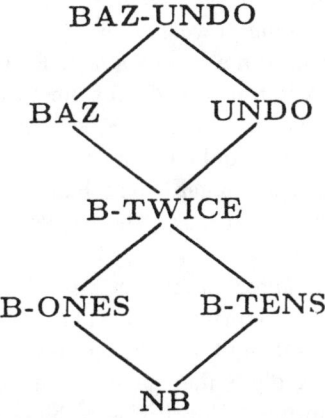

Figure 3 A coarse-grained taxonomy and prerequisite structure for three-digit subtraction. The nodes are categories of problems, defined in the text. A problem type at the lower end of an arc should be mastered before the student attempts the type at the upper end of that arc.

who needs to be backed up to its prerequisites. We have tentatively dealt with this problem by some introductory exercises on the prerequisites. We will discuss some general ideas on prerequisites below. When the mode of interaction is problem-solving, the tutor must be able to select suitable problems and, within a problem, know when to interrupt and what to say when doing so. We considered the latter problem earlier under the rubric of HG. We now consider how to choose a problem, in two steps. First, determine whether to continue on the same (sub)topic within the syllabus, and know how to choose a new one when appropriate. Second, choose a good problem on the topic.

Given the kind of taxonomy described in the syllabus section, be it in subtraction or some other subject, how does the tutorial component navigate it? We now address this question of the choice of topic within any syllabus. The areas or topics in a syllabus may have various individual attributes and pairwise relations that are tutorially significant. These attributes and relationships are the key to determining a good choice for the next topic. Useful attributes of a topic are its importance, its degree of difficulty, and the extent to which the current student has apparently mastered it. One relation that would seem indispensable is the notion of a prerequisite. Part-whole relations may also prove helpful and possibly too the notion that one topic should immediately precede another. Some of this information, like the prerequisite relation, may be regarded as imposing firm constraints on the ordering of subject matter, while others, like difficulty, might indicate only a preference. Notice that predecessor implies prerequisite and that prerequisite must be acyclic if it is to be enforced.

If the foregoing kinds of information can be supplied by a domain expert, it is possible to assemble automatically from that information an order in which topics occur. Suppose that topic A is a prerequisite of B and B is a prerequisite for C. Let it also be specified that A should be an immediate predecessor of C and further that C is very important. We will first consider how a planner might proceed and then look at a myopic forward reasoner. To formulate an order among the topics, one might first consider the important topic, C, and put it into the plan. If the predecessor relation is taken as a firm constraint, then the new plan becomes ((A C)), where the outer list can tolerate insertions anywhere, but the inner list only at its ends. Neither permits reordering. Checking prerequisites, we must add B before C. We amend the plan to (B(A C)) which is less constrained than ((B A C)) and yet meets the requirements considered so far. Another check of prerequisites leads to the final plan (A B(A C)).

Now consider a forward reasoning process. C is rejected as the initial topic because its prerequisite and predecessor have not been studied. B also has a prerequisite, so A becomes the first topic. When A is finished, C, for which it is predecessor, is considered, but the prerequisites of C are not met, so B becomes the next topic. After that, only C remains unmastered, but it cannot be taken up unless A immediately precedes it. If the predecessor constraint is again regarded as binding, then A is repeated, after which the conditions for studying C are met. The sequence is the same as that found earlier by the planner.

The use of problem solving requires that the tutor select an appropriate new problem from time to time. In our earlier structuring of the subtraction syllabus, the topics were defined in terms of relative size of digit values, so once a topic is chosen, finding a problem is a matter of generating digits that satisfy those relationships. The choice of particular digits influences the difficulty level of a problem within the particular topic. Partly on the basis of findings by Clapp [24], we assume certain generalizations about the relationships between numerical properties and the resulting difficulty. Actual use of the system will enable us to check these assumptions. For example, we assume that the difficulty of a one-digit subtraction fact increases monotonically with the minimum of the subtrahend and result, for a given minuend, making 9-7 harder than 9-8. Getting a problem at the right level of difficulty within a topic can be achieved with a repertoire of three methods, one to generate an easy problem and two others for transforming a problem into one that is harder if the student is doing well, or easier otherwise.

The Nature of the Material

In addition to breaking topics into partially ordered subtopics, there must be a decision about the nature of the material. Should an arithmetic operation like multidigit subtraction be regarded, and also taught, as something that is like production rule programming, procedural programming, or something else? In other words, in what form, possibly a computational paradigm, shall we express the material to be learned? Presumably one wants to choose the form in which the material will be most readily learnable or "communicable" (in the sense of Wenger [18]), best understood and easiest to use. Learnability and usability in turn depend on fit with cognitive capabilities. New material fits to the extent that its parts are equal to or closely analogous to what is already known or can easily be figured out by types of reasoning that are relatively effortless for a person.

Subtraction is a good domain in which to examine the choice of expressive form. It is typically taught as an algorithm but can also be expressed as a set of rules or even as a set of principles. We next spell out these three possible viewpoints for subtraction, commenting also on the nature of the comparison relations, greater/less than. After that we comment on the facts or data used in conjunction with the algorithms, rules, or principles.

First, as an algorithm, subtraction of multidigit numbers can be expressed as an outer loop over the columns from right to left. For each column, there is an if-then structure that branches according to whether borrowing is needed. Borrowing itself involves two loops, one to find the nearest leftward non-zero digit in the minuend, and one to carry out the required succession of regroupings (possibly only one regrouping). One can also treat comparison as an algorithm, such as the following: If one number is longer it is larger; for numbers of equal length, start at the leftmost column and search rightward for an instance of unequal digits in corresponding columns; etc.

Notice that the search loop in this comparison algorithm and the main processing loop in the subtraction algorithm move in opposite directions. For this reason, their joint use might tend to obscure the similarity of the concepts they embody. After all, comparison is the idea of one number being larger than another, whereas subtraction reveals by how much. In the Coinland environment it is possible to emphasize this similarity by creating a comparison exercise directly analogous to the subtraction exercise. Specifically, the student can be given two trays to compare and allowed to make value-preserving exchanges in either until one has at least as many coins of each denomination and more of at least one denomination.

As a set of rules, subtraction can take the form provided in [25]. For example, as adapted in [18], the rule for processing a column that has no digit in the subtrahend is:

   IF number-a is in column-a and row-a
      AND row-a is above row-b
      AND your are processing column-a
      AND there is no result for column-a

AND you are focused on column-a
AND there is no number in column-a and row-b
THEN write number-a as the result for column-a.

One benefit of rules like these is that they break up the requisite knowledge into chunks that are mutually independent and are of a suitable size for learning. A related aspect that proponents of rules appreciate is that they can be individually deformed by small perturbations in their conditions, given in the "if" part of the rule, on when to apply the operation specified in the "then" part. For example, one of the conjuncts might be omitted. In this way the system can generate malrules or bugs that may provide an account of a given student's incorrect performance.

Without the explicit sequencing that procedures impose, rules need extra conditions to keep from being used at inappropriate times. These conditions would always be satisfied (and not need to be checked) at the right position in the ordering of a procedure. Using excessive rule conditions in this way violates the dictum that one should use a rule system only for processing that is inherently unordered. An apparent compensation for the excessive use of explicit conditions is that the explicit loops found in the algorithm become unnecessary. However, there must be looping in the control structure of the system that executes the production rules.

Coinland, in contrast to the six-condition rule given above, places no firm conditions on the applicability of operations. Its operations of shifting and exchanging are applicable at all times, in the sense that they never violate the semantics of the domain. A learner may have inappropriate conditions for the use of these operations, but that is a matter of inefficiency rather than incorrectness, a weaker strain of bug. By moving subtraction into a general problem-solving framework, the Coinland approach generalizes the problem of seeking the conditions that favor the use of an operation. Here an operation is to be used when it appears to move toward a goal state, perhaps by achieving some subgoal or getting to a state with a high heuristic value.

Third and last, having looked at subtraction as an algorithm and as a set of rules, consider what it might mean to regard subtraction as a set of principles. In particular, we will look at a specific (partial) set of subtraction principles. These principles would come later in the syllabus than—and can therefore depend upon—an understanding of small positive integers, counting forward and backward, and the principles of addition. The first principle is just definitional.

1. A subtraction problem includes two numbers, called the minuend and the subtrahend, and demands an answer which is the former reduced by the latter. ("Reduce" is defined in 2.–4.).
2. Reducing a number by 1 yields its predecessor in the counting sequence. (This relies on knowing how to count backward).
3. A reduction may consist of several component reductions by amounts that add up to the (composite) reduction amount.
4. The order among component reductions (as in 3.) is immaterial. (This follows from (3.) and the fact that order is immaterial in addition).

Clearly such principles are to be conveyed far more subtly than simply by stating them. What may be less obvious is that to ascribe knowledge and use of such principles to a child is not to claim that the child can articulate them, any more than we demand that a child state the rules of grammar before we concede that she speaks her native tongue.

Having introduced these sample principles, we are in a position to say something more concrete about the issue raised above, prerequisites and granularity. As for prerequisites, first note that 2. must precede 3., for the latter to make sense. Second, principles 1.–4. do not include place numeration, so it would seem that the principles of place numeration, collectively, do not have a prerequisite relationship to 1.–4. in either direction; that is, place numeration can be learned earlier or later than these principles of subtraction. A child possessing principles 1.–4, and able to use them, could solve subtraction problems with small integers, without having knowledge of place numeration (much less a complete algorithm for subtraction). The principles support the solution of 7 − 2 by counting backward from 7 to 6 to 5. Having memorized that result, a child might us it to solve 7 − 3 in just two steps, going from 7 to 5 to 4.

Such a knowledge of subtraction potentially extends in applicability to new numbers as soon as the counting order is learned. For example, without having full command of subtraction a child might nevertheless be able to deduce that "two years ago" in 1989 refers to 1987. Once a further principle concerning the subtraction of 10 is mastered, it becomes conceivable to be able to subtract 11, 12, 21, etc., by principle-based exploration, a possibility for which we believe that Coinland can provide a basis.

One last form of knowledge, simple but essential, is the fact. It is important that when a student is working on a problem in some domain the relevant facts of that domain are indeed facts for that student. If instead they are not facts but actually little problems that the student has to work out, they will presumably thereby interrupt the flow of thought in the proper domain. This discussion hints a kind of "super-prerequisite" relationship, one in which an easier topic should be not just mastered as a procedure but compiled into a set of facts before the related harder topic is taken up. Even in such a situation, however, the facts need not be treated monolithically. That is, not all the super-prerequisite facts of an entire domain need be known before work can begin in that domain. For example, it may be possible to teach multiplication to a student who does not know the whole multiplication table, if problems are restricted to those digitwise products with which the student is familiar.

Future Directions, Testing, Acknowledgments

We have implemented subtraction and a variety of related games designed to foster learning of both the Coinland interface itself and earlier topics in the curriculum, including the notion of a place value system of expressing numerical quantity. We also plan to extend the curriculum upward at least as far as division by small integers. As the curriculum becomes more thoroughly covered, it will be possible to make a more meaningful selection of where within it a particular student ought to be. It will then make sense for us to refine our currently primitive diagnostics and tutoring decisions. The system is written in C and is being developed on a PC/AT, but can be delivered on a PC/XT.

We earlier conducted extensive testing in grades 1–3 in a local elementary school and earlier did some pilot testing. Though even the recent tests were informal, we found the results very encouraging. Students found Coinland comprehensible and engaging, and some went on to use it to check or instruct themselves in Numberland. A few ultimately reached an apparent knowledge of subtraction aspects beyond what the teacher believed they knew. Such testing is not only part of the design cycle but will ultimately serve as evaluation of the system, including transfer to paper-and-pencil arithmetic.

Testing of the system with students in grades 1–3 is in progress as this chapter goes to press. The initial results are very promising. Six of the children have completed the entire sequence of pretest, use of the system, and posttest. Every one of these students improved from pretest to posttest, and the improvement in the mean was from 45% to 83% on a 20-problem test. The improvement is even sharper if attention is restricted to the 14 problems that call for borrowing. On these problems the improvement in the mean of the correct scores for the six children increased from 26% to 76%. It is also worth noting that two of the students did not have time to work on the full range of problems that the system can present, to assist with learning how to do problems with borrowing. The children also showed improvements on every component of the skill, the mean improvement being 26%. The mean of successful borrows on the pretest was 32%, whereas on the posttest it was 82%. The mean improvement on subtraction facts, where the minuend was greater than 9, was 41%. The pretest and posttest were also administered at the same times to six other children who did not play the games. For these children the mean improvement for problems done correctly was 2%.

The authors will continue to seek the valuable advice and assistance of Dr. Thomas Nuttall, an education specialist in Fairfax County, VA, who recently concluded three years on the faculty of George Mason University. We gratefully acknowledge Joseph Psotka's insightful comments on an earlier draft of this paper.

References

1. V. J. Glennan and L. G. Callahan, *Elementary school mathematics: A guide to current research (3d ed.)*, Washington, DC: Association for Supervision and Curriculum Development (NEA), 1968.
2. J. D. Holland, E. L. Hutchins, and L. Weitzman, "STEAMER: An interactive, inspectable simulation-based training system," *AI Magazine*, 5:2, 15–27, 1984.
3. H. Hamburger and A. Lodgher, "Exploration in Coinland," *Proceedings of the International Conference on Computer-Assisted Learning*, Dallas, TX, April, 1989, Berlin: Springer-Verlag, 1989.
4. T. P. Carpenter, "Learning to add and subtract: An exercise in problem solving," in E. A. Silver (ed.), *Teaching and learning mathematical problem-solving: Multiple research perspectives*, Hillsdale, NJ: Lawrence Erlbaum Associates, 1985.
5. A. Collins, J. S. Brown, and S. E. Newman, "Cognitive apprenticeship: Teaching the craft of reading, writing and mathematics," in L. B. Resnick (ed.), *Knowing, learning and instruction: Essays in honor of Robert Glaser*, Hillsdale, NJ: Lawrence Erlbaum Associates, 1989.
6. L. B. Resnick and S. Omanson, "Learning to understand arithmetic," in R. Glaser (ed.) *Advances in instructional psychology, vol. 3*, Hillsdale, NJ: Lawrence Erlbaum Associates, 1987.
7. M. Lampert, "Knowing, doing, and teaching multiplication" *Cognition and Instruction*, 3:4, 305–42, 1986.
8. J. S. Brown and R. R. Burton, "Diagnostic models for procedural bugs in basic mathematical skills," *Cognitive Science*, 2, 155–192, 1978.
9. J. S. Brown and K. Van Lehn, "Repair theory: A generative theory of bugs in procedural skills," *Cognitive Science*, 4, 379–426, 1982.
10. S. Papert, *Mindstorms: Children, computers, and powerful ideas*, New York: Basic Books, 1980.
11. H. Hamburger and S. Crain, "Plans and semantics in human processing of language," *Cognitive Science*, 11:1, 1987.

12. L. B. Resnick, "Syntax and semantics in learning to subtract," in T. P. Carpenter, J. Moser, and T. Romberg (eds.), *Addition and subtraction: A cognitive perspective*, Hillsdale, NJ: Lawrence Erlbaum Associates, 1982.
13. T. W. Malone, "Heuristics for designing enjoyable user interfaces: Lessons from computer games," *Proceedings of the ACM conference on human factors in computer systems*, 1982.
14. CERI (Centre for educational research and innovation), *Information technologies and basic learning: Reading, writing, science and mathematics*, Paris, France: OECD (Organization for Economic Cooperation and Developmnent), 1987.
15. B. Y. White and J. R. Frederiksen, "Qualitative models and intelligent learning environments," in R. Lawler and M. Yazdani (eds.), *Artificial intelligence and education*, Norwood, NJ: Ablex Publishing Co., 1987.
16. R. R. Burton, "Diagnosing bugs in a simple procedural skill," in D. Sleeman and J. S. Brown (eds.), *Intelligent tutoring systems*, New York: Academic Press, 1982.
17. J. A. Self, "Bypassing the intractable problem of student modelling," *Proceedings of the conference on intelligent tutoring systems*, Montreal, 1988.
18. E. Wenger, *Artificial intelligence and tutoring systems*, Los Altos, CA: Morgan Kaufman, 1987.
19. M. K. Singley, "The effect of goal posting on operator selection," *Proceedings of the Third International Conference on Artificial Intelligence and Education*, Pittsburgh, PA, 1987.
20. J. R. Anderson, "Cognitive psychology and intelligent tutoring," *Proceedings of the Cognitive Science Society Conference, Boulder, CO*, Hillsdale, NJ: Lawrence Erlbaum Associates, 1984.
21. H. Ikeda and M. Ando, "A new algorithm for subtraction?" *The Arithmetic Teacher*, 21, 716–719, 1974.
22. W. J. Clancey, *Knowledge-based tutoring: The GUIDON project*, Cambridge, MA: MIT Press, 1987.
23. M. C. Volpel, *Concepts and methods of arithmetic*, New York: Dover Publications, Inc., 1964.
24. F. L. Clapp, *The number combinations, their relative difficulty and the frequency of their appearance in textbooks*, Madison, WI: Bureau of Education Research, University of Wisconsin, 1924.
25. P. Langley, J. Wogulis, and S. Ohlsson, "Rules and principles in cognitive diagnosis," in N. Fredericksen (ed.), *Diagnostic monitoring of skill and knowledge acquisition*, Hillsdale, NJ: Lawrence Erlbaum Associates, 1987.

Chapter 8

Transfer, Adaptation, and Use of Intelligent Tutoring Technology: The Case of Grace

WAYNE D. GRAY
MICHAEL E. ATWOOD

NYNEX Science and Technology Center
500 Westchester Avenue
White Plains, NY 10604

Abstract *The Grace Tutor, an intelligent tutoring system (ITS) for teaching COBOL, is part of the ACT* [1, 2] family of tutors. The Grace Tutor and the student interact in a mixed-initiative dialogue. The tutor's side of the dialogue is controlled by four components: a cognitive model (or simulation) of the ideal student, an "overlay" model of what the student does and does not know (knowledge tracing), a curriculum specification, and an interface component. For the Grace Tutor the ACT* tutor technology was transferred from the university research laboratory to an independent, corporate development laboratory. In this chapter we discuss our first year of work on the Grace Tutor as a case study in how an ITS architecture, developed at a university as a research project, was transferred, adapted, and used by a corporation. A second theme that we interweave with the first is that of ITS as CHI—or ITS as a good domain in which to study and explore issues in computer-human interaction.*

Introduction

The Grace Tutor is an intelligent tutoring system (ITS) for teaching the COBOL programming language. Unlike most efforts to build tutors, we did not start the Grace Tutor as a research project but rather as an advanced development effort. Our goals were to acquire in-house expertise in tutoring technology while getting three lessons of approximately two hours each, up and running for a large-scale trial in a NYNEX classroom in New York City. Meeting these goals would give us the experience required to plan and implement a complete COBOL curriculum.

Our focus on "advanced development" rather than "basic research" meant that what we built had to fit into a classroom in the real-world of corporate training. Equally important, rather than having the luxury of building our own system from the ground up, we had to go out into the basic research world to identify an architecture that we could transfer to our laboratory, adapt to our situation, and use in a corporate training center. The transfer, adaptation, and use (TAU) of such an architecture forms the basis for this chapter.

Although our group had never built its own tutoring system, we are a group of computer scientists and psychologists with a history of involvement in the field of computer-human interaction (CHI). Like many people who work in the field, we have been frustrated by its slow and fragmented pace of progress. We knew enough about

---

Present address for Wayne D. Gray is Graduate School of Education, Room 1008, Fordham University, 113 W. 60th Street, New York, NY 10023.

ITSs to agree with Anderson that "work on intelligent tutoring can be classified as a special case of research on human/computer interaction" [3]. We were eager to acquire enough expertise to either verify or refute Newell and Card's claim that the ITS "just might be the path to major advances in hardening the psychology of the interface" [4]. Hence, as we discuss our experience with the transfer, adaptation, and use of tutoring technology we will offer our insights and assessment of ITS as CHI.

In the next section we provide an overview of the Grace Project, how we found a corporate champion, why we adopted the ACT* tutoring architecture, a view of our target classroom, and a discussion of the look and feel of the Grace Tutor from the student's perspective. The following section discusses features of an ACT* tutor. The remaining sections discuss the transfer, adaptation, and use of ACT* technology from Carnegie-Mellon University via our laboratory to a NYNEX classroom. We focus on our successes and our mistakes with the hope that the ensuing lessons learned may be valuable to other groups.

Overview

The Grace Project

The Grace Tutor is part of a suite of tools that is being developed for the student and novice COBOL programmer. The other tools include the Grace Critic, the Grace Hypertext, the Grace COBOL Editor, and the Grace Mainframe Interface. The idea is to provide a special purpose tool for each of the many activities a COBOL student performs and to migrate the tools along with the student to the programmer's workplace. The Grace COBOL Editor will provide a structured editor for COBOL programming. Most of the syntactic details of COBOL will be handled by the Editor. Interfaces to the Grace Tutor and the Grace Critic will be application-specific variations of the interface to the Grace COBOL Editor.

The Grace Tutor provides students with focused practice on well-defined COBOL skills. When the student has mastered a concept and is working on a case problem (or the novice programmer is working on a program) then the Grace Critic could be used. In contrast to the Grace Tutor, the Grace Critic allows the student to work on any COBOL problem and will permit the student to code any syntactically correct COBOL expression (even if it is semantically incorrect). The Grace Critic stays in the background, almost indistinguishable from the Grace COBOL Editor, until some student action triggers a critic to provide advice or the student asks for criticism (or praise) [see ref. 5 for a discussion of critics].

The Grace Hypertext is being built upon the Symbolics Document Examiners™ hypertext facilities and will be completely integrated with the Grace Tutor, the Grace COBOL Editor, and the Grace Critic. If the student does not understand a concept that is used in the problem description, help, or feedback messages, the student can select the concept with the mouse and go directly to a written discussion and example. Likewise, in the Grace Tutor when knowledge tracing (discussed below) reveals that students fall below a certain threshold on a COBOL concept, then they will be immediately transferred to a Grace Hypertext window for reteaching on that concept. Finally, the Grace Mainframe Interface will permit work done in the Grace Suite to be compiled and run on the mainframe. Although work is proceeding on the Grace Critic, Grace COBOL Editor, and Grace Hypertext, our main thrust and the primary focus of this chapter is the Grace Tutor.

Getting Started

Finding a Champion. The record of getting educational projects out of the laboratory and into the field is dismal. A study of 293 educational programs [6] found no case in which the program was implemented unchanged. All 293 projects were either not implemented at all, or if implemented were being used in ways unanticipated by the developer. Awareness of this record caused us to look for a corporate champion (a friendly training center with enthusiastic supporters among the staff and management) who would support ITS development and, after development, champion the implementation of the tutor within its own courses.

With an estimated 5 million student training days per year, NYNEX has many training centers conducting training in many disparate domains. In searching for a champion we adopted the following criteria. First and foremost, the management and staff of the training center had to have a progressive and high-technology outlook. Rather than feeling that the way they were taught was the way that things should be taught, they had to be open and receptive to outside ideas. Equally important, they had to believe that advanced technology, although unproven, could play a significant role in the training process. Second, since studies showing the cost-effectiveness of ITS have not yet been conducted, our champion training center had to believe that there were positive, if somewhat intangible, benefits to being perceived as the sort of organization that supports leading-edge technology. Third, although we did not expect the training center and staff to be familiar with artificial intelligence (AI) or high-powered workstations, we looked for a center that was more than just "computer literate." Ideally the staff would be immersed in computer technology and see it as an integral part of training and the job being trained. These criteria led us to Barry Howard and his staff at New York Telephone's Computer Technology Learning Center (CTLC).

Having found a champion in the CTLC, we still had to identify a niche within the champion's environment. This search required three additional criteria. First, the course for which the tutor was targeted had to be recognized "corporately" as an important course and had to handle a large number of students each year. Second, to lend itself to current tutor technology, the material covered by the course had to be well structured. Third, and finally, to provide us with a wealth of material the course had to be of moderate (2 to 6 weeks) duration. These criteria were met by CTLC's 10-week COBOL course (Classroom factors affecting tutor development and implementation are discussed later.)

Finding an Architecture. Rather than develop the Grace Tutor from scratch, we sought to transfer, adapt, and use an existing tutoring technology. In selecting an existing technology we had several criteria. First, the technology had to have been used to tutor a complex, cognitive skill. This criterion ruled out many of the declarative knowledge tutors that abound in the literature [e.g., 7]. Second, believing that not every approach was appropriate for every domain, we looked for a technology that had been applied to our target domain, introductory programming. This ruled out most of the rest of the tutoring systems [see ref. 8 for several excellent chapters on completed tutoring systems]. Third, we wanted a technology that had been used in the classroom in support of an entire course (as opposed to just a few hours). This criterion ruled out various part-course tutors such as Proust [9] and Bridge [10]. Fourth, we wanted a living technology, one that was still being used and undergoing continual development. This ruled out the classic programming tutor, BIP [see Chapter 6 of ref. 7]. Fifth, we wanted a technology that was a demonstrated success. Since we were working under a tight schedule, we could afford to adapt, but not substantially alter, existing approaches.

Sixth, and finally, the group that had developed the technology had to view transfer as an opportunity to extend and validate their approach while recognizing that transfer entails a loss of control and ownership. Once another laboratory had their technology, changes would inevitably occur. Presumably some of these changes would be improvements that the originators would want to adopt. However, since university developments are driven by theory whereas industry developments by pragmatism, the very essence of the system could change in ways in which the originators would not approve. All in all, these criteria are stringent and it was not obvious beforehand that they could be met.

These criteria led us to the ACT* tutors [11, 12]. We were familiar enough with ACT* tutors to know they met our first five criteria. At the time we contacted them (August 1988), Anderson's group at Carnegie-Mellon University (CMU) had been using the LISP Tutor to teach introductory LISP since 1984 [13, 14]; the Geometry Tutor [13, 15] had been built and used in the Pittsburgh Public School System; other tutors had been built and used as research vehicles.

During the summer of 1988 a new generation of tutors was emerging using a production system specialized for building tutors, PUPS-TUTOR [11]. Anderson's research interests were again focusing on programming as a complex, cognitive skill and his interest in transfer led to the decision to build three programming language tutors (PASCAL, PROLOG, and LISP) using one tutoring architecture. Hence when we contacted them, Anderson and his group were beginning to wrestle with the problem of using one tutoring architecture to build three different tutors. The prospect of using the PUPS architecture to build a fourth language tutor immediately intrigued them. The architecture was being designed with three specific programming languages in mind yet the hope was that it would be general enough to handle any programming language. Was this true? Could PUPS-TUTOR handle COBOL? The response of Anderson's group met our sixth criterion with the result that, with both a champion and an approach in hand, the Grace Tutor became an official project.

The Grace Tutor in Context

Approximately 300 programmers a year graduate from CTLC's 10-week Programmer Basic Training (PBT) course (a course with no computer science prerequisites).[1] Literally, the day after graduation these new programmers start work as NYNEX programmers joining the over 4,000 programmers currently employed. The PBT is a full-time, 40+ hour-per-week course in which introductory COBOL takes 6 weeks of the 10-week curriculum. More students take the PBT than any other programming course (and have been for well over 15 years).

In service of implementation and curriculum analysis [16, 17], two of our team went, as students, to portions of the PBT. The experience was invaluable. The current course is approximately 60% workshop and 40% classroom instruction. Workshops center on a series of five case problems. Each case problem is programmed on the IBM mainframe using the tools and environment [JCL (job control language), TSO (the mainframe editor used by NYNEX COBOL programmers), and debugging packages] that are used on the job. Classroom instruction is focused on the COBOL, JCL, TSO, and

---

[1] Included in this number are those who graduate from CTLC's New England Telephone counterpart, NET Educational Technology Development.

debugging packages that students need to know for the current case problem. Any innovation that diminishes the role of the case problems or of the tools used in the work environment will never be accepted and implemented.

Unlike our closest model, the LISP Tutor, the Grace Tutor is being built for a course we do not "own"—that is, we cannot modify the course to suit the structure of our tutor. Our goal is for the Grace Tutor to become a modular tool for tutoring COBOL skills. After a concept (e.g., iteration, conditionals, and so on) is introduced in lecture, we envision the instructor sending the students to practice on the tutor. Time spent on the Grace Tutor will come from time now spent on in-class practice plus savings that result from needing less time to do the case problems. (We are assuming that practice with the Grace Tutor will reduce the amount of thrashing, and hence the time spent on case problems.) This concept of use supports a strategy of "incremental implementation" by which the Grace Tutor can be expanded one COBOL skill at a time with four to six exercises per skill. In this way, the Grace Tutor can be used initially without having to cover the entire COBOL curriculum. As it expands to tutor additional COBOL skills, these additions can be easily inserted without massive disruptions or revisions to the course.

Look and Feel

The Grace Tutor tutors COBOL programming as a complex cognitive skill. We refer to it as a practice tutor. The student acquires COBOL concepts (declarative knowledge) through lecture or reading and then uses the tutor to practice applying these concepts in a programming exercise. A screen dump taken during the middle of an exercise is shown in Fig. 1. The code window dominates the screen. The small window beneath the code window accepts keyboard input and provides a place for feedback to appear. The window along the right is reserved for menus. The code window contains completed COBOL code (indicated by uppercase letters) and placeholder symbols (indicated by bold, lowercase type and enclosed by angle brackets). Placeholder symbols indicate subgoals that must be accomplished to complete the exercise.

In Fig. 1 the student or the tutor has selected the node ⟨?more–data–items?⟩ which caused a particular menu to appear on the right. Essentially the student is engaged in variable declaration and will have to choose a "dataname" and a "picture clause" (PIC) that declares the type (e.g., alphanumeric or numeric) and number of bytes in memory for each variable.

The Computer–Student Interaction Cycle. At the beginning of every computer–student interaction cycle, the computer selects and highlights a placeholder symbol (subgoal). (This selection is based on a top-down, left-right algorithm.) The student may elect to work on this default selection or may override the computer's choice by selecting any other placeholder symbol. In either case, after the placeholder symbol has been selected, the student picks a template from the menu-window (on the right) or, if the placeholder symbol represents a terminal goal (one that is not decomposable into 2 or more subgoals), types the code symbol into the feedback/type-in window. (Placeholder symbols are selected and menu entries are chosen by clicking on the item with the left mouse button.)

If the student's action was correct then the effects of the action appear on the screen. If the student chose the wrong menu entry for the selected placeholder symbol or typed in the wrong code symbol, then error feedback is provided. In either event, the Grace

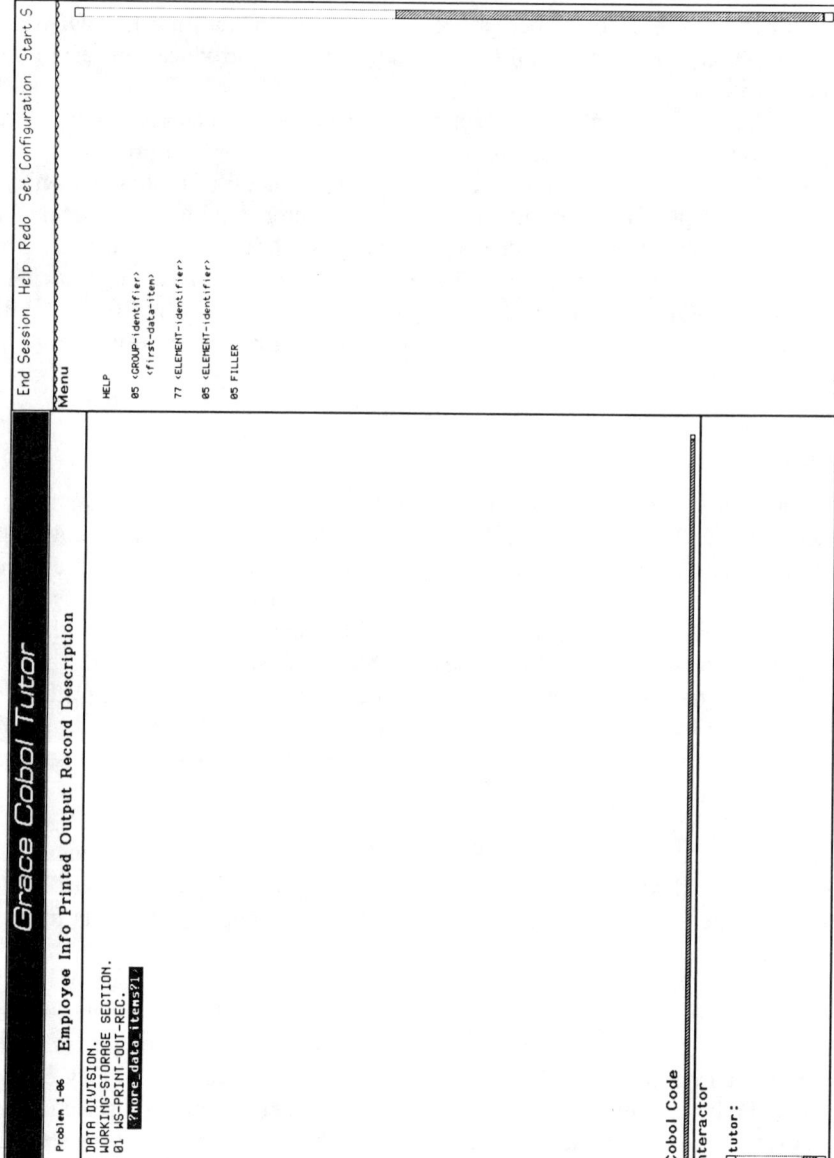

Figure 1. The Grace Tutor interface.

Tutor's response represents the end of one computer–student interaction cycle. If the student's action was correct then the tutor begins the next cycle by selecting a new placeholder symbol. In contrast, if the student's action was incorrect then the tutor reselects the last cycle's placeholder. (In either case the student has the option of working on the placeholder selected by the computer or selecting another one.)

Summary: Look and Feel. The Grace Tutor can solve the COBOL exercise in most of the right ways that an ideal student can. As long as the student is pursuing a valid path through the problem space then the tutor stays in the background and acts as a structured editor. As soon as the student tries to apply an illegal operator (one that is wrong in the context of the currently selected goal) then the Grace Tutor intervenes. The student's attempt is blocked and feedback is provided. At any time the student may request help on any goal (placeholder symbol) that is on the screen or on any item that appears in the menu-window. (Help is obtained by clicking on an item with the middle mouse button.)

When students complete one exercise the Grace Tutor selects another exercise for them until they complete the current lesson. Note that implicit in this simple description of the Grace Tutor are a myriad of modeling and interface details. Some of these details will be discussed as we go along.

Features of an ACT* Tutor

There are four aspects of an ACT* tutor that are crucial—three have been considered in past work and one has been largely ignored. The three that have been considered extensively are: the ideal student model, knowledge tracing, and curriculum specification. The area that has been largely ignored is interface design. In this section we describe these four aspects. In the following sections we discuss their transfer, adaptation, and use.

The Ideal Student Model

The ideal student model knows about as much of the domain as a good student. For example, since there are no computer science prerequisites in NYNEX's curriculum, neither our good student nor our ideal student model are assumed to bring any knowledge of computer science to the course. Like our good student, all that our ideal student model knows about computer science is what is taught in the COBOL course. The ideal student model can solve any COBOL exercise in *most of the right ways* that a good student can. This statement deserves emphasis since it is both a feature and a limitation of ACT*-based tutors.

For any COBOL exercise, the ideal student model can code the exercise in many of the right ways that have been taught up to that point in the curriculum. However, it cannot code the exercise in ways that come from beyond that point. For example, an instructor could easily think of more expert solutions than the Grace Tutor could accommodate. Likewise, a student with a strong computer science background can also try solutions that the tutor cannot handle.

The ideal student model is more than just an expert system of a good student: it is a cognitive model, or simulation, of that student. The claim here is that not only can the ideal student model come up with all the same answers that a good student can, but also that it solves the problem in the same way. Hence, the knowledge in each production corresponds in size and content to a knowledge structure (chunk) that the student has;

furthermore, the number and organization of the productions correspond to the number and organization of the student's knowledge structures.

Knowledge Tracing

As the student completes an exercise the tutor follows along by tracing a solution through its ideal student model. If a production can be found for a student action, then that action is judged correct and the judged probability of the student having acquired the knowledge in that production is incremented. If a production cannot be found then the tutor selects the production that it would use in its ideal solution of the exercise and decrements the judged probability that the student knows the production. The body of weighted productions is a trace of the student's knowledge; the process by which these weights are obtained is referred to as knowledge tracing.

Individual exercises are organized into lessons where each lesson covers a certain to-be-learned topic. (For example, in the Grace Tutor one lesson consists of exercises on "variable declarations" whereas another consists of exercises on "tables," that is, CO-BOL arrays.) At the level of the ideal student model, what distinguishes one lesson from another are the new productions. Knowledge tracing occurs only for these new productions. Knowledge tracing and the ideal student model provide ACT* tutors with a very fine grain and very thorough approach to mastery [18] instruction. Immediate feedback is provided whenever the student makes an error, and the student stays in a lesson until mastery of all new productions has been demonstrated. Immediate feedback is also provided when the student attempts something that the ideal student model does not know about. Providing immediate feedback and preventing students from deviating from the ideal student model is a hallmark of ACT* tutors [11] and is enabled by knowledge tracing.

Every action that a student takes is mapped onto one of the ideal student model's productions and leads to an updating of the judged probability that the student has learned that production. At the end of a fixed set of exercises the tutor decides whether the student has mastered the knowledge required to solve those exercises. If the judged probabilities for learning all new productions is greater than some criterion (typically 0.95) then the student can go on to the next lecture or to the next section in his or her textbook. If there are productions that the student has not mastered, then the tutor picks remedial exercises for that student. The remedial exercises provide the student with the opportunity to acquire and practice the correct knowledge. This process continues until mastery on all productions in a module is achieved.

Curriculum Specification

Curriculum specification includes both the exercises within a lesson and the lessons within a course. The lesson is the most basic curriculum unit. Although the exact span of a lesson is somewhat arbitrary, it tends to cover a coherent set of concepts that an instructor might spend two to four hours introducing. Both the ideal student model and knowledge tracing are built to handle the lesson, not higher levels of curriculum organization. Sequencing from one lesson to the next is pretty much handled by traditional means of curriculum organization (namely, instructor intuition). Some guidance, however, is provided by the ideal student model used in conjunction with the identical elements theory of transfer [19].

Interface Issues

The purpose of every component of a tutor is to facilitate the interaction of the student with the domain knowledge. From this perspective an ITS can be thought of as having a deep or cognitive interface as well as a more traditional surface or physical interface. The cognitive interface structures and organizes the domain's knowledge in ways that are accessible to the student. The physical interface is the plane (screens, menus, help messages, mouse gestures, typing, and so on) on which the cognitive interface and the student communicate.

The cognitive interface of an ACT* tutor is primarily embodied by the ideal student model and knowledge tracing. The physical interface is handled in a language-independent fashion with a minimum of guidance from ACT*. ACT* tutors have software hooks to connect various miscellaneous functions to the PUPS-TUTOR production system on one side and the interface on the other. Examples include mechanisms for help messages, feedback on wrong menu selections, judging of keyboard input, error messages for keyboard input, and an approach to prettyprinting (or formatting) code in the code window. At present, the implications of ACT* theory for physical interface design have not been elaborated as much as we believe they could be. This topic is an aspect of ITS as CHI and will be discussed in more detail later in this chapter.

Each of the next four sections discusses the transfer, adaptation, and use (TAU) of one key feature of an ACT* tutor: the ideal student model, knowledge tracing, curriculum specification, and interface issues. *Transfer* entails both what was transferred and how it was transferred. *Adaptation* involves adapting ourselves to the ACT* technology as well as adapting the technology to the specific needs of the Grace Tutor. *Use* also has a dual focus. Use entails our experiences in using the technology to do what it is supposed to do, as well as feedback from instructors, students, and others. Guiding many of our discussions is an underlying concern for defining and assessing the claim of ITS as CHI.

TAU of the Ideal Student Model

Transfer

Transferring the ideal student model involved exchanging code for the PUPS-TUTOR production system and knowledge of how to use that code to construct an ideal student model for COBOL. The initial transfer issue was one of software compatibility. The CMU version of PUPS-TUTOR was written in InterLISP-D and ran on the Xerox 1108 series of computers, whereas our delivery vehicle, the Symbolics MacIvory™, required Common LISP. Although some automatic translation programs existed, they were less useful than we had hoped. The problem was that PUPS-TUTOR used a lot of LISP macros and the translation programs translated these macros into very low-level LISP constructs. Although such code would run, it was almost impossible to maintain or enhance.

Basically two solutions emerged to this problem. The first involved our working closely with the CMU group to rewrite InterLISP-D macros into Common LISP. The second involved going up a level and transferring algorithms rather than code. Knowing what PUPS-TUTOR was supposed to do enabled us to take full advantage of the rich object-oriented coding environment that Symbolics provides.

PUPS-TUTOR is a flexible production system that is specialized for developing tutors. The basic data structure of a production system is the working memory element

(WMELT). WMELTs in PUPS-TUTOR are framelike slot and value combinations. For example, a WMELT that describes a COBOL data element could be written as:

```
(WMELT    first-dataelement
   ISA                input-element
   HELP-INFO          ''the first variable''
   dataNAME           dataNAME-first-dataelement
   hasPARTS           (first-dataelementPIC first-dataelementVALUE)
   superORDINATE      (input-GROUP1))
```

In this example the name of the WMELT is "first-dataelement" and its description consists of five slot-value pairs. These provide us with a flexible means for describing parts of a COBOL program. What the slot names are and how many is not fixed by PUPS-TUTOR but is completely determined by how we represent COBOL. Likewise the role that the slots play is also completely flexible. For example, the HELP-INFO slot is used by one of the interface functions, "defhelp," to provide a help message to the student. The value of the dataNAME slot is itself another WMELT. The hasPARTS slot and superORDINATE slot both take lists of WMELTs.

The productions, as in any production system, consist of conditions and actions. The conditions consist of one or more WMELT structures with slot-value pairs that must be matched for the production to fire. The actions are one or more WMELT structures. Some of these structures are modified versions of existing WMELTS, others are newly created by the production. Fairly sophisticated pattern-matching macros enable some of the items in a slot-value pair to be selected and either revised or cut and pasted into a new slot-value pair in a new WMELT.

As with many other production systems, there is easy access to the host language, in this case, LISP. This access permits much of the "keeping track" functions of the productions to be accomplished in LISP, while freeing the productions per se for more cognitive modeling purposes.

Finally, unlike the original LISP Tutor that used the GRAPES production system, tutors built using PUPS-TUTOR do not need to precompile solutions [20]. This arrangement allows for real-time knowledge tracing.

Adaptation

There was much that we had to learn about PUPS-TUTOR and it took a good deal of time to learn it. PUPS-TUTOR was developed as a research tool, not as a commercial product. At the time we began using it there were not more than three or four people in the world who had used PUPS-TUTOR. All of these people were at CMU and all had played some role in its development. Because it had only been used by those who had developed it there were no user guides, no manuals, and no documentation.

Adapting Us to PUPS-TUTOR. While PUPS-TUTOR was being translated into Common LISP, we began learning how to use it by studying an early version of the PASCAL tutor running on the Xerox 1108. In retrospect, this time was largely wasted. The PUPS-TUTOR production system in the context of a fairly advanced PASCAL tutor was much more complex than PUPS-TUTOR by itself. Indeed, when the chance came to sit down with one of the CMU group and discuss our problems, we realized that 90% of our effort had been spent struggling with how the PASCAL tutor was implemented and not PUPS-TUTOR per se.

Around February 1989 our Common LISP translation was sufficiently advanced that we abandoned the 1108s and moved our work onto the Symbolics MacIvories. At this time we began writing productions for the "variable declaration" lesson. These productions have now been rewritten many times with at least three complete "fresh starts." In retrospect these rewrites have progressed through fairly discrete stages.

In stage one we tried to do all of the work in the productions and took the narrow approach of writing productions for one exercise at a time, incrementally adding new productions for additional exercises as needed. This worked well enough to get us through a May 1989 empirical trial with students. However, there were problems that soon emerged.

The first stage covered three very simple exercises and enforced a fairly dogmatic solution. Students and instructors alike complained about being forced into one solution (actually many more than one final solution was acceptable, but the complaints did not acknowledge this). The "clincher" was when we tried to expand our system to cover a more complex exercise. At that time we realized that our existing productions were handcrafted to deal with a very limited range of exercises and could not be easily scaled up. The first stage ended when we realized that the hard part of writing productions was not in getting them to solve the exercise but in building into the ideal student model all the flexibility that real students require [21].

The second stage focused on giving students the flexibility to code an exercise in many right ways. In building in this flexibility we came to understand that our productions were doing two very different jobs: student modeling versus "keeping track." Student modeling was the job for which production systems were well suited. "Keeping track," however, was another matter.

In giving students the ability to code an exercise in various idiosyncratic ways we exacerbated what until then had been a fairly small tracking problem. At each step, every right choice that students make rules out a number of alternative right choices. As students proceed through an exercise their degrees of freedom become increasingly restricted by the decisions they have already made. The tutor needs to keep track of the students' decisions and to propagate these decisions via changes to working memory elements. As a simple example, if a student decides to name a variable "WS-NAME-IN-REC" then the next time the student refers to that variable he or she has to use the same label.

PUPS-TUTOR allows such tracking to be accomplished via LISP code on the action side of a production. Such resort to LISP allows us to write fewer productions. For example, in stage one we required six productions to handle "data groups"; in stage three we had just two. Our heuristic is that if a distinction is important for student modeling then we write a new production. If the distinction is not important then we do it in LISP. (See ref. 20 for a discussion of "keeping track" of "mutual constraints" with the GRAPES-based LISP Tutor.)

Stage three came when we tried to extend stage two productions to deal with optional grouping. At that point we realized that our existing productions were too narrow in scope and we had to rewrite everything.

Our lesson learned from this experience is to first define the scope of the lesson, that is, the complete range of exercises and the allowable variations in their solution. After the scope is determined then the set of new productions is planned (with attention to existing productions). Finally, production writing can begin.

Adapting PUPS-TUTOR to Us (and to COBOL). Somewhere between stage two and stage three we began adapting PUPS-TUTOR to us rather than vice versa. Part of the adaptation process entailed defining a style of production writing. Although we are hesitant to try to define exactly what our style is, it manifests itself in the decisions we make regarding how to do what.

A different type of adaptation has been the development of two complementary interfaces, one for the student (COB) and one for production writers (PSI) along with the ability to easily (two keystrokes, two seconds) switch from one to the other. An exercise can be started in COB to enable us to see it as the student sees it. If a problem (bug) occurs we can switch to PSI and load the exercise from COB in its current state. PSI gives us larger windows for type-in and feedback. It also allows complete access to LISP and to a variety of tools for tracing productions and examining the state of the production system as a whole or of individual WMELTs. Likewise, PSI can host a number of distinct production systems. If two or more production writers are editing productions, they can each have their own production system in which to work out problems, modify existing productions, and add new productions. In this way changes that one writer makes to a production in his production system does not affect the use of that production by other people. At the same time PSI allows anyone to load in the current productions and WMELTs from any subsystem. This flexibility greatly facilitates development. At the implementation level we have "flavorized" (used Symbolics' object-oriented features) large portions of the underlying code. At this point any new addition is written completely in flavors based on algorithms, not code, obtained from CMU.

Use

The crux of the issue is building a tutor that allows students to think about COBOL the way they are taught. In our experience this involves much work with both the ideal student model and the physical interface. Even when the productions are right the interface can force students to think in unaccustomed ways and actively interfere with their ability to solve exercises.

Writing productions that follow student thought is essentially a grain-size issue. Our stage one tutor had much too fine a grain-size as evidenced by student and instructor complaints. For example, starting a variable declaration required three complete computer-student interaction cycles. Students first had to decide to declare an anonymous "data item" then decide to make it a "data element" (rather than a "group item") and finally decide whether or not the "data element" should be initialized (requires a COBOL VALUE clause). Achieving the same result in the current system requires just one cycle.

A lot of trial and revision has occurred to get us from the stage one grain size to the present one. The basic realization was not simply that stage one involved too many cycles but that it required students to think at too small a grain size. Rather than make three separate decisions, our students simply decide to "code a data element with value."

These changes ran contrary to our abstract notions of elegance and top-down coding, and we resisted them at every stage. Fortunately for the Grace Tutor, we are committed to empirical test and evaluation. Although without such testing we would be much further along than we are at present, what we would have built would be a system that nobody would want to use.

The TAU of Knowledge Tracing

Knowledge tracing involves having the tutor follow the student through an exercise by tracing a solution through the ideal student model. If a production can be found that accounts for a student's actions, the probability that the student knows that production is incremented; if not, the judged probability that the student knows the production is decremented and feedback is provided.

Transfer

With knowledge tracing, what was transferred includes algorithms only, no code. The version of PUPS-TUTOR that we transferred left CMU in December 1988. At that time knowledge tracing had not been added. (Knowledge tracing was an important feature of the original GRAPES-based LISP tutor and was planned all along for the PUPS-TUTOR tutors.) By the time that CMU added knowledge tracing and we were ready for it our code had so evolved that it seemed a better idea to implement the algorithms ourselves rather than integrating two different styles of coding.

Adaptation

Adapting ourselves to knowledge tracing has required writing productions with knowledge tracing in mind. Ideally we would have one production for every skill in the ideal student model. In reality we tend to have too many productions and have to fight the tendency to have too few. Of the two types of sins, having too many versus having too few productions, too few is worse. If the knowledge that must be tutored is not discretely represented then it can be neither directly taught nor remediated.

The sin of having too many productions is often a necessary side effect of using productions to "keep-track" or to implement the structured editor. The problem is essentially one of having two or more productions that the ideal student model views as equivalent. The knowledge tracing problem in this case can be solved by the invention of *metaproductions*.

A metaproduction is essentially a label followed by a list of equivalent "real" productions. When any production in the list is used correctly (or not used when it should be), the weight for the metaproduction is revised. If the criterion on the metaproduction falls below the 95% level, then the problem selection algorithm picks a problem that requires one production from the metaproduction's list for its solution.

Use

Knowledge tracing works well as a fine-grained tool for implementing mastery learning of procedural knowledge. We have discussed extending knowledge tracing to remediate deficits in declarative knowledge. With a little effort, knowledge tracing could tie into an additional tutorial mechanism that puts the student into the Grace Hypertext. The hypertext would provide a declarative description of the production and, possibly, examples of its use. Issues that must be resolved include determining what judged probability for not knowing a production should trigger the hypertext tutorial, how or when to interrupt the student, and whether (and how) the student should demonstrate that he or she has acquired the declarative knowledge before returning to the exercise. This hypertext link will represent the first extension of an ACT* tutor to remediate deficits in declarative knowledge.

The TAU of Curriculum Specification

Transfer

Although curriculum specification may be the "make it or break it" step in successfully implementing tutoring technology, there are no formulas or theory to guide the process. Rather, any curriculum we develop has to simultaneously fit the constraints of an existing course, the instructional style imposed by industrial training (as opposed to academic education), the specific knowledge demands of the job of a NYNEX COBOL programmer, and of course, COBOL itself.

Adaptation

The curriculum specification process can be characterized as adapting the Grace Tutor to the constraints imposed by the classroom environment. In doing so we adopted a broad range of strategies. Two of our group learned COBOL by attending the existing COBOL course as students. We have studied the NYNEX-produced instructor and student materials as well as the current textbook. We have examined exams and case problems from a variety of students. We have created our own exercises and have studied the variety of solutions that students (and instructors) have provided. Simultaneous with our beginning the Grace Tutor, NYNEX's New York Telephone and New England Telephone companies decided to design and adopt a common COBOL course. A member of our group (S. Dews) worked extensively with curriculum developers in both companies. Throughout our curriculum specification process we have strived to maintain ties with the classroom instructors.

Our job to date has been simplified in that we have not had to specify a curriculum for the entire COBOL course. For our first deadline we only have to develop three lessons. One of these lessons we chose, the other two chose us.

Our first lesson introduces students to COBOL variable declarations, or in COBOL terms, the COBOL "DATA DIVISION" with an emphasis on the "WORKING-STORAGE SECTION." This lesson occurs near the beginning of COBOL instruction in both the current and revised course.

Lessons two and three cover single-dimension arrays, or in COBOL terms, "TABLES." Tables involve indexing, subscripting, redefining existing variables, and much more that students find confusing. These lessons chose us in that surveys of supervisors consistently indicate that new programmers have problems with tables. These are the lessons that the instructors and curriculum developers thought would be the best test of the Grace Tutor.

Although picking the three lessons to develop was relatively easy, deciding on the content of those lessons was not. The basic problem is one of grain size. Neither the instructors nor the curriculum developers are accustomed to thinking at the grain size required by the ideal student model. Like most experts their expertise represents compiled knowledge with the formerly individual components pretty much inaccessible.

It is interesting to speculate that the different perspectives on grain size from differences in the task and time frame that confronts a classroom instructor versus a tutor builder. The classroom instructor instructs a group of individuals and his or her main focus is covering a certain body of knowledge in a 10-week period. Class after class, most of the students acquire most of the concepts before graduation. A student who is trying to learn a microissue has access to multiple sources of knowledge (including the instructor covering the same concept in a slightly different way in a later lesson) over a prolonged period of time (at least hours, usually days, and sometimes weeks). Very few problems with microissues prevent students from doing any work until they completely understand that issue. Hence the problem with individual microissues may never become explicit.

In contrast, the tutor builder focuses on tutoring one individual at a time within the time frame of one exercise. Any failure to know a particular microissue could easily stop the student from completing an exercise. For the tutor the ideal student model serves to immediately identify any problem as resulting from a lack of knowledge about one or more microissues. Hence for the tutor the student's problem is not only immediately obvious but also must be remediated during the current lesson. This remediation limits the amount of time that students spend "thrashing" and improves instruction by limiting unproductive student activities.

Use

Although we have said that having to prepare just three lessons has simplified our curriculum specification, in a way it has complicated it. In a complete curriculum we would know that if students reached lesson n then they have acquired all the skills taught by lesson n-1, n-2, and so on. The only skills we would have to worry about tutoring would be the skills introduced in lesson n. With our discontinuous curriculum, these skill acquisition assumptions cannot be made. It is not the case that all students will have demonstrated acquisition of all microissues considered by the tutor as prerequisites to our second lesson (tables).

This is a problem that we are trying to design around. Each table exercise may require three types of COBOL knowledge: knowledge that was taught by the tutor in the "variable declaration" lesson (lesson 1), knowledge required for the exercise but not taught in lesson 1 (such as iteration), and knowledge specific to COBOL tables. In each exercise, students will have to complete the parts that require either lesson 1 or "table" knowledge. COBOL knowledge that the tutor has not taught will be supplied as a series of partially completed templates. This approach should maximize the acquisition of table productions (microissues) by allowing students to concentrate on table-specific concepts rather than on concepts such as iteration. In the long run there will be a penalty for this approach. When our complete COBOL curriculum is specified then the table exercise problems and productions will need to be rewritten (or at least reedited).

The TAU of Interface Design

At its most basic, an ITS is a computer tool for achieving the task of learning a particular problem domain. The ideal tutor would be one in which students thought only about learning the domain and not about learning the tutor; that is, the language for communicating with the tutor would be the same as the language for working in the domain. This is a stringent usability requirement, but one which we believe that tutoring systems for computer languages can approach.

The Grace Tutor's side of the computer–student interaction cycle is controlled by the tutor's cognitive interface and expressed through its physical interface. The cognitive interface is composed of the ideal student model and the knowledge tracing mechanism. These components are theory-driven and largely, but not completely, embodied in the production system architecture. The cognitive interface communicates with the student through a physical interface embodied in a multimedia system of windows, text, hypertext, mouse gestures, menus, placeholder symbol selections, typing-in, and so on.

Viewed from the perspective of the cognitive and physical interface, our design goals become one of ensuring that the physical interface provides a language and medium that facilitates communication between the student and the ideal student model. These adaptation and use issues are discussed here, after a discussion of the interface hooks we obtained from CMU.

Transfer of Interface Hooks

In ACT* tutors the ideal student model and knowledge tracing components are essentially independent of the particular machine and physical interface in which they are implemented. A number of hooks are provided to connect these tutoring components to the windowing and input/output aspects of a particular hardware and software environment.

Example hooks include a variety of LISP functions for getting the value of slots of particular WMELTs. These functions can be used by other functions to update the physical interface contingent on the current state of the production system. This functionality is used to make menus context specific. Other examples include sets of functions for taking keyboard input from the students, judging whether the input meets certain criteria, and if it does then passing it onto the appropriate production. A parallel set of functions exists for handling menu selections.

Particularly useful was a set of functions for formatting or prettyprinting COBOL code. These functions are based on an approach developed by Pugh and Sinofsky [22]. They enable us to specify the spacing, indentation, and so on required for properly formatted COBOL. The functions look at the "CODE" slot on the action-side of the production that fires. The value of the slot is a list and each element of the list is parsed according to the rules for that type of COBOL statement. (The action is recursive in that an element in a list may itself be a WMELT with a CODE slot that is itself parsed.) The resulting formatted COBOL appears in the CODE window (see Fig. 1).

A basic "help" mechanism also was transferred. Students can ask for help on a specific placeholder symbol. The help they receive is based on a combination of the production that the ideal student model would select at that symbol and the particular exercise on which they are working.

Adapting the Physical Interface to the Ideal Student Model (and Vice-Versa)

Our first physical interface was built quickly to enable us to collect usability data. Out of several small trials we have derived two usability criteria to guide interface development and evaluation. First, students should be able to master the interface after less than an hour of use. Second, coding COBOL with the tutor should be easier than with the media currently available. Students and NYNEX COBOL programmers typically use one of two methods to code COBOL: writing with pencil onto coding sheets and then typing this code into the mainframe using the TSO™ editor, or skipping the coding sheets and typing directly into TSO. Although TSO provides some support for COBOL programmers, it is far from being a structured editor.

After reflecting on why the existing physical interface did not meet these criterion we achieved the crucial insight of "ideal student model as cognitive interface" and came to two conclusions based on this insight. First, the physical interface forced communication in ways that were out of sync with both the ideal student model and the "ideal student." Second, compromises required by the physical interface were having an unacknowledged effect on how we wrote productions. Hence the ideal student model was being influenced by factors that had nothing to do with how students thought about COBOL.

In designing a physical interface for an ITS we now view the basic usability problem as one of adapting the physical interface to the cognitive interface. In what follows we first elaborate the concept of "ideal student model as cognitive interface" and then provide an abbreviated example of problems and solutions in designing menus for the Grace Tutor.

Ideal Student Model as Cognitive Interface. The ideal student model is more than an expert system of a good student; its intent is to provide a cognitive simulation of a good student. If our simulation is accurate, then the ideal student model will solve the COBOL exercises in much the same way as a good student does. There are several claims embedded in this statement. First, at the most macro level, the solutions that the ideal student model achieves are similar to solutions that real students will achieve. Second, the intermediate states along the solution path are similar for the students and the model. If we interrupted the student part way through an exercise, what we would find in the code window would be reproducible by the ideal student model by applying some sequence of productions to the initial state of the exercise. Put another way, given a partial solution we could not determine whether it was achieved by a good student or by the ideal student model. Third, the ways that students get from one intermediate state to another (the operators they apply to transform one state into another) are similar to those used by the ideal student model. Again, we could not determine whether a given sequence of states and operators was generated by a good student or by the ideal student model.

Given the similarity between computer and student problem-solving, the goal of the physical interface should be to support communication about COBOL programming between the two problem-solving systems: the student and the ideal student model. We propose three heuristics to support this process.

The first heuristic is to be clear about the task that both problem-solvers are performing. Although the task may seem obvious, the obvious answer is often wrong. Neither the student nor the ideal student model are using a tutor. Rather, both problem-solvers are coding COBOL exercises using a structured editor. Conceptualizing the primary task as using a structured editor means that a host of physical interface issues can be divorced from tutoring per se and treated primarily as issues in designing a structured editor. In the example that follows, Menu Design, we make good use of this perspective.

The second heuristic is one-to-one correspondence: firing one production in the ideal student model (tracing one operator in the student's head) should require one choice or selection (e.g., selecting a menu entry or typing something in) at the physical interface.

The third heuristic is to use a common language to describe the choices that are to be made or the feedback provided. For this heuristic the burden is on the tutor builder to program the student's language into the ideal student model. Unfortunately, the prerequisite step of determining the student's language is more of a problem than it may seem. For example, Bellcore studies [23, 24] show distressingly little agreement among the public at large for terms that describe common text editing operations. The goal for the physical interface designer is to design a communication system—language and media (gestures, windows, menus, and so on)—that is easy for students to learn and facilitates the exchange of ideas. Our three heuristics provide guidelines, not fixed principles, that experience (not an experimental test) shows will enhance such communication.

Menu Design. By the end of the June 1989 trial, we knew that our menus were one aspect of the Grace Tutor that did not meet our usability criteria: not only did students dislike the menus, but even after an hour on the system they had problems using them.

The problems fell into three areas. The biggest source of overt errors was that to type something in, such as a variable name, the "type-in" entry had to be selected from the menu. Students and instructors (as well as ITS developers) had a strong tendency to skip the menu and begin typing as soon as the placeholder symbol was selected. A second area was a dislike of the hierarchical menu structure. To make a change on the screen a student might have to go through as many as four menus. Finally, the labels used for menu entries had to be explained and reexplained throughout the course of the trial. Even though we got the labels from the textbook, students did not think of COBOL in these terms.

All three problems arose because the physical interface violated these heuristics—with the result that it was out of sync with the ideal student model. The current menu system is completely revised. The menus are context-sensitive; that is, different menu entries appear depending on what placeholder symbol is selected. All menus are flat; that is, nonhierarchical. The menu entries are templates; they are small versions of the COBOL code and new placeholder symbols that will appear in the code window when the correct menu entry is selected (Fig. 2). If the correct action at a placeholder symbol is to type in something, then no menu appears when that placeholder symbol is selected. Rather, the type-in window flashes, the prompt "Please enter:" appears, and the cursor

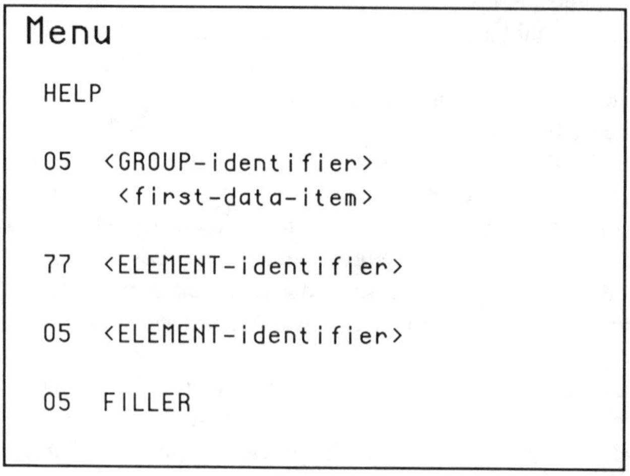

```
Menu

HELP

05    <GROUP-identifier>
      <first-data-item>

77    <ELEMENT-identifier>

05    <ELEMENT-identifier>

05    FILLER
```

Figure 2. Example of templates used in a context-sensitive menu.

starts blinking. How our heuristics for enhanced communication led to these changes is the basis of the rest of this section.

We had earlier considered context-sensitive menus but quickly became lost in a maze of "tutoring" issues that we could not resolve. If not all choices appear on the menu, how should we determine what to include? How many and what "distracter" items should appear? Although the tutoring and educational literature did not offer guidance, our first heuristic did. As we indicated earlier, the student's primary task is to use a structured editor to code COBOL exercises.

From the perspective of a structured editor, context-sensitive menus should contain all legal actions; that is, all actions that could legally occur at that point in any COBOL exercise. Since COBOL is a structured language with four major divisions and two or more sections per division, the number of legal actions within a particular section of a particular division is limited (especially when compared to languages such as LISP). Currently, menu entries are largely determined by the division and section in which the placeholder symbol occurs.

The "structured editor" perspective also sheds light on the "type-in" issue. From a tutoring perspective, it seemed that putting students into a type-in mode immediately after selecting a placeholder symbol was too much of a give away. However, for a structured editor, this automatic type-in would not be a give away but is exactly what should occur. The important thing, from the structured editor perspective, is to have a principled basis for deciding when a template (and hence a menu) is required versus when something should be typed in. At present, we have divided placeholder symbols into two types: those that are replaced by more than one COBOL and placeholder symbols (i.e., templates) and those that are replaced by one (contiguous) COBOL symbol. The former lead to menu selections whereas the latter lead to type-in's.

For the type-in's there are as many potential things to type-in as there are different types of terminal nodes. Hence, letting students know that a given placeholder symbol requires a "type-in" action is less of a hint than limiting their menu entries to templates from the current COBOL division. If we define choice as the students' hitting the carriage-return after they have typed in their input, then in both the type-in case and the menu case the students are only one "choice" away from getting to a "fireable" production.

Having context-sensitive menus does not require having flat or nonhierarchical menus; rather, flat menus follow from the one-to-one correspondence heuristic. If the grain size of our ideal student model is correct (each production corresponds to a chunk of knowledge that a good student would have), then making the student select three or four menu entries before getting to a "fireable" production is functionally equivalent (from the student's perspective) of dividing that production into three miniproductions. With our current flat menu structure every menu entry leads directly to the instantiation of one production or to an error message.

Neither context-sensitive nor flat menus necessarily result in having templates as menu entries (see Fig. 2). This revision derives from our third heuristic, using a language that the students understand. Before templates, our menu entries were verbal-conceptual descriptions. For example, one entry might be "Workarea Element with Value" whereas the corresponding template would be "77 ⟨ELEMENT-identifier⟩." This is an interesting case since both the syntactic (template) and semantic (verbal-conceptual) descriptions are inherent in the production. By iterative trial and evaluation we have convinced ourselves that COBOL students think about COBOL in terms of the syntax, not the semantics; so our menus contain templates, not descriptions.

Summary. The aforementioned example shows how the cognitive interface can and should influence the physical interface. This influence is possible because of the close correspondence between how the computer and students solve the problem. Although such a similarity in processing is not unique to ITS, it is a hallmark of these systems.

Conclusions and Summary

In going from inspiration to proof-of-concept we had to master ITS technology, build a complex software system, acquire knowledge of COBOL, acquire knowledge of an existing COBOL curriculum, and build six hours of instruction. As our trial date approaches we have met or exceeded most of these goals and believe that we now have the experience required to plan and implement a complete COBOL curriculum.

In retrospect, we naively underestimated the task of building an ITS. Such success as we have had resulted from much hard work on our part combined with the transfer, adaptation, and use of the ACT* tutoring technology, the support and enthusiasm of its developers, and our corporate champion, the CTLC.

Although in many ways we still feel like a White Plains outpost of the ACT* tutor project, as we have mastered the technology we have begun to adapt it to the unique needs of the NYNEX COBOL training center and to place the Grace Tutor in the broader perspective of an emerging Grace Suite of student and programmer tools. Likewise, because of our backgrounds and research interests in CHI, we were attracted from the onset to the notion of ITS as CHI.

Lessons Learned

Concept of Use. A tremendous advantage of most of the CMU tutors is that the ITS developers also controlled the classroom and, in the case of the LISP tutor, the textbook. Such control makes it easier to structure the curriculum around the lessons that the ITS covers. With the Grace Tutor, we did not control the classroom and could not structure the curriculum around the tutor. Rather, we had to develop a concept-of-use for the tutor that would make it an integral and important tool for an existing classroom.

Getting the Structured Editor Out of the Productions. As one of our heuristics, we discussed the importance of defining the task that the student is doing. In our case we realized that the student's primary task was not interacting with a tutor, but using a structured editor to code COBOL exercises. In our current system such confusion is inevitable. Our structured editor is not a discrete, software entity, but is completely implemented in the same productions that define the ideal student model. Each production does a double duty by implementing part of a structured editor along with a chunk of COBOL knowledge.

This double duty has several consequences, all of which are negative. First of all, writing productions that serve two purposes is more difficult than writing productions for one. Second, combining both functions in one production makes it difficult for production writers to keep student modeling independent of the structured editor. Finally, many more productions have to be written than are required for student modeling. Although essentially the same COBOL knowledge may be used in different places, the requirement to implement a structured editor means that several productions may be required that differ only in, for example, where on the screen the code appears. The solution, that has emerged in discussions with CMU, is to create a structured editor as an independent software entity. The productions of the ideal student model could then be written, almost exclusively, from the perspective of how the ideal student would interact with the structured editor to produce COBOL code.

Class Instructors Use a Different Grain Size than ITS. Instructors are the best source of expertise on what the ideal student model should include. They prefer, however, to work at a higher level than required by the individual productions of the ideal student model. We need to find better methods for eliciting this expertise from instructors and determine whether the ITS would benefit from incorporating the concept groupings recognized by instructors.

We Need a Suite of Tools, Not a Single Tool. The ACT* tutors have been shown to be effective in introductory programming courses. Although getting students started out appropriately is certainly important, we do not wish to abandon them after the first few weeks of instruction. Rather, we want to continue to support advanced students and to provide support after they leave the classroom. In addition, we need to adapt our tools to the class and work environments.

For these reasons, we are working on the Grace Suite. In addition to the Grace Tutor, other tools include the Grace Critic, the Grace Hypertext, the Grace COBOL Editor, and the Grace Mainframe Interface. The idea is to put a suite of tools at the student's disposal and to migrate the tools along with the student to the programmer's work place.

ITS Technology Transfer from Lab to Lab. The transfer of ACT* technology was successful. We were able to take software, ideas, and intuitions and produce a functioning tutor. By adapting this technology to a corporate training center, we are learning lessons that should transfer from our lab back to the ACT* group.

Transfer from the Classroom to the Job Site. Based on work with the ACT* tutors, it is clear that training transfers within the classroom from one lesson to the next. However, the most important issue is how well classroom training transfers to the job site. Although this is not an ITS issue per se, it is an important question to ask of any educational innovation, especially those intended for job training. Despite its obvious importance, transfer of training to the job site is seldom assessed. The difficulties in conducting these types of longitudinal studies are tremendous. There are, however, some advantages to pursuing these questions in a corporate setting. Unlike university education or military training, the day after our students graduate from the COBOL course they report to work as COBOL programmers. As COBOL programmers their job sites and work environment are well-known and fairly stable. For these and other reasons we are optimistic about collecting longitudinal data on transfer of training.

Real-World Trials. Throughout this chapter we have alluded to a trial that we will be conducting to provide a proof of the Grace Tutor concept. Currently the CTLC is revising the existing COBOL course. After its revision, the Grace Tutor will be used for the three lessons that we have completed. (For this trial special arrangements have been made with Symbolics to provide each student with their own workstation.)

Recently, the Metropolitan Life Insurance Company has contacted us about conducting trials of the Grace Tutor at their training center in Manhattan. Given the major differences in course length (three versus six weeks of COBOL) and student population (experienced programmers versus novices) at the two corporate training centers, we view the Met Life trial as an important test of the tutor's generality. Has the Grace Tutor been "hand-crafted" to fit into CTLC's curriculum or can the concepts and approach taken be easily adapted to other situations?

Assuming successful trials, we will revise our software platform while working on a complete COBOL curriculum. It is our expectation that by the time our curriculum is completed, New York Telephone, New England Telephone, and Metropolitan Life will have purchased sufficient numbers of workstations to integrate the tutor into the classroom on a routine basis.

ITS as CHI

From the perspective of CHI a key feature of the ITS is that it is a knowledge-based system that models limited domains. Key features of users of ITS are their lack of indepth domain knowledge and their single-minded purpose in interacting with such systems; namely, to acquire domain knowledge. These features provide powerful constraints on the range and variability of factors that affect most computer-human interactions.

Throughout our project we have been impressed with continuing discoveries that more and more of the physical interface can be prescribed by the cognitive interface. Our evolving understanding of the close connection between the physical interface and the knowledge in knowledge-based systems leads us to go beyond the cautious enthusiasm of Newell and Card [4] and Anderson [3] to proclaim the ITS as the drosophilae of CHI.

In most software systems the computer and user speak completely different languages and approach the task from completely different perspectives. The function of a physical interface in such systems is to provide a way of translating "German to Japanese" (and vice versa). In contrast, with an ITS the languages and approaches are very similar. The function of a physical interface in these systems is to provide a place where two members of the same culture can leave messages regarding a common task. This similarity of language and approach enables us to better explore the relationship between the optimal physical interface and the user's goals.

The benefits of this approach are not limited to knowledge-based systems. As we gain more experience with ITS as CHI we may begin to understand the problems in specifying an "ideal" interface when the computer and user speak different languages and approach the task from very different perspectives—that is, for nonITS applications.

In conclusion, a hallmark of an engineering discipline is having a principled basis for knowing what can and cannot be done (e.g., how long a suspension bridge can be or how hot a tire can get before it explodes). Up until now the CHI community has focused on how to build the "optimal" interface, and the unspoken attitude of the field is that a less than optimal interface is in some way a failure of the field. Perhaps this is not true, and perhaps by advancing ITS as CHI we will be in a better position to discover our potential as well as our limitations.

Postscript

Close to six months after the aforementioned information was written we were given the opportunity of writing a postscript. Much has happened. The main event of the past six months has been the preparation, conduct, and analysis of a trial at the Metropolitan Life

Insurance Company's (Met Life) New York City training center. The trial was a "usability" trial; that is, rather than being a test of the tutor's effectiveness, it was a trial to determine whether the target population could use and would use the tutor.

Pretrial

The tutor that was trialed was basically the tutor described earlier. Three lessons were completed and implemented. The biggest change was that the lesson 1 productions were rewritten again (stage four in our rewrite chronology). Although the stage three productions could handle the scope required by lesson 1, they did not easily match the style required by lesson 2. The result was another change in grain size to something smaller than stage three but larger than that of stage two.

Trial

The trial was conducted onsite at Met Life's training center. Twelve students and several instructors participated. The results were encouraging. First, the tutor passed the "smile test"; the students liked it. This result was shown both by comments made during the session and by the questionnaire administered afterward.

Second, the tutor did no harm. This statement is a cautious way of introducing our pre/post tutor results. Each of our students had already completed Met Life's COBOL course. To determine whether the tutor had anything to teach such students, we introduced a written pretutor and posttutor COBOL exercise. We scored the written exercises as if we were the tutor; that is, we looked for "correct productions." This scoring method yielded a pre/posttest difference of 65 to 84%. (A second and blind scoring, by a Met Life instructor who used his own criteria, resulted in an even larger pre/post difference.) Although the lack of any control group prevents us from making claims about the tutor's effectiveness, we feel safe in saying that we are "on the right track."

Third, the tutor passed our first usability criteria; namely, students were able to master the interface after less than an hour of use. Many of our improvements (including the new menu structure) worked. A major caveat is that although students (and instructors) had clearly mastered the interface they still had complaints, many of which we believe are justified. For example, it took students a while to get used to using the bottom window for keyboard input. They all wanted to be able to type in the code window.

All in all, the trial was a great success. We obtained feedback on every aspect of our system from writing better exercise descriptions, to what our help message should contain, to where our feedback messages should appear.

Current Events

We have entered phase two of our tutor development and are advancing on two fronts. One front involves the design of a new, integrated "bull's eye" architecture for Grace. The "bull's eye" architecture has a structured editor as its center, the critic in the next ring, with the tutor on the outermost ring. The structured editor will be a type-in editor. One reality that we have finally accepted is that COBOL programmers want to type and do not like to use menus. Any editor that is going to win the hearts and minds of the COBOL community has to allow type-ins. The critic will be built on top of the editor and will extend the editor's ability to check syntax into semantics and logic. The tutor will be built on top of the critic and will have access to all of the critics rules as well as additional rules of its own.

The second front involves using the current architecture to build a full curriculum tutor (FCT). The FCT will cover most of the lessons in a full COBOL curriculum. There are three reasons for working on the FCT. First, we expect that all of the curriculum, lessons, and exercise descriptions will transfer completely to the new architecture. We hope that at least the grain size of the productions also will transfer. Hence a majority of the curriculum development work will transfer. Second, we need to collect data comparing the tutor with the standard COBOL classroom. Such effectiveness data is required to obtain continuing support for the project within NYNEX and without. Third, we expect to use the FCT as an environment for testing out ideas about tutoring. For example, one of our "lessons learned" from the Met Life trial is that we cannot count on students acquiring the prerequisite declarative knowledge before using the tutor. A possible solution to this is a "master/apprentice" mode in which the tutor solves the first exercise in each lesson as the student observes. As each rule fires the tutor pops up a text window that explains the rule in the current context. Hence, the declarative knowledge required for the current lesson is taught in the context in which it it needed. The master/apprentice mode is just one of the ideas that are easily implemented within our current architecture.

Acknowledgments

Building an ITS is a large enterprise involving many people at many different levels of involvement. Some of the people who deserve thanks include: Carnegie-Mellon University: John Anderson, Albert Corbett, Jon Finchman, Raymond Pelletier, and Lael Schooler; New York Telephone: Susan Brand, Jane Ellison, Thelma Gattis, Gary Henninger, Barry Howard, John Ott, Joan Pipkins, Ken Warburton, and Rosemary Willsen; New England Telephone: Marvin Cruse, Gus Martinson, Pat Neville, and Marie Tansey; Metropolitan Life Insurance Company: Romia Bull, Robert Flast, and Marcia Hearst; NYNEX Science & Technology: Bart Burns, Shelly Dews, Anders Morch, Robert Radlinski, Edmond Thomas, John Thomas, and Athea Turner.

Thanks also to John Anderson, Bart Burns, Anders Morch, and Robert Radlinski for their comments on earlier versions of this chapter.

References

1. J. R. Anderson, *The architecture of cognition*, Cambridge, MA: Harvard University Press, 1983.
2. J. R. Anderson, "Skill acquisition: Compilation of weak-method problem solutions," *Psychological Review,* 94:192–210, 1987a.
3. J. R. Anderson, "Methodologies for studying human knowledge," *Behavioral and Brain Sciences,* 10:467–505, 1987b.
4. A. Newell and S. K. Card, "The prospects for psychological science in human-computer interaction," *Human Computer Interaction,* 1:209–42, 1985.
5. G. Fischer, A. C. Lemke, T. Mastaglio, and A. Morch, "Using critics to empower users," *Proceedings of CHI,* Seattle, WA: ACM, 1990.
6. P. Berman, "The study of macro- and micro-implementation," *Public Policy,* 26:157–84, 1978.
7. E. Wenger, *Artificial intelligence and tutoring systems.* Los Altos, CA: Morgan Kaufmann Publishers, 1987.

8. J. Psotka, L. D. Massey, and S. A. Mutter (eds.), *Intelligent tutoring systems: Lessons learned,* Hillsdale, NJ: Lawrence Erlbaum, 1988.
9. M. L. Johnson and E. Soloway, "PROUST: An automatic debugger for PASCAL programs," *Byte,* 10(4):179–90, April 1985.
10. J. G. Bonar and R. Cunningham, "Bridge: Tutoring the programming process," in J. Psotka, L. D. Massey, and S. A. Mutter (eds.), *Intelligent tutoring systems: Lessons learned,* Hillsdale, NJ: Lawrence Erlbaum, 1988.
11. J. R. Anderson, C. F. Boyle, A. Corbett, and M. W. Lewis, "Cognitive modeling and intelligent tutoring," *Artificial Intelligence,* in press.
12. A. T. Corbett and J. R. Anderson, "The LISP intelligent tutoring system: Research in skill acquisition," in J. Larkin, R. Chabay, and C. Scheftic (eds.), *Computer assisted instruction and intelligent tutoring systems: Establishing communication and collaboration,* Hillsdale, NJ: Lawrence Erlbaum, in press.
13. J. R. Anderson, C. F. Boyle, and B. J. Reiser, "Intelligent tutoring systems," *Science,* 228:456–62, 1985.
14. J. A. Anderson and B. J. Reiser, "The LISP tutor," *Byte,* 10(4):159–75, April 1985.
15. J. R. Anderson, C. F. Boyle, and G. Yost, "The geometry tutor," in *The Proceedings of IJCAI,* Los Angeles, CA.
16. S. Dews, "Developing an intelligent tutoring system in a corporate environment," *Proceedings of the 33rd Annual Meeting of the Human Factors Society,* Denver, CO.
17. W. D. Gray, "A role for evaluators in helping new programs succeed," in J. S. Wholey, M. A. Abramson, and C. Bellavita (eds.), *Performance and credibility: Developing excellence in public and nonprofit organizations,* Lexington, MA: Lexington Books, 1986.
18. B. S. Bloom, "The 2 sigma problem: The search for methods of group instruction as effective as one-to-one tutoring," *Educational Researcher,* 4–16, June/July 1984.
19. M. K. Singley and J. R. Anderson, *The transfer of cognitive skill.* Cambridge, MA: Harvard University Press, 1989.
20. A. T. Corbett, J. R. Anderson, and E. J. Patterson, "Problem compilation and tutoring flexibility in the LISP Tutor," *Proceedings of ITS-88: The International Conference on Intelligent Tutoring Systems,* Montreal, Canada, 1988, 423–9.
21. M. K. Singley, J. R. Anderson, J. S. Gevins, and D. Hoffman, "The algebra word problem tutor," in D. Bierman, J. Breuker, and J. Sandberg (eds.), *Artificial intelligence and education,* Amsterdam: IOS, 1989.
22. W. W. Pugh and S. J. Sinofsky, "A new language-independent prettyprinting algorithm," Ithaca, NY: Cornell University, Computer Science Department, 1987.
23. G. W. Furnas, T. K. Landauer, L. M. Gomez, and S. T. Dumais, "Statistical semantics: Analysis of the potential performance of key-word information systems," *Bell System Technical Journal,* 62:1753–806, 1983.
24. G. W. Furnas, T. K. Landauer, L. M. Gomez, and S. T. Dumais, "The vocabulary problem in human-system communication," *Communications of the ACM,* 30:964–71, 1987.

Chapter 9

Design, Development, and Implementation of an Intelligent Tutoring System for Training Radar Mechanics to Troubleshoot

LAURA C. KURLAND
ROBERT ALAN GRANVILLE
DAWN M. MACLAUGHLIN

BBN Systems and Technologies Corporation
10 Moulton Street
Cambridge, MA 02138

Abstract *The Maintenance Aid Computer for the HAWK-Intelligent Institutional Instructor (MACH-III) intelligent tutoring system (ITS) contains a qualitative radar simulation and an expert troubleshooter that can both demonstrate the correct troubleshooting procedures or monitor, advise, and evaluate the student's performance. The student or instructor can select three different modes that allow the student to: (1) troubleshoot with assistance or not, (2) see and explore how devices contribute to the system function, and (3) see an expert performance. Our ITS provides the functional decomposition that a radar mechanic needs to troubleshoot by providing selectable, multiple views of the system's functional circuits. These circuits show the testable and untestable components and connections in a physical orientation. Correct device and system models are provided by dynamically showing the power and information signals' paths of influence. One unique aspect of MACH-III is that it decomposes relevant troubleshooting knowledge into a (selectable) functionally organized, hierarchical tree that shows the symptoms, the malfunctions that contribute to those symptoms, and the components that contribute to those malfunctions. Because the ITS is being used as a trainer at Fort Bliss, development has focused on creating a flexible instructional tool, an environment, and a tutor with instructionally relevant features that can be easily selected. The use of these features and the supportive ancillary materials are critical for successful implementation.*

Introduction

In the last few years, ITSs have emerged from the laboratory for use in real-world environments. This chapter has three major topics: First, it describes how the MACH-III (Maintenance Aid Computer for HAWK-Intelligent Institutional Instructor), an intelligent simulation and tutoring system, addresses the real-world task of teaching novice mechanics to troubleshoot a complex electronic system, the HAWK HIPIR radar (ANPQ-57). Second, this chapter describes the process by which MACH-III is being integrated into a well-established curriculum. Third, the chapter addresses research investigating how MACH-III was received by students and instructors, and observations as to what they did with it.

The MACH-III consists of an underlying model-based simulation of the radar and an expert troubleshooting system that can either demonstrate the correct troubleshooting procedures or monitor, advise, and evaluate a student's troubleshooting performance. The model-based simulation actually consists of two domains, the receiver and the transmitter subsystems. The expert troubleshooting system contains two distinct modules: a knowledge base that represents the functional hierarchy of troubleshooting knowledge (the troubleshooting tree) and an inference engine that interacts with the troubleshooting knowledge tree, the model-based simulation, and the student. Development was done on a Symbolic LISP machine. Each stand-alone workstation consists of a Symbolics 3600 computer with a monochrome monitor, a color monitor, a mouse, and a keyboard. Three workstations are currently in use for instruction at the HAWK Department, Fort Bliss Army Air Defense School, El Paso, Texas.

The MACH-III was mandated to be a "training system" that had to be integrated into the curriculum and utilized in a school setting. Its functions had to be flexible, friendly, and tailored for both instructor and student use. To meet these multiple demands, MACH-III had to serve three different roles: As an *instructional tutor*, students are supported throughout the troubleshooting session with advice and feedback. As an instructional tool, instructors may use MACH-III features selectively, given their teaching style or particular student need. As a multimedia instructional environment, instructors or students can use MACH-III's multiple capabilities as an animated textbook (e.g., dynamic diagrams), a dynamic workbook with selectable exercises for troubleshooting, an explanation database that provides explanations about devices or reasons about why devices pertain to a symptom, or just as a simulated radar where they can break things and run the built-in tests to see the resulting symptom indicators.

The design of the MACH-III ITS was based on observations of troubleshooting sessions on the radar and protocol analyses of interviews with experts (instructors, radar engineers) and mechanics with a range of experience from student to expert. During development, the design has been formatively evaluated with individual instructors and students, individually and in pairs. Research on embedding the MACH-III system in the present curriculum and on training instructional staff to effectively utilize system capabilities is reported in this chapter, along with preliminary results of the system's effect on student learning.

Front-End Research: Cognitive Task Analysis

The radar mechanic's job is to maintain and troubleshoot the radar. If, while running the daily checks and specialized built-in monitoring tests, a particular fault indicator illuminates, the mechanic needs to perform the steps in the *Fault Isolation Procedures (FIP) Manual*. This consists of continuity checks on cables, reading meters, setting switches, and replacing devices. It should be noted that although these mechanics do take a brief basic electronics class and see resistor-level detail in their wiring schematics manual, these smaller items are usually contained by the devices they test or replace, such as circuit cards, or larger, more complex assemblies containing many components.

From our research [1, 2] and other similar studies on technicians [3], a number of related difficulties hinder the students and junior mechanics from understanding how the radar works and from troubleshooting successfully and efficiently. From a cognitive viewpoint, student problems fall into four categories:

Information Overload

The critical information in the voluminous and cumbersome documentation (the schematics and the FIP) that is needed for troubleshooting is often too detailed, too general and too difficult to comprehend, or the information is simply missing.

Unsuitable Hierarchical Knowledge Organization

Novice mechanics lack a hierarchically organized, cognitive framework that is suitable for troubleshooting. They need a system-level model that supports the acquisition of relevant functional and procedural information and allows this knowledge to be easily recalled. The fault isolation procedures (FIP) manual does not offer any high-level information about why one is testing for a voltage or replacing a device in the procedural steps. Because the FIP has a branch and flow organization that directs the reader to another seemingly arbitrary, often redundant step, a mechanic who misreads, mismeasures, or skips a relevant step can get hopelessly lost.

Inadequate System and Device Models

Novice mechanics lack a sufficiently detailed and accurate model that organizes and integrates knowledge from a functionally based, troubleshooting point of view. This view needs to include how the radar works at the system, circuit (functional feedback loops), and device levels. Students have great difficulty acquiring these accurate models because the schematics are organized by an engineer's taxonomy of functional blocks and show an overly detailed view of circuits (e.g., amplifiers and resisters), and yet, at the same time, do not show complete system- or device-level information that a troubleshooter needs. For example, if a device has four inputs, one functional block may only show two of the signals. Also, the functional feedback loops, like idioms in language, are entities that perform specific functions. A loop's purpose or existence cannot readily be discerned from just looking at the schematic detail. One must know it exists and what its purpose and constituent components are a priori; because if one component within the loop is malfunctioning, it affects all the other components within that loop. Furthermore, if such a broken component belongs to multiple loops, all those loops are impaired, producing multiple faults.

Lack of Strategies

Novices lack a functional understanding of why the fault isolation procedures require them to perform particular tests or replacements. This lack of procedural understanding, along with their inadequate system and device models, leaves them unable to cope with unexpected results or mishaps. Because the FIP does not provide any high-level insight into its troubleshooting procedures, mechanics do not understand that it may implicitly be employing high-level functional strategies, such as separating specific loops or investigating a device's various power sources.

The purpose of our ITS is to provide scaffolding that will strengthen and integrate the student mechanic's functional understanding of the radar and its documentation, and to organize this knowledge to help the students troubleshoot in a more logical manner.

Understanding the documentation is important, since these manuals are the only cognitive tools that mechanics will have available on the job. They must learn to utilize them effectively while troubleshooting, rather than just trace circuits in the schematics without comprehending their function or perform test and replacement steps without understanding the reason for each step.

Based on our research, the MACH-III must:

Show the problem space at various levels of detail for each functional area;
Provide troubleshooting practice on problems that increase in difficulty;
Provide a range of procedural help, such as physical location, switch setting, correct sequence where necessary, and correct use of the documentation;
Provide supports to make strategies explicit; and
Provide general and specific explanations about purpose, location, and behaviors of devices, signals, functional circuits, and the system.

Addressing Instructional Needs with MACH-III

To address these problem areas, we developed an ITS with the following constructs:

An underlying simulation that qualitatively simulates the radar's internal/external components and signals and their dynamic properties;
Diagrams (needed for troubleshooting) that provide multiple views of the system at different levels of detail;
A troubleshooting tree that organizes the procedures into a functional hierarchy;
A troubleshooting advisor that guides mechanics through the process of troubleshooting the simulated subsystems through these views; and
An explanation system that provides device background information and high-level information (reasons) about the procedural steps outlined in the troubleshooting tree.

The constructs, described in the next discussion, were implemented to take advantage of the computer's dynamic and interactive capabilities and were designed for maximum flexibility to adjust to student and instructor needs.

What the User Sees

Diagrams—Multiple Views of the System

To bridge the gap between a complex machine, the schematic documentation, and the student's limited, possibly inaccurate understanding, animated diagrams depicting multiple views of the radar system were developed: purely physical, purely functional, and a mixture of physical and functional at different levels of detail.

Purely physical diagrams show only those parts that are testable and replaceable as indicated in the technical manual, along with all built-in test equipment, such as meters, indicators, and push buttons. Purely functional diagrams group labeled boxes representing components (including adjusts) into color-coded, high-level functional groups (e.g., "makes signal"). Physical-functional diagrams show an overview and detailed functional circuits (loops) in a physical context (see Fig. 1). These diagrams attempt not only to show signal flow but also functionally important devices that tell the "functional story" within a physical context and orientation.

High Frequency Noise Degeneration Loop

The high frequency loop is used to cancel high frequency noise in the XMTR signal being created. To do this, it isolates the high frequency noise in the degen bridge (through the AM and FM loops), which is then sent to the isomodulator to cancel the noise in the signal that is simultaneously being created.

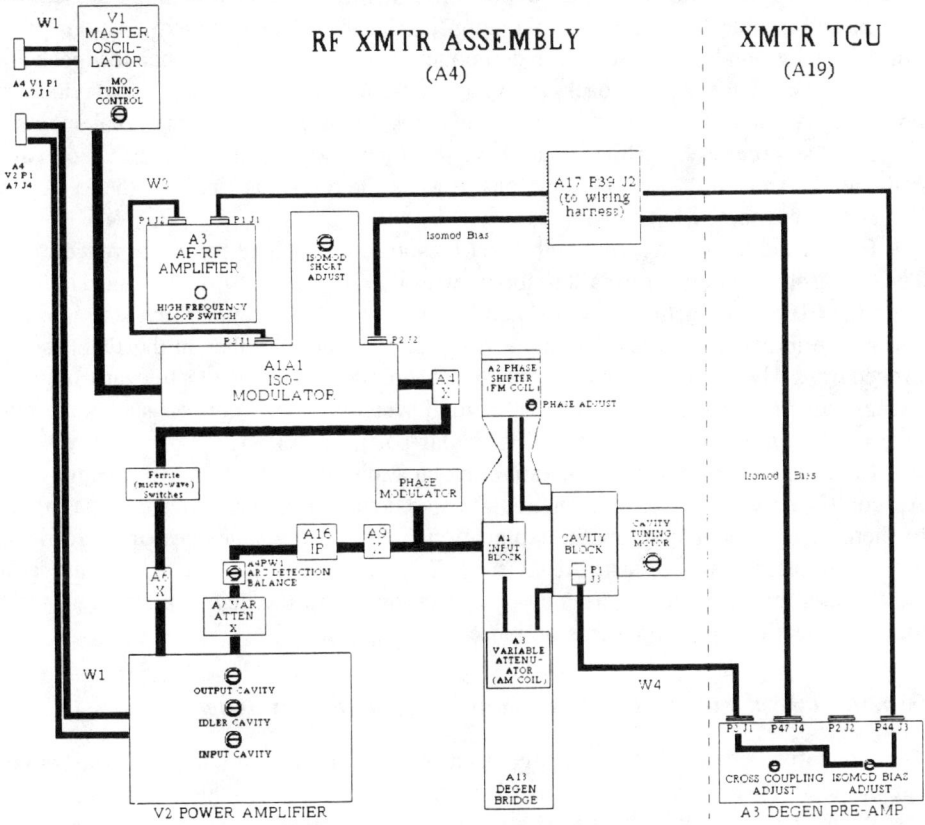

Figure 1. A physical functional loop.

These three types of views attempt to summarize the schematic wiring diagrams in selective ways: power is distinguished from information (controlling signals), their paths of influence are shown with animated "flow," functional loops are specifically labeled and drawn at the large component level, and device models are provided, both visually and qualitatively (at the simulation level). With these various views of the radar system, students can: troubleshoot as if on the real system (using physical diagrams); develop a high-level functional understanding of how major subsystems (like the transmitter) work, and comprehend what and how malfunctions affect them (using functional diagrams); and comprehend how certain functions are performed by components organized within functional circuits (mostly loops), how faults affect these circuits, and how to fix the faults (using physical-functional diagrams).

Explanation Systems

MACH-III embodies two distinct types of knowledge available for explanation. The first type is static background information we call *canned explanation*, which provides background functional information necessary for functional troubleshooting. The second is dynamic, context-sensitive information, which we call *generative explanation*, needed for the troubleshooting advice and critiques generated during troubleshooting, described in the section, Generative Explanation: Troubleshooting Tree and Expert System.

There are two kinds of canned explanation facilities in this system are device explanations and troubleshooting tree explanations (reasons). The device explanations provide critical background information for components, wires, and functional feedback loops (that act like devices). This background device information includes: the device's purpose, the larger context in which it participates (such as loops), its signal inputs and outputs, the process by which it does its job ("the how"), and the effect of various behavioral states on the system functions (e.g., if the device is "noisy," then x happens; if it is broken, then y happens).

In the troubleshooting tree, high-level reasons are provided for each procedural test. These reasons contain information about: why the step is performed (which is missing from the FIP); what action is performed by that step, if any; and which step is relevant by cross referencing the relevant table, step, and substep number in the Fault Isolation Procedures. The reasons connected with each entry (node) in the tree are important, because this information helps develop a good system model. Figure 2 shows the reason why a procedure is performed for a particular part, the A3W1.

Both the canned and the generative explanation systems use a hypermedia format. All bold-faced words in the explanation are also devices that have device menus attached to them. The student may obtain explanations of these other devices or perform other actions on them (replace, test, break, etc.). The interactive tree is a visual navigational aid through the procedures, grouped by symptoms. Both the device menus and the tree menu connect to the components on the color diagram.

Generative Explanation: Troubleshooting Tree and Expert System

The generative explanation knowledge used in MACH-III is organized in a hierarchical data structure, which we call the troubleshooting tree. Because the documentation is all that the mechanics will have available to them on the job overseas (fresh copies, not the ones they used at school) and because they have difficulty understanding why steps are performed and thus "branch and get lost" (their term), a troubleshooting tree was designed for each subsystem (Fig. 3). The troubleshooting trees summarize the possible problem space in a manner organized for flexible, functional troubleshooting. The troubleshooting tree essentially provides a functionally organized overview of the Fault Isolation Procedures, much like the diagrams provide an overview of the schematics.

For each subsystem (e.g., the receiver), the functional troubleshooting tree displays the problem space associated with each malfunction symptom, as determined by the Fault Isolation Procedures, and high-level troubleshooting knowledge organization. The trees start with a node called "Perform SPB Test" or "Perform XMTR Auto Test." These nodes represent the built-in test (a preprogrammed test that the student activates by a pushing the *Test* button) by which the mechanic collects the symptoms. The tree then moves "down" to the next level representing the visible symptoms (fault lights). Under each symptom, possible high-level malfunctions (hypotheses) are listed (e.g., "Incorrect Monitoring"). The next level organizes the technical manual's procedural tests into very high-level functional tasks.

The malfunction and functional task groupings are important improvements over the paper-based FIP, since these overarching organizers of troubleshooting information are generally missing in the technical documents. Levels under the functionally organized

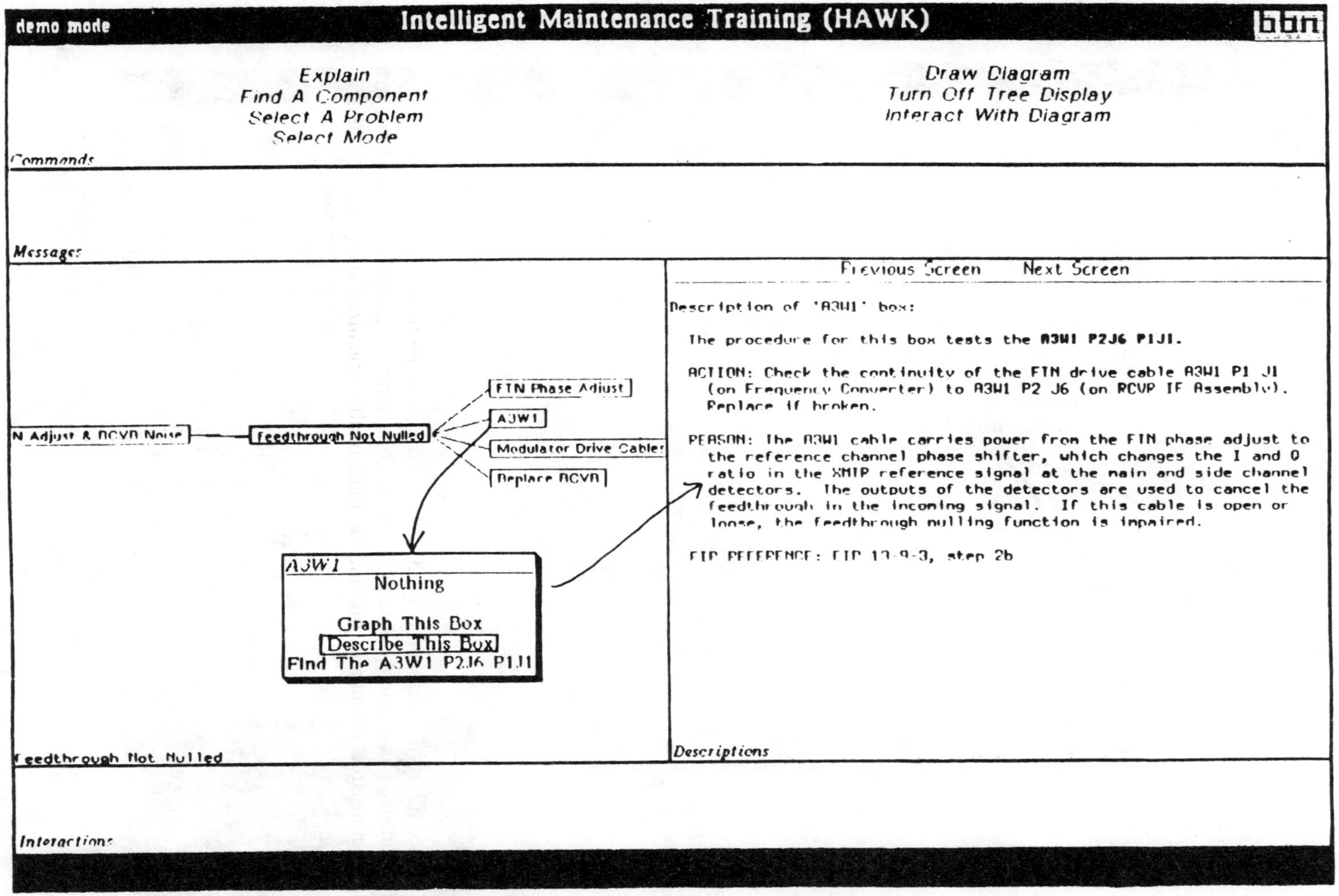

Figure 2. A troubleshooting explanation.

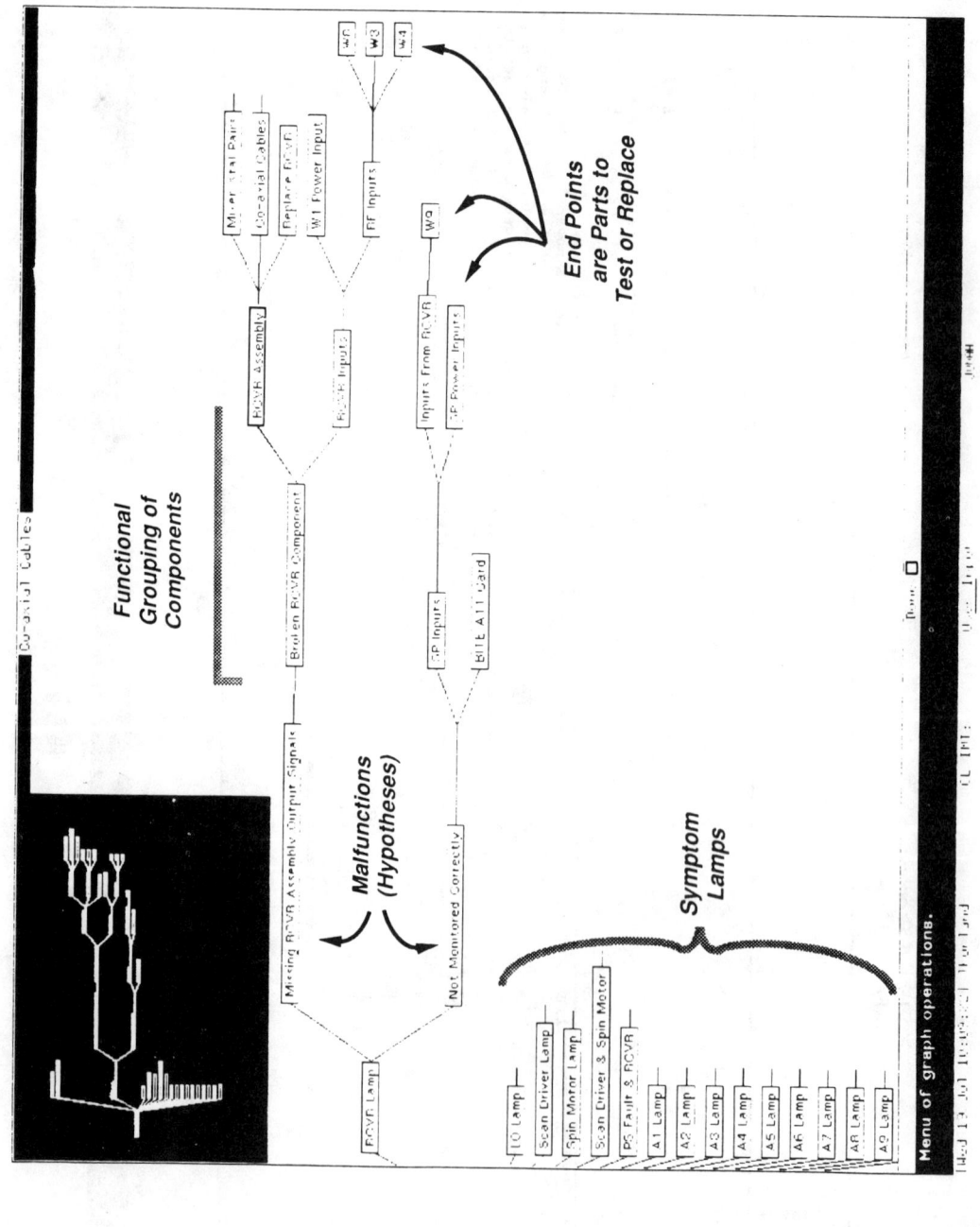

Figure 3. A troubleshooting tree.

procedures are tests pertaining to specific items, such as "Check Input X." As an item is tested, it "greys out," giving the student a quick overview of what items in the problem space have been completed and what items remain.

A troubleshooting expert system guides mechanics through the process of troubleshooting by providing supports to make strategies explicit and provide a range of procedural help in the context of troubleshooting. These articulate hypotheses about malfunctions describe which parts remain unchecked and have been ruled out, give increasingly specific hints, alert the student to inconsistencies, and specify what procedural actions are absolutely sequential. More is said about the expert systems' functions in the troubleshooting modes section below and the following technical section.

Modes of Troubleshooting Interaction

Troubleshooting occurs in three modes: Magic Mode, Real-Life Mode, and System Demonstrates. In Magic Mode, all "signal flow" is shown, so if a component is broken and produces no signal, there is no flow emanating from it. Also, normally inaccessible components are testable (but to fix them, the real replacement unit is given as the only option by request of the instructors). This mode does not require the student to perform the required sequential procedural actions as found in Real-Life or System Demonstrates Modes.

In Real-Life Mode, no dynamic flow is shown in the diagrams until the fault is located and only real-life items may be tested or replaced. In Real-Life Mode, the *Advice* and *Critique* functions may be selected separately, together, or not at all. The Advisor helps the student decide what to do next (e.g., based on what functionally grouped procedural tests have been left undone) and provides feedback about whether there have been previous requests about the same point. Advice is graduated according to where the student is within the troubleshooting tree structure—that is, it takes the student down one level of specificity.

Advice at the topmost level generally involves giving a functional hypothesis (e.g., "perhaps feedthrough is not being removed"). At the next most general level, *Advice* may suggest a very large area to check ("test the receiver parts") with an option provided for more specific information ("click HERE for more information about receiver parts"), which may result in a list of specific parts (e.g., "W9 cable") or a functional grouping of parts (e.g., "the receiver input cables") that also has the option of further specification.

The Critiquer gives feedback about the "reasonableness" of the student's last step, specifies whether items in a functional group have been skipped or do not pertain to that symptom, and provides hints and other context-specific information about a particular step. The text is "generated" according to the part, where it lies in the tree, and the student's previous actions.

In System Demonstrates Mode, like Real-life Mode, signal flow is not displayed. Here the student requests the expert to perform whatever next action it has determined is best at this point. At each step, the expert explains what next action it is considering and why, what the result of that action is, and what it means. The expert highlights the part where it will perform an action, changing the diagram if necessary. As in Real-Life Mode, the parts that have been done are greyed out on the troubleshooting tree.

Technical Overview of MACH-III

This technical overview of MACH-III describes MACH-III's constituent software components, the component interactions, and the functional hierarchy of troubleshooting knowledge (the tree) and the benefits of such a structure. MACH-III is composed of two integrated software modules: the Model-Based Simulation and the Troubleshooting Expert System. The Troubleshooting Expert System can further be broken down into two independent components, the Knowledge Base, consisting of the Troubleshooting Tree, and the Inference Engine, known as the Conduit Program. Figure 4 shows the interactions between these modules.

The Model-Based Simulation

The Model Based Simulation is the "radar program" of MACH-III. This is the component that models the behavior of the HAWK. It consists of a Simulation Engine and a System Model. Each of these is described here.

The Simulation Engine. The Simulation Engine controls how devices are simulated and how signals are propagated throughout the system. The simulation runs off a queue: each step in the simulation involves simulating the device at the front of the queue by executing its behavior; if any of the device's outputs change, the devices that receive those outputs are placed onto the end of the queue so that they will be simulated. The simulation begins in one of two ways: either by placing all "producers," which are devices with no inputs, onto the queue when the entire simulation needs to be run, or by placing a specific device onto the queue when the state of the device is changed (e.g., it

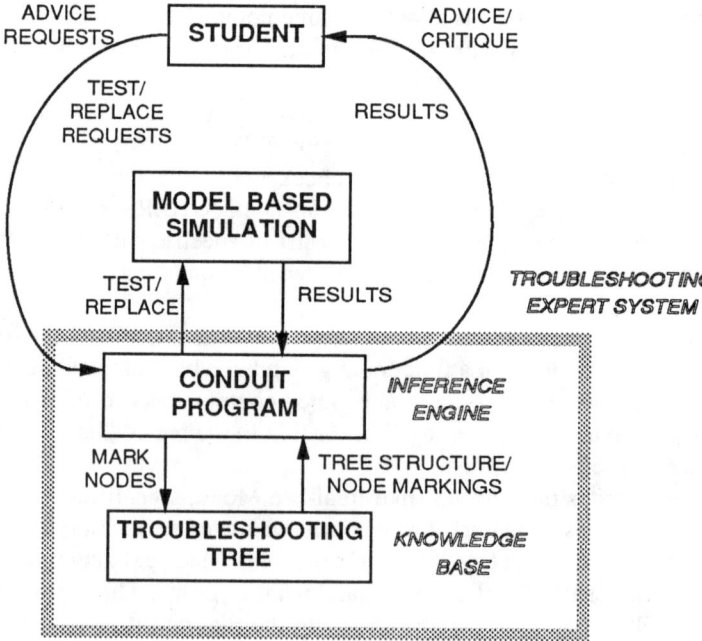

Figure 4. MACH-III software modules (Real-Life Mode).

is repaired, turned on or off, or set to a new position). The simulation terminates when there are no more signal changes to be propagated (the queue is empty).

Loops often create problems in simulation systems, since signal propagation is not as instantaneous as in the real electrical world. The feedback signal is not available when the loop is initially simulated, but the device that joins the feedback signal with the loop's input (the *join* device) needs the feedback signal to create its output and get the loop going. In MACH-III, we experimented with two different solutions to this problem:

> Solution 1: The simulation engine makes an assumption about the feedback signal and then simulates the loop with this assumed signal. If this assumption proves to be wrong, perhaps because of a fault somewhere in the loop, then the corrected feedback signal will be propagated through the loop again;
>
> Solution 2: The behavior of the join device, where the feedback signal joins the input of the loop, is defined in such a way that it can produce an output signal without requiring all its inputs (including the feedback input) to be present. In this way, the join device is simulated, producing an output that effectively represents the situation where there is no feedback. When the feedback signal propagates around, the output of the join device will be refined to incorporate the feedback signal.

Both solutions have proven to be effective. However, Solution 1 places more burden both on the person who is building the system model and on the simulation engine. The modeler must both define the properties of the signal to be assumed and explicitly define the join device to be part of some functional loop device. The simulation engine uses this information to know both when to make a signal assumption (if it is trying to simulate a device that is defined to be part of a loop and that device is missing an input) and what signal to assume (it creates a signal with the properties specified by the modeler).

At first glance, Solution 2 seems to imply that more care must be taken in defining the behavior of the join device to account for the case where there is no feedback signal. However, in a system that is modeling faulty behavior, this case will inevitably arise because of a fault somewhere in the loop, and thus would have to be accounted for anyway. Also, Solution 2 intuitively seems closer to represent the real world.

The System Model. The System Model defines three things: the types of devices that are modeled, the specific device instances that are to be simulated, and how these instances are connected together to form the system. Device types and device instances are described in more detail here.

Device types. A device type represents a class of devices, of which there may be several different instances in the system model. Examples of device types are WIRE, BUTTON, and COMPLEX DEVICE. A device type defines the inputs, outputs, and parts of the device, as well how instances of the device behave in the simulation under both normal and faulty conditions, and what actions can be performed on that device type (e.g., wires can be tested for continuity and buttons can be pushed). The device types are organized into a taxonomy, as exemplified by the items in dark boxes in Fig. 5. In this figure, each box represents a device type. The bold label in the box is the name of the device type, and the text under the bold lists some operations that apply to that device type. The links between nodes are ISA links (a BUTTON is a BASIC DEVICE), and operations that apply to the higher nodes are inherited by the lower nodes (both BIDIRECTIONAL WIRE and CABLE WITH CONNECTORS inherit the "test for continuity" operation). The taxonomy is rooted in BASIC DEVICE, which defines the basic

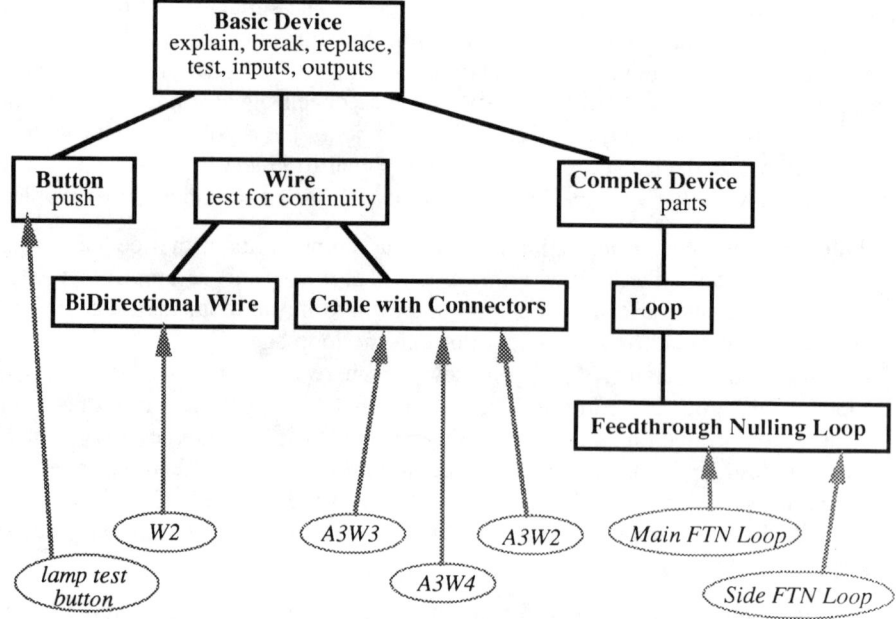

Figure 5. Device taxonomy with instances of device types.

operations that apply to all devices (e.g., explain, break, replace, and test), as well as methods for manipulating, for example, the repair state (e.g., good, broken), internal state (e.g., on, off), replaceability, and testability of devices.

A device type that consists of defined parts is built on a special device type called COMPLEX DEVICE, which has associated with it some special operations for manipulating parts. Complex devices may represent physical replaceable assemblies (e.g., the receiver, the transmitter control unit) or aggregations of parts into functional loops (e.g., the main feedthrough nulling loop, the high-frequency noise degeneration loop). The inputs and outputs of a complex device are associated with the inputs and outputs of their parts (e.g., the main channel output of the receiver is the output of the main band pass filter that is part of the receiver). Although complex devices are not simulated (their parts are), they are still manipulated in the same manner as other devices (they can be explained, replaced, etc.).

The behavior of a device, defined in a special rule language developed for MACH-III, specifies how the device behaves when it simulates. Specifically, the behavior defines how to derive the output signals from the inputs and the repair and internal states of the device (e.g., good, noisy, off). For example, the WIRE device type is defined to have one input, one output, and no parts. The behavioral definition of a WIRE is such that under normal conditions, the output is simply the same as the input; but when the wire is open, the output is absent, no matter what the input is.

Device instances. A device instance represents a specific component in the radar. It is an instance of a device type. The device instance (or simply, device) has a unique part name and is uniquely connected in the system. Figure 5 shows some instances added (represented as gray circles) below the taxonomy. In this figure, A3W2, A3W3, and A3W4 are three different instances of CABLE WITH CONNECTORS; each is testable for continuity (this operation is inherited from WIRE). The Lamp Test Button is a button that can be pushed. The Main FTN Loop and Side FTN Loop are instances of FEED-THROUGH NULLING LOOP, a complex device with operations for manipulating its parts.

The MACH-III system models. The MACH-III contains two distinct system models for the receiver and transmitter subsystems of the HAWK HIPIR radar. These two models do not interact: they are simulated separately, and no signals are passed between the two. A session with MACH-III involves working with one model or the other; one does not normally switch back and forth between the two. The system enforces this separation by not allowing information about one model to be accessible from the other model.

A total of 158 device types have been defined to model both the receiver and the transmitter. The taxonomy formed by these device types has been found to be wide and shallow: the device types in the MACH-III system are dissimilar enough that they are not built from more general device types; they do not inherit many characteristics from other device types. Each device type tends to define all the necessary characteristics by itself.

The receiver model contains 96 devices (instances), whereas the transmitter model contains 147. The receiver models 137 potential faults (all the devices in the receiver can be broken a total of 137 ways, although some may produce the same symptom, e.g., receiver noise lamp on), whereas the transmitter models 111 faults.

The Troubleshooting Expert System

The Troubleshooting Expert System is MACH-III's "expert troubleshooter program." The Troubleshooting Expert System knows how to troubleshoot the HAWK, and is capable of either demonstrating the correct procedures or advising and critiquing students as they attempt to troubleshoot a problem. The Troubleshooting Expert System is made up of two distinct independent modules, a Knowledge Base consisting of a Troubleshooting Tree, and an Inference Engine consisting of a Conduit Program.

The Troubleshooting Tree forms the Knowledge Base of the Troubleshooting Expert System. It represents a hierarchical organization of the troubleshooting knowledge that not only tells the student what to do, but also explicitly has knowledge as to why a particular action should be taken. It does this by representing a hierarchy of finer and finer hypotheses on what could be causing the current problem.

The Conduit Program forms the Inference Engine of the Troubleshooting Expert System. This module is the "workhorse" of the Troubleshooting Expert System, performing three distinct tasks by which information flows between the student, the Troubleshooting Tree, and the Model-Based Simulation.

The first task is to use the Troubleshooting Tree as a database to advise and critique the student. The Conduit Program accesses the tree structure and the markings set on nodes to determine the best course of action in the current situation and to judge the actions currently being taken by the student. It also marks nodes in the Troubleshooting Tree depending on results of actions. The second task is to directly interact with the Model-Based Simulation. The Conduit Program performs actions in the simulation and collects the reported results of those actions. Thus, the Conduit Program can be seen as applying the knowledge of the Troubleshooting Tree to the problem represented in the Model-Based Simulation. The third task is to provide the appropriate interface between the Model-Based Simulation and the student, depending on the selected mode of interaction. The Conduit Program performs actions in the Model-Based Simulation by the student and reports back the results of these actions to the student, as well as any advice or critique required by the mode of interaction.

The Conduit Program is designed to be strictly domain independent, and therefore is a completely separate module from the Troubleshooting Tree and the Model-Based Simulation, which are both strictly domain dependent. A direct result of this is that to apply the MACH-III technology to other domains, the Conduit program would not have to be changed or modified at all. Although the Model-Based Simulation and the Troubleshooting Tree would obviously have to be developed for a new domain, the Conduit program as it now exists would run equally well with these new modules as it does with the current ones for MACH-III. In fact, MACH-III actually does consist of two domains, one for the HAWK Receiver Assembly and one for the Transmitter Assembly. Each domain has its own Model-Based Simulation, and each has its own Troubleshooting Tree. The one Conduit Program is used with equal success for both sets of Model-Based Simulations and Troubleshooting Trees.

Component Interactions. As we have described, MACH-III has three modes of interaction with the student. These modes affect how the student interacts with the software components and how these components interact with each other. In Magic Mode, only one software component, the Model-Based Simulation, is in use. Students interact directly with this simulation, testing and replacing pieces of the HAWK and having the results of their actions reported directly to them.

In Real-Life Mode (see Fig. 4), the Troubleshooting Expert System is added. Rather than interacting directly with the Model-Based Simulation, the student acts through the Conduit Program (although the interface continues the illusion of direct interaction). The student issues action requests to the Conduit Program, as well as advice requests, depending on the feedback mode. The Program then performs these actions in the Model-Based Simulation and collects the results itself. It also interacts with the Troubleshooting Tree at this time, using the tree structure and marked nodes to judge the appropriateness of the requested actions and marking nodes in the tree based on the performed actions and the collected results from the Model-Based Simulation. The Conduit Program then reports back to the student the results of the actions, as well as any advice or critique based on the feedback mode. Thus we can see that in Real-Life Mode, all interactions between the student, the Model-Based Simulation, and the Troubleshooting Tree must flow through the Conduit Program.

In Demo Mode, the Troubleshooting Expert System behaves very similarly to that in Real-Life Mode. The chief difference is the interaction between the student and the Conduit Program. Rather than issue requests for advice or actions determined by him or her, the student can only request the Program to perform the next action it determines is best at this point. The Conduit Program then interacts with the Troubleshooting Tree exactly as it did in Real-Life Mode to make this determination. It then performs this action by interacting with the Model-Based Simulation exactly as it did in Real-Life Mode, again collecting the results. Nodes are marked in the Troubleshooting Tree the same as they were in Real-Life Mode, and the action and its results are reported to the student. And just as it was in Real-Life Mode, all interactions among the student, the Model-Based Simulation, and the Troubleshooting Tree flow through the Conduit Program.

The Functional Hierarchy of Troubleshooting Knowledge. Functional hierarchy is a knowledge-base organization paradigm based on the procedural abstraction methodology of Liskov and Guttag [4]. Procedural abstraction focuses on the behavior of a procedure while hiding its implementation to achieve a high degree of modularity. To quote, "Thus the behavior—'what' is done—is relevant, while the method of realizing that behavior—'how' it is done—is irrelevant" [14].

Following this principle, we arrange the knowledge base into a hierarchy of definitional rules. Each individual rule defines the performance of an individual action by describing the subactions that must be performed without defining how those subactions themselves are performed. Instead, the subaction definitions are only to be found in the rules specific to those subactions. For the rule at hand, these subactions can be thought of as calls to procedures that are defined elsewhere, and may themselves contain procedural calls, and so on. We can continue in this fashion, breaking down procedures into more and more basic actions until we get to the set of actions actually modeled, the primitive actions of the simulation. The rules than form a hierarchy where the children of each rule are those procedures directly called by the rule definition. In this manner we build our knowledge base as a modular hierarchy of procedural abstractions.

A concrete example from MACH-III is shown in Fig. 6. The root node of this tree, labeled LO LAMP, defines how to determine why the LO Lamp is lit, indicating a fault. In other words, this node defines how to isolate an LO problem. According to the tree, isolating the LO fault is done by checking for an LO malfunction (the left child of the LO LAMP node) and checking to see if the LO function is being monitored correctly (the right child of the LO LAMP node).

It is important to note that at this root node neither subaction is defined. That is, the rule defined by this node contains calls to the procedures to check for an LO malfunction and to check to see if the LO function is being monitored correctly, but does not contain definitions for these procedures.

Continuing our example, checking for an LO malfunction (the MALFUNCTIONING LO node in the tree) consists of checking LO inputs (the left child) and checking the LO itself (the right child). Since checking individual devices, such as the LO, is a primitive action in the MACH-III simulation that needs no further definition, the CHECK LO node has no children. The CHECK LO INPUTS node, defining how to check LO inputs, has three children, and so checking LO inputs consists of three steps: checking the LO power cable W1, checking the LO plug P12J3, and checking the RF signal input cable W6. Since these actions are again primitive in the simulation, they need no further definition, and their corresponding nodes have no children.

Problems with production rules. Any system that is being designed for tutorial purposes, as was MACH-III, must have adequate explanatory capabilities to instruct student users. Although the production rule paradigm is well known for expert systems, its problems, particularly those that adversely affect explanation generation, are also well known:

- Production rules can hide knowledge in their design. For example, it may be important how the constituent elements of a particular rule are ordered, since production rules invoke their tests and actions in the order in which they are written. However, looking at the rule does not reveal whether the order of its conditional clauses is deliberate or circumstantial. Furthermore, if a particular ordering is deliberate, the reason why this ordering is critical is also missing and unavailable for explanation. Thus, rules can be very hard to understand, and explanations based on them are unsatisfactory.
- The creation of a control structure is needed to decide which production rule to apply when, since the rules themselves cannot determine how to choose among

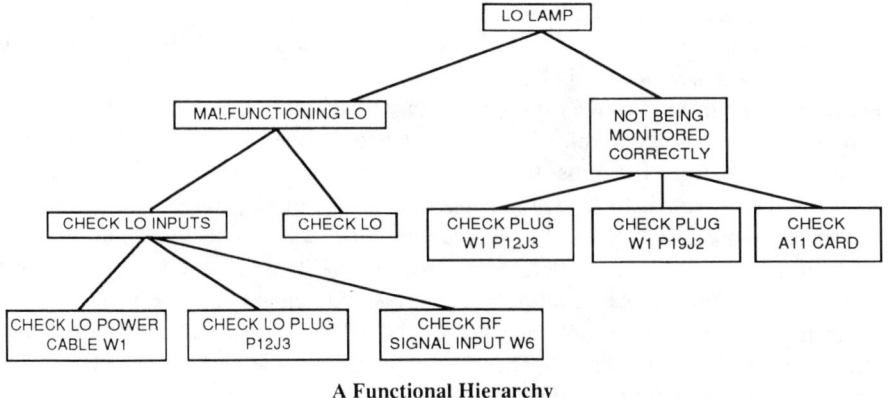

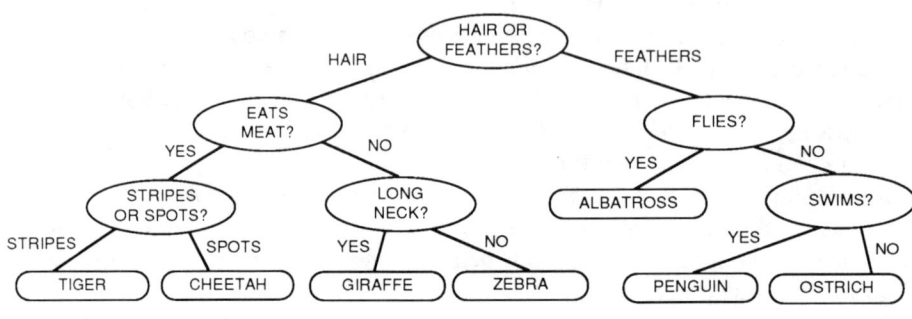

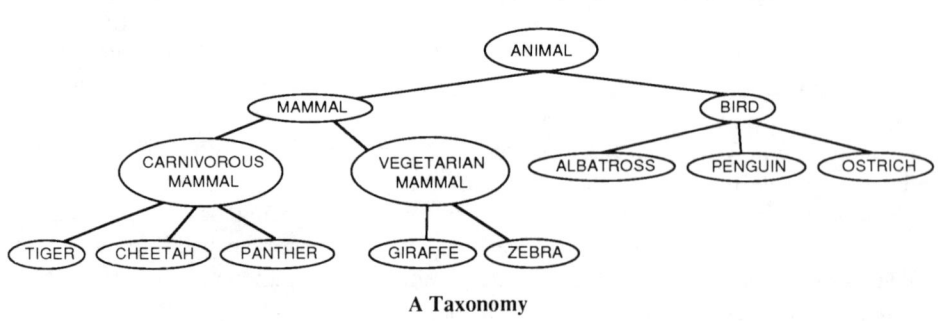

Figure 6. A functional hierarchy, a decision tree, and a taxonomy.

the several possible applicable rules. These control strategies tend to be simple, such as selecting the first applicable rule or selecting the applicable rule with the highest priority number assigned by the programmer. In each case, it is the programmer who assigns the priorities to the rules (by rule ordering, by explicit numbering, or whatever), based on knowledge that is nowhere represented in the system. Thus the only possible system-generated explanation for choosing one rule over another is that a particular rule has a higher priority over another, without any reason why this is true.
- Since each production rule is designed to represent a single level of knowledge

[5], there is nothing explicit in any given rule that describes how it interacts with other rules in the system. Therefore, it is extremely difficult to determine how the changing of one rule will affect other rules and therefore the performance of the entire system. Furthermore, since the control knowledge is not represented in the rules, changing the rules may have unpredictable effects on the control structure operation. Thus, building and modifying a production rule system can be extremely difficult.
- The organization of a production rule system, a collection of conditional rules with possibly ordered clauses and a separate control structure, is difficult for programmers to understand, much less end users. Furthermore, this organization has little relationship to the way people cognitively organize knowledge, since it does not provide a single organizing framework (a mental model) that can be decomposed into its smaller, more detailed constituent units. If the goal of a system is to teach a body of knowledge and make a specific organization of that knowledge explicit to the student, then the production rule paradigm is not suitable for this system.

Although much research effort has been spent on addressing the aforementioned problems [6-8], these efforts, mostly consisting of the addition of various mechanisms to restore lost knowledge, have not been entirely satisfactory. This had led to the development of alternative representations, one of which is functional hierarchy, described here.

An alternative to the production rule: The functional hierarchy. The functional hierarchy paradigm explicitly incorporates relevant design and interdependency knowledge into the rules themselves and the knowledge that would be hidden in the production rule control structure. The hierarchy then arranges this knowledge into the organization the student needs to learn. This allows us to avoid the pitfalls inherent in the production rule paradigm.

The functional hierarchy clearly and concisely shows the knowledge the system explicitly contains. If some facet of the system such as clause ordering is important, rules defining how to perform this subtask can easily be incorporated into the hierarchy, making this knowledge available for explanation with no additional effort. (It should be noted that such a clause ordering was not incorporated into the knowledge base for MACH-III, solely because it was not relevant for this task. There is nothing in principle that would have prevented the inclusion of this knowledge into the troubleshooting tree if it had been needed.)

The rules in a functional hierarchy define all the actions the system must take, not just describe conditions under which changes to the database must be performed, as do production rules. The control structure knowledge is explicitly incorporated into functional hierarchy rules, which are treated no differently than other rules. Thus, no separate control structure is required, and the relevant knowledge for control decisions is explicitly in the knowledge base and available for explanation.

The interdependence of rules are strictly defined by the hierarchy, so the effects of rule changes are relatively easy to predict. This high degree of modularity has all the benefits of procedural abstraction [4], including the facilitating of building, adapting, and maintaining large systems.

The functional hierarchy allows us to organize and present the relevant knowledge in the framework we desire the student to acquire. No additional mechanisms are required to convert this knowledge into a form compatible with the mental model the student should have, since the functional hierarchy already represents this mental model to the desired level of detail. Since the system bases its reasoning on this framework, it is much more capable of demonstrating proper procedures and evaluating student performance with respect to the correct knowledge organization.

The functional hierarchy is superior to production rules in an ITS because it contains all the relevant knowledge explicitly, clearly shows the relationships and interdependencies of the various pieces of knowledge, and organizes this knowledge into the correct framework.

Functional hierarchy in an ITS. Using a functional hierarchy offers several advantages from an instructional perspective. Although a production rule describes the circumstances under which a particular set of actions may be appropriate, functional hierarchy rules are definitional in nature, describing how a particular task is performed. Therefore they are more easily understood out of context. The knowledge is a cognitively compatible organization that is designed explicitly for instruction and can therefore help the students obtain this organization for themselves. In MACH-III, the visually displayed troubleshooting tree structure directly reflects the instructors' organization of the functional knowledge that underlies the Fault Isolation Procedures.

Because implementational details are hidden from each rule, but are available by examination of subsequent rules, students can be provided with information at the appropriate level of detail, with the ability to delve deeper for more explicit information if desired. Thus, the instructional capabilities of an expert system are enhanced by the organization of the knowledge base into a functional hierarchy.

In MACH-III, the advice and critique tutoring functions were also facilitated by the use of a functional hierarchy knowledge base—its troubleshooting tree. The tree is organized first by nodes representing symptoms, which branch to children nodes representing the possible malfunctions or hypotheses. This decomposition continues until we reach functional subtasks supported by the simulation, including testing and replacing groups of devices or the devices themselves.

At any given moment, the system assumes and encourages the student to work on one hypothesis at a time, represented by one node in the troubleshooting tree. A request for advice is interpreted as a question about how to perform the task at hand, which is checking the current hypothesis. Since the children of a node represent how to perform a task, advice then consists of suggesting any one of the children of that node, as long as the child has not already been done.

For critiques, the system provides feedback from the student's actions in four possible ways that depend on the tree structure. First, if the node representing the student's action is part of the hypothesis branch currently being worked on (that is, the node is a descendant of the node representing the current hypothesis), then the action is judged as being reasonable. Second, if the action is not part of the current branch (the action node is not a descendant of the hypothesis node), but the action does belong to a relevant branch (the action node and the hypothesis node have a common ancestor), then the action is judged reasonable, but the student is encouraged to stay within one branch (pursue one hypothesis at a time) and stop "skipping around." Third, if the action does not pertain to this symptom at all (the node is not in the part of the tree representing this problem), the student is told that the action is not relevant to the problem, and the system briefly lists the problem(s) to which this action is relevant. Finally, if the action has already been performed while working on the current problem, the student is simply reminded of this. (This last critique is merely a bookkeeping matter, rather than a trait of the functional hierarchy.) Thus the system exploits the fact that the troubleshooting tree reflects the correct organization of troubleshooting knowledge and encourages the student to learn and use this organization through advice and critique.

In addition to these instructional benefits, there are additional programming advantages to the functional hierarchy paradigm in how it uses procedural abstraction. These advantages are well documented [4] and need only brief mention here. Systems based on procedural abstraction are easier to write, since the programmer need only be concerned with the high-level problem of accomplishing a task, without worrying about the implementational details before they are appropriate. These systems are also much easier to maintain and modify, since the interactions of the rules are explicitly specified in the hierarchy.

Functional hierarchy versus other tree-based knowledge organizations. At this point, it may be useful to contrast the functional hierarchy paradigm with other tree structure knowledge organizations to clarify their differences.

The first such organization is the *decision tree*, where each node represents a question to be asked. Each possible answer is represented by a link to another node, for another state and another question to be asked. Figure 6 shows an example of a simple decision tree that determines the type of an animal. First, we ask if the animal has hair or feathers. If the answer is hair, we move to the node that says EATS MEAT?, and if the answer is feathers, we move to the node labeled FLIES? Assuming we have an animal with hair, we next ask if it eats meat, the question in the node. If yes, we move on to the node that asks STRIPES OR SPOTS? If the animal does not eat meat, we go to LONG NECK? We continue in this way until we get to an answer.

As this example shows, the purpose of a decision tree is to organize the possible tests into an order that eventually reaches a conclusion while eliminating unnecessary tests. In our example, once we know the animal has hair, we do not care whether it flies or swims. However, the decision tree lacks any information as to why this ordering makes sense, why results from certain tests can eliminate the need for other tests, or why a certain set of test results indicates one solution as opposed to another.

In contrast, nodes in a functional hierarchy represent definitions of action to take, such as the definition of how to check the LO inputs that we saw earlier. The organization reflects more naturally both how the tasks are composed of subtasks, and the "bigger picture" as to why low level subtasks, such as testing specific parts, are done in the order they are, or even why they are needed at all. Decision trees reflect only this ordering of low-level tasks, without the reasoning behind it. The Fault Isolation Procedures manual for the HAWK radar is actually an example of a decision tree. Each step consists of a test to perform on the equipment. Based on the results of that test, the troubleshooter is instructed as to which step to perform next. The MACH-III troubleshooting tree reflects the ordering given by the FIP decision tree, and through its use of a functional hierarchy, also teaches the student *how* and *why* this ordering was derived by organizing the relevant background knowledge into a hierarchy of hypotheses and steps on how to pursue these hypotheses.

Taxonomies. Another knowledge organization that makes use of a tree structure is a *taxonomy*, where nodes represent classes of objects, and the hierarchy denotes subclass/superclass relationships by links between nodes. Because any member of a class is also a member of the superclasses of that class, these links are frequently called IS-A or AKO (*A Kind Of*) links. Figure 6 shows an example of a small taxonomy. The link between the TIGER node and the CARNIVOROUS MAMMAL node represents the fact that all tigers are carnivorous mammals (a tiger IS-A carnivorous mammal, or a tiger is A Kind Of carnivorous mammal). Similarly, the link between CARNIVOROUS MAMMAL and

MAMMAL represents the fact that all carnivorous mammals are mammals. The simple syllogistic conclusion that all tiger are mammals is represented by the fact that the TIGER node is a descendant of the MAMMAL node.

The taxonomy structure can also be used to make assertions at the appropriate levels of abstraction. Rather than say that tigers have hair, cheetahs have hair, panthers have hair, giraffes have hair, and zebras have hair, we can state that all mammals have hair. Then by virtue of their being mammals, all these animals must have hair. Since having hair is a distinguishing feature of mammals, this is really where this knowledge should be represented. This feature of taxonomies is known as *inheritance*. A taxonomy also permits statements about what is usually true about members of a class. For instance, we could say that most birds fly, and then make explicit exceptions at the PENGUIN and OSTRICH nodes. This is generally known as *default reasoning* or *reasoning with exceptions*. A discussion of the advantages and implications of inheritance and default reasoning is beyond the scope of this chapter. The interested reader may begin investigating this field with Woods [9], Brachman [10], and Touretzky, Horty, and Thomason [11].

A functional hierarchy is not a taxonomy because the functional hierarchy does not use the node parent/child relationship to represent the generalization/specialization relationship of superclasses and subclasses. Rather, the parent/child relationship represents a decomposition of a task into lower level subtasks that define how to perform the original task. Because this is not a taxonomy, concepts such as inheritance and default reasoning do not apply to this structure.

To make clearer the distinction between a functional hierarchy of procedures and a taxonomy of classes of objects, we refer to the functional hierarchy example in Fig. 6. The root node LO LAMP has two children, MALFUNCTIONING LO and NOT BEING MONITORED CORRECTLY. If this procedural tree were to be interpreted as a taxonomy, the positioning of these nodes would mean that anything that was a "malfunctioning LO" was also an "LO lamp," and that anything that was a "not being monitored correctly" was also an "LO lamp." Similarly, trying to interpret a taxonomy as a functional hierarchy yields equally bizarre results. If the taxonomy in Fig. 6 were a functional hierarchy, it would be saying that "bird" is an action that is comprised of three subactions, "albatross," "penguin," and "ostrich." Clearly, taxonomies and functional hierarchies represent different kinds of knowledge, even though their physical structures are similar.

Integrating MACH-III into the Existing Curriculum

Many classroom innovations, however good they may be, do not get integrated into the existing curriculum. Recent studies in instruction by cognitive scientists suggest that cognitive changes at the individual level are highly influenced by the existing social organization of instruction [12–15]. What follows is a description of the multiple approaches taken with the MACH-III project.

Cognitive and Formative Research to Design MACH-III Features and Interface. From our protocol analysis and troubleshooting observations, and from our collaborations with a knowledgeable radar engineer and experienced troubleshooters/instructors, we constructed various types of MACH-III diagrams. We did this to help bridge the gaps between the physical radar, the detailed and abstract schematics, and the confusing procedural manual, as well as to provide students with the appropriate functional system (loop) and device models. Our formative research, iteratively conducted throughout the project (three years) with instructors and students, indicated that these diagrams and troubleshooting trees were clearly meeting an instructional need.

The user interface was also iteratively evaluated through formative research. This not only included screen design, feedback message placement, scrolling, hypertext features, menu design and actions, and pop-up meters, it also included making all three buttons on the mouse do the same action. As a result, little training is needed to use the system. Its ease of use has been a strong point in getting people to come over and "mouse around" without great fear of "breaking the computer."

Although MACH-III was still under development at the time, we observed how students used MACH-III in a classroom situation. Real-Life Mode's Advise and Critique options were selected. Formative research revealed that: (1) some students used MACH-III like the real radar, following the FIP closely to test parts and ignored the tree; (2) some students ignored the documents, if given the option. This also happens when they are on the real radar. Similarly, they also ignored the tree, and either just tested/replaced everything that highlighted, or if somewhat familiar with the FIP, first checked the relevant parts they could recall; or (3) some students ignored all documents and the functional grouping of the tree. Instead, they just viewed the parts displayed at the end of the tree as a check list. The last item was viewed as a problem by instructors and designers, so the troubleshooting tree display functions were modified such that students would be required to "expand" each box to see the lower levels of the tree (child nodes). Although less user·friendly, this treatment of the tree makes it less likely that students would simply run down the ends of the tree as a parts list. The first two items were viewed as under the instructor's control and discretion, and although viewed as problematic by the designers, no changes were made to the system.

Team Collaboration with Instructors. Collaboration with the instructional staff was key to designing and integrating MACH-III. The prototypes were always available for all the instructors to peruse at this discretion throughout the development cycle. One expert instructor with 30 years experience served as our resident advisor for functional information (he was not that familiar with this most current version of the radar), and eventually became our on-site computer technician. Although he had relatively little prior computer experience, he became the resident computer expert: he loaded tapes, distributed software across the network connections, brought up the computers everyday, reported bugs, and so on. Another expert troubleshooter/instructor familiar with the current radar helped us understand where the students had difficulty by letting us watch him and other mechanics of varying levels of experience troubleshoot. Many of his comments and suggestions helped us design and structure the troubleshooting trees. What also helped create a collaborative atmosphere between the designers and the instructors were frequent visits for formative purposes and many telephone calls for help (in both directions).

Categorization of Problem by Curriculum Area and Difficulty. To support an instructional situation, all possible faults in MACH-III were placed into a problem database, and a menu item called "Select a Problem" was installed. When a problem fault was inserted, the menu item toggled to become "Reveal Broken Component," which students selected if they wanted to "give up" or get a new problem. Note: Instructors could still break a specific device and choose whatever mode (e.g., Magic Mode) they desired for troubleshooting. Category areas were first defined by instructors to match up to the curriculum topics as closely as possible. The problems were then categorized by the instructors. (This included all possible ways to break devices, e.g., device x could be broken, noisy, shorted, etc.) For the receiver simulation, the problem areas neatly matched up to the curriculum's small functional circuits that comprised a major loop (e.g., signal circuits). However, for the transmitter simulation, the circuits were so complex and interrelated that many parts belonged to overlapping multiple functional circuits. Instructors chose five very broad functional areas rather than specific circuits.

Problems within a problem area were then categorized by the instructors into three levels of difficulty: easy, medium, or hard. Difficulty levels seem to depend on a combination of factors: (1) how hard it was to determine what the part was (e.g., if the light indicated the exact problem, such as a circuit card, this would be easy); (2) how hard it was to trace "back" in the circuit in the schematics (if the part is next to the item indicated by the symptom light, this would be easy); (3) how likely the problem would be (frequent problems are easier since the students specifically learn these and, if possible, are given the opportunity to practice replacing these problem parts); and (4) how easy it is to fix (parts that are often faulty have been made easy to replace).

Exercises. After a "dry run" with an instructor and a real classroom schedule, a selectable exercise mode was created that had the same curriculum areas as before, but presented only a small instructor-selected subset or problems from that curriculum category. This was an important feature to add to reduce the instructor's load for answering student questions, since some problem areas contained more than 30 problems. This became more important when the instructors had to divide their attention between three or four workstations.

Workshops. Instructors who were interested in MACH-III signed up for one of four workshops held during a two-day time period. Each workshop covered the MACH-III functions and provided support materials. The instructor's responses to the overall system ranged from very enthusiastic to mildly enthusiastic. The diagrams were given very high praise; they were easy to understand and had been seen by mostly everyone, since the system prototypes had been around for quite some time. The majority of instructors thought the troubleshooting tree was much better than the FIP because it concisely organized the troubleshooting procedures into an easily understandable framework and gave reasons. These instructors wanted the paper version of the tree to take with them right away. A few instructors (generally less experienced) viewed the tree more cautiously for several possible reasons: its purpose was not understood, it was thought to be too helpful, and/or they thought the student should learn to use the FIP and nothing else (the crutch problem). The teaching schedule and turnover are major roadblocks for the integration process. Nearly all of the instructors (a total of 16) attended the workshops, but six of them left for overseas duty within two months of the workshop. Some instructors have all night duty or staggered schedules, starting at 4 a.m. These problems and other demands led to the development of many off-line support materials.

Curriculum Support Materials. An ITS needs to be supported with a broad spectrum of off-line curriculum materials that ties into the existing curriculum and social organization. Curriculum materials can aid curriculum development by the instructors, who have severely limited resources and time. In our case, we developed four device and tree explanation manuals, a user manual, color and black-and-white viewgraphs, and posters of the trees. We also provided suggestions on how to use both the viewgraphs with the lectures or schematics and the trees with the Fault Isolation Procedures. These materials will be incorporated into their curriculum.

Initial Field Test Results

The first field test of MACH-III with an actual class at the Fort Bliss School took place during the first two months of 1989. Although the field test was intended as a "first pass" look at how MACH-III and its support materials would be utilized as this very early stage of integration, a traditional experimental paradigm was set up for initial evaluation purposes. Two classes were involved in the evaluation. A control class preceded the MACH-III treatment class by approximately three weeks. Both classes had the same number of students ($N = 11$) and the same instructor for lecture. This instructor, appointed by the school, had received the previous year's "Instructor of the Year" award. Although not ideal, the treatment and control classes had two different sets of school-appointed practical lab instructors to teach hands-on troubleshooting on the radar. To compare the two classes, various performance measures were taken after the transmitter and receiver portions (a total of 2½ weeks) of the eight-week radar course. The performance measures reported here include the school's normalized written tests and questionnaire responses.

The evaluation took place within the constraints of the school's classroom schedule and traditional instructional format. In both groups, the lecture and practical lab times were equivalent. For the control class, one pair of students was on the radar with the instructor, whereas the rest of the students were "paper troubleshooting" three to five problems. This consisted of giving students a symptom and direction as to where to start in the daily checkout manual. They were then expected to find the correct fault isolation procedure and figure(s) in the schematics and then trace the circuit in the schematics. Students were to keep asking the lab instructor for the results of testing a component until they found the fault. In the treatment class, the treatment (troubleshooting sessions on the MACH-III system and use of ancillary materials) had to be "squeezed in," without adding any extra time. Thus, the normal paper troubleshooting time was spent doing only one to two paper troubleshooting problems, troubleshooting on the MACH-III system, and answering the MACH-III written exercises that required them to read the device and tree manuals and look at the diagrams. The treatment class spent approximately one third of their lab time on the radar, one third troubleshooting on the MACH-III system, and one third answering their MACH-III written exercise materials for paper troubleshooting.

Student and Instructor Questionnaires

At the end of the treatment period, questionnaires were given to the treatment class and to the instructors. Both questionnaires asked the rater to indicate how helpful MACH-III's features and support materials were, their general feelings about MACH-III, usefulness and perceived problems with the system, and whether they would recommend the MACH-III program to others. In addition, the instructor questionnaire also asked them how they would use the various features to teach and how much time they had spent with MACH-III.

Student Questionnaire Results. Although we expected that most of the students would like MACH-III's features, we were concerned that some students would give unfavorable ratings because they might not like computers, might not think simulations could substitute for the real thing, or some other reason. Students' responses were somewhat, although not entirely, surprising. (See Table 1 for Student Questionnaire Results.) Their reactions to MACH-III's individual features or support materials were very favorable, as expected, particularly for the tree, the diagrams, the explanations, and selecting problems by specific area.

Student responses to the questions regarding their general feelings about MACH-III and whether they would recommend MACH-III program to others unexpectedly diverged into two distinct groups, "fairly enthusiastic" or "didn't like it." Those who liked it cited the features, ease of use, and value as a training device. Those who did not like it thought that it "took time away from the radar," which was not true. MACH-III took time away from paper troubleshooting, which is what they usually do when they are not on the radar. Students did not know that they would still only get to solve one fault on the actual radar per lab session, regardless of whether they had the MACH-III trainer or not. The strong urge to keep "hand-on" the "real radar" is not surprising, given that novices tend to be physically oriented, as our research and others' [3] have found. There were also a few complaints about the unexpected bugs and breakdowns in the system.

Instructor Questionnaire Results. Eleven instructors were given the questionnaires to fill out during the latter part of the field trial It was expected that the instructors would be favorable toward MACH-III and its support materials, although the project staff were aware that they had had a horrendous teaching schedule during the fall and winter and had little time to try out the system. However, we wanted to know about which features they thought most positive, how they had used the system, how much time they had spent, whether they had tried any students out on the system, and how they intended to use it in the future.

Rating MACH-III feature and support materials. As expected, the instructors' general feelings about MACH-III were strongly favorable (see Table 2). This matched their ratings of specific MACH-III features and support materials, where virtually every MACH-III item received the strongest favorable rating. It was very encouraging to note that their enthusiasm had not decreased since the workshop in early fall of 1988.

How instructors personally spent their time on MACH-III. Because of the instructors' teaching and hectic work schedules, they had had little time (<10 hours) to explore MACH-III since the workshop. However, most of them made a point of apologizing profusely to the project staff when they turned in their questionnaires. They stressed that MACH-III was an excellent teaching system, and for some junior instructors, a means for learning a lot more.

For those instructors who did use the system, their pattern of use is shown in Table 3. They usually preferred to have the computer randomly select problems within specified curriculum problem areas. They also preferred to read about the diagrams more than about the loops or chains, and the least about specific devices. This differentiated reading emphasis appeared to reflect the instructor's level of experience. Less experienced instructors read about the system and complex devices more than the more experienced instructors. When instructors themselves practiced troubleshooting, they spent more time in Magic Mode than in Real-Life Mode, and did not use System Demonstrates Mode at all, although it was unanimously rated "very helpful." Magic Mode may have been more popular because MACH-III first comes up in Magic Mode by default and/or possibly because the signal flow is more dynamic and therefore more interesting.

Instructors' plans for using MACH-III to teach. Although nearly two thirds of the

Table 1
Student Ratings: MACH-III Features and Documentation

	Don't Know	Not Helpful	Sort of Helpful	Very Helpful
Diagrams				
Computer diagrams	7	60	33	
Color viewgraphs	10		70	20
Problem selection				
Instructor picks problem by specific area		20	20	60
Computer picks problem from any area	44	22		33 ($N = 9$)
Modes				
Magic Mode	10	10	40	40
Real Life			50	50
System Demos		10	50	40
Advisor	10	30	50	10
Critiquer	10	50	40	
Explanations				
Device, written	10		50	40
Device, computer		10	20	60
Tree, written	10	10	40	40
Tree, computer		10	40	50
Trees				
Tree posters	20	10	40	30
Tree on computer		10	50	40
Exercise				
(Questions and answers)		50	40	10

General Feeling about MACH-III

Don't Know	I Don't Like It	Lukewarm	Fairly Enthusiastic	Very Enthusiastic
	40	10	40	10

All results reported in percentages.
$N = 10$.

instructors reported actually having used MACH-III with students, this use (an average of 5 hours) had been informal, such as with study hall or as a visual aid in teaching foreign students. After we attempted a "dry run" using MACH-III (in the late fall of 1988) as a classroom trainer, one instructor independently had his class use MACH-III to practice troubleshooting during the transmitter portion of the course. Although this was still a more causal attempt than we would have preferred, the school administrator and a few key instructors seemed convinced that MACH-III would fit in as originally

Table 2
Instructor Ratings: MACH-III Features and Documentation

	Don't Know	Not Helpful	Sort of Helpful	Very Helpful
Diagrams				
Computer diagrams		27	75	
Color viewgraphs			44	56
Problem selection				
Select a problem by specific area	10		30	60
Computer picks problem from any area	20		30	50
Modes				
Magic Mode			30	70
Real Life	10		30	60
System Demos				100
Explanations				
Device, computer			33	67
Device, written			33	67
Tree, computer	10		30	60
Tree, written	10		50	40
Trees				
Tree posters			33	67
Exercises				
(Questions and answers)	14		29	57

General Feeling about MACH-III				
Don't Know	I Don't Like It	Lukewarm	Fairly Enthusiastic	Very Enthusiastic
10			30	60

All results reported in percentages.
$N = 10$.

envisioned. Thus, the field test was the first time MACH-III had been fully utilized to teach both the transmitter and receiver portions of the curriculum.

Table 4 shows how much instructors have used or planned to use MACH-III's features to teach. When we compared these results to their ratings of the individual features, the results diverged in ways that possibly reflected some pragmatic aspects, their reservations about how to use this "new technology," as well as some philosophical differences about instruction. For the computer diagrams and signal flow, their comments corresponded to their highly favorable rating on the MACH-III's features, which was not surprising since the diagrams were easily comprehensible. However, despite a highly favorable rating, just over half of the instructors were planning to use the viewgraphs of the diagrams. This disparity may be pragmatic, since there is only one work-

Table 3
Instructor's Personal Use of MACH-III

How much have you used MACH-III for yourself alone?					
Hours	<1	1–5	6–10	11–15	16+
% instructors	9	45	36	0	9

Which part of MACH-III have you used and roughly how much?	
Subsystem	Instructor Time
Transmitter	4.3
Receiver	4.8
TOTAL hours	8.1

Note: Several instructors just marked these categories with an X, only those entries with numbers were used to calculate the averages. The average is probably lower.

Problem Selection (*N* = 10)

% Time	
30	I just break devices.
43	I select a category and let computer randomly pick problem *within* the category.
27	I let the computer select a problem from *any* category.

How do you use MACH-III? (*N* = 8)

Use	Not at All	A Little	Fairly Often	Almost Always
Read device explanations (total averages)		41	30	29
Specific devices		56	33	11
Loops or chains		36	36	27
Diagrams		30	20	50
Read tree explanations	14	43	29	14
(86% preferred to use both written and computer explanations)				
Practice troubleshooting (*N* = 4)		50		50
Magic Mode		43	29	29
Real-Life Mode		43	43	14
Ask system to demonstrate	38	38	13	13

Table 4
MACH-III Use with Students

How much have you used MACH-III with how many groups of students?

Students groups	0	1	2	3	4	5	6
Percent	27	27		27	9		9*

Which subsystems of MACH-III have you used and roughly how much?

Subsystem	Time Spent with Students on MACH-III
Transmitter	3.3
Receiver	1.9
TOTAL hours	5.2

Note: Several instructors just marked these categories with an X, only those entries with numbers were used to calculate the averages. The average is probably lower.

How do/will you use MACH-III instructionally? ($N = 9$)

Use	Not at All	A Little	Fairly Often	Almost Always
Show computer diagrams and flow		20	30	50
Use color viewgraphs	20	20	30	30
Show troubleshooting tree	33	33	33	
Practice troubleshooting				
with FIP only		56	33	11
with FIP and Advisor	22	11	11	56
with Advisor only	44	44		11
without anything	50	33	17	
			($N = 6$)	

*This instructor interpreted this to mean the number of sessions for one student group, while others indicated how many different student groups.

ing overhead projector and practical "hands-on" instructors teach in a cavernous, noisy radar "garage" without walls or outlets.

Differences in instructional philosophy were reflected in the instructors' more reserved ratings as to how they would use the troubleshooting tree and the Advisor/Critiquer to teach, in contrast to the enthusiastic ratings as individual features by the majority of instructors. For the tree, over half of the instructors would use it to teach. Some instructors were concerned that students would not use the documentation or thought that the tree made troubleshooting "too easy." Similarly, for the Advisor, the FIP, or a combination of both, there was a preferred ranking. When mapped out for individual instructors about two thirds of the instructors clearly preferred the Advisor-FIP combination over either one alone. Nearly half would use the FIP alone often. The majority strongly believed that the Advisor function should not be used alone. These reserved ratings contrast with what actually happened during the field trial, as described here.

Field Trial Observations

Members of the project observed the MACH-III interactions from the beginning of the transmitter classes through the beginning of the receiver classes. One pair of students at one of the three workstations was videotaped and extensive field notes were taken while they were troubleshooting.

During lecture, the instructor showed the viewgraphs of the screen diagrams, before and after going into detail with the schematics. Thus, he chose to use them simply as orientation materials for the official documentation, rather than as the primary means for telling "the functional story."

In the computer room, students started out using the cumbersome official documentation (checkout manual, the FIP, and/or the schematics), while troubleshooting on MACH-III. This use rapidly decreased over time, which is also typical of students on the radar. When troubleshooting, students heavily used the tree's menu option, "Find the Component," available for each component. This resulted in highlighting the component on one of the colored diagrams, and thus practiced their "parts location" skills, one of the curriculum goals.

The instructor had difficulty simultaneously working with three pairs of students troubleshooting three randomly selected problems from a very large problem set. (Some categories have as many as 58 problems.) Instead, he would work intensively with one pair while the other two pairs were troubleshooting alone. He would then check the other two pairs or just monitor all three pairs. It is an overwhelming task for an experienced instructor, much less a junior instructor, to recall every step for every possible problem, particularly since some problems have very long, complex procedures. Thus, when we reminded the instructor of how to preselect which small subset of problems would be presented for each area, he was delighted and immediately selected a handful of problems for that day's sessions. It was clearly important to the instructor to control which problems should be presented initially.

We were pleasantly surprised to find that students' use of the troubleshooting tree kept increasing until they were continually using it. This increase may be attributed to students' highly favorable reactions to the tree's concise organization and availability. When first introduced to the tree, one student remarked, "It's like the FIP, only better and easier." Also, when one student started to take out the FIP, his partner stopped him with the command, "Use the tree. It's easy."

Aspects of MACH-III that involved reading text explanations were not used much, for several reasons. First of all, students seem a bit "macho" about asking for the Advice option or reading the text appearing in the explanation window. Second, the instructor initially approached MACH-III only as a radar simulation. When pressed for an explanation for not using these features, he (and another key instructor) voiced the opinion that students should "work up to" the Critiquer and become "familiar with the circuit first" by what was essentially "discovery learning."

The two instructors were then specifically asked by the project staff to use the Critiquer option and explanations during the last lab sessions. When the instructor explicitly requested students to read the material, this seemed to make it "okay" for students to do so. However, because the instructor did not continue to remind them and did not capitalize on the tutorial or explanations, most of the students eventually resumed ignoring the explanatory text generated by the Tutor option and just looked for the test results or what to test next—until they were really "stuck." Those few who did continue to pay attention to the information given by the Advisor and Critiquer options, usually the slow students, appeared to solve the problems much faster and easier than before.

It was not entirely surprising that the instructor did not reinforce the tutor-provided information about how the part caused the fault. We have observed that this is not done very often or consistently by the instructors, since the goal of the school is only to familiarize the trainees with the documents and to find the chassis containing the fault. Since this type of mechanic is not allowed by doctrine, and thus is not trained to locate a fault down to the component level—only the battery replaceable unit (chassis)—the emphasis has not been on what specifically caused a fault. However, since it is hard to understand how the system works, much less breaks, without this information, the MACH-III diagrams, tree and tree explanations have been tailored to show and "tell the story" down to the functional component within the battery replaceable unit. Furthermore, the Critiquer continually gives feedback about the "reasonableness" of a student's actions and describes the relationship of that part to other parts and their functional relationship via the tree structure. In short, MACH-III's Critiquer tells the "why" and largely contains the tutorial components of the system in its feedback. By turning off the Critiquer and not encouraging use of the Advisor, the instructors turned off the AI tutor and made the system behave according to their philosophy at the moment in time.

It appears that what really saved the day was the interactive troubleshooting tree. It took the place of the FIP and the Advisor, although this was not anticipated but in retrospect, was logical, natural, and desirable. Thus, in spite of the instructor's reserved predictions of how they would use the tree, this observed use actually met the instructor's desired combination of FIP and Advisor, since the tree was designed to combine the information contained by cumbersome FIP and is the visual instantiation of the Advisor, which derives its information from the functional hierarchy. However, the instructors have since decided to use both the Advisor and the Critiquing function—but only the latter at the end of each troubleshooting problem (an option we have since inserted), rather than after every step. Thus, the instructors appear to have overcome the notion that the Critique function is "too helpful" and cautiously accept that it may be a valid and useful instructional tool.

Student Performance on Exams

Preliminary evidence that MACH-III is effective appears in student performance on the transmitter written exam. MACH-III was used throughout the transmitter portion of the course, but for the receiver portion, MACH-III was used for less than an hour, and thus was not representative of a "treatment." Since there was no apparent difference between the two groups' mean performance on either the receiver written exams (control = 89.1; treatment = 88.6) or the practical (hands-on) exams for the transmitter (control = 99.6; treatment = 98.1) or the receiver (control = 98; treatment = 96.7), no further statistics were calculated. The nearly perfect practical exams essentially had a ceiling effect. The instructors scored students on their ability to: locate parts and then provide their functional definitions—using handouts, correctly perform the daily checks and find the correct symptoms, find and correctly perform the right FIP procedure(s), trace schematics, find the fault (yes or no), and correctly observe safety precautions. It was interesting to note that nearly all the MACH-III students chose to use the MACH-III manuals to provide functional definitions instead of the school's handouts.

When comparing the transmitter written scores for the ITS group (mean = 85.9) to those of the control group (mean = 78.6), the difference (7.3) is significant. The ITS group scores are one standard deviation higher ($t = 1.88$; $df = 20$; $p = .04$) than the control group scores. Although these results are preliminary, they are intriguing. The range for the control group is broader (55 to 95, standard deviation = 11.0) in comparison to the treatment group range (75 to 95, standard deviation = 6.6). Thus, not only does the treatment group score higher, but also as a group performs more consistently.

Conclusion

From our perspective, MACH-III is a definite technological success. The knowledge representation and system design has succeeded in covering the complex transmitter and receiver subsystems of the radar. Whether MACH-III will prove to be an instructionally effective and a cost-effective solution will take more time to find out. The Army is currently evaluating the MACH-III system at the school.

The results of this field trial occurred without any modification to the current curriculum, and no adjustments were made to reflect the fact that students were doing two more things (MACH-III troubleshooting and the written exercises) in the same amount of time. However, there were some very encouraging indicators.

The school has requested the development of the radar's other subsystems, so that the majority of the subsystems are contained by MACH-III and can be used for teaching throughout most of the course. The school has been mandated to revamp its program so that the material is taught in a "functional context." The instructors and the head of the school believe that MACH-III and its support materials, particularly the written device and tree explanations, fit neatly into this program. Plans are underway to utilize the development team's suggestions on how to employ MACH-III effectively.

The school has continued to use MACH-III and its support materials in the classes following the field test. Students clearly get more troubleshooting practice, however tutorless it appears to be. With the real radar, each pair gets one problem per lab on the average. The rest of the time, they are paper troubleshooting. With more subsystems, they will clearly get more troubleshooting experience. It remains to be seen how the instructors and students choose to use the system's features—just as a simulation or with the simulation along with the tutor, tools, and instructional environment.

To integrate any technical innovation into an existing curriculum requires multiple steps and stages that involve not only the designers and instructors, but the social organization at different levels and times [16]. Most of the lessons recently relearned from integrating software into the public schools also apply to integrating AI software into training programs. As found in public schools, the teacher will be the key factor [17] in determining how successful MACH-III will be, as with so many other previous technological innovations.

Finally, our work has revealed several key points that we feel may help future ITS builders successfully integrate their systems into classrooms. Designers should:

- Think of an ITS not only as a tutor, but as a set of teaching tools, and as an instructional environment around which students and instructors work;
- Support an ITS with a broad spectrum of off-line curriculum support materials

that tie into the existing curriculum and social organization. In our case, these are the four device and tree explanation manuals, user manual, viewgraphs, posters of trees, and suggestions on how to use both the viewgraphs with the lectures or schematics, and the trees with the Fault Isolation Procedures;

- "Sow seeds" early and continually throughout the development cycle with key personnel (head of school department, inside expert instructor) about how the ITS should be used. Therefore, as it develops, these key personnel have a chance to evaluate the appropriateness of the ITS functions and interface and also prepare their curriculum to incorporate its capabilities. There are often many "heads of state" who will never see your system, but control its fate, nonetheless;
- Develop a collaborative relationship with the instructors to build a teaching tool and environment designed around their needs. Furthermore, designers should be the first to try teaching with the system to model how the system and support materials should be used. By teaching with it, designers also can see what features or ideas on how it should be used are lacking or unrealistic. The instructors should then be successively coached through several passes in how to best utilize the system's capabilities, according to their preferences and restrictions (i.e., a teaching apprenticeship). In this way, designers learn what is feasible by being in the instructor's shoes and instructors assimilate what the designers have been describing about the ITS. This collaborative and iterative process results in a more accurate and integrated teaching tool/tutor/environment, as well as off-line support materials (which, oftentimes, only the designers are in the position to produce);
- Allocate sufficient and succinct development and integration time. The process described previously requires a minimum of a year beyond software development time for the instructors to begin to understand the capabilities of the software and to see how they can use it to teach. This means they need time to integrate their teaching styles with their understanding of the ITS system to structure an appropriate class schedule, generate appropriate curriculum materials (edit those made by the designers or make their own), see which instructors use the medium well and feel most comfortable with it, and find out where the ITS is best used.

What designers and educators need to consider for the future is how instructors can best utilize an ITS's interactive tutoring capabilities as tools and as an instructional environment. This is a political issue as much as a curriculum design issue. It is not realistic to think that students can be "parked" in front of an ITS and left to read or listen and just learn it all. Someone needs to monitor that learning situation to make sure that a good match exists between the students and the technology.

Acknowledgments

The MACH-III was developed under contract number MDA903-85-C-0425 from the Training Research Laboratory of the U.S. Army Research Institute for the Behavioral and Social Sciences (ARI), Alexandria, Virginia. We would like to thank Dr. Joe Psotka for his gracious support. Many people worked on this massive project and their endeavors have helped shape and create MACH-III and its ancillary materials. At BBN, they were L. Dan Massey, Stephen McDonald, Yvette Tenney, Denis Newman, Jos deBruin, Bruce Roberts, and Pam Schmitt. Significant assistance in the development of the HAWK MACH-III system was provided by the instructional staff of the HAWK Department, U.S. Army Air Defense Artillery School, Fort Bliss, El Paso, Texas. At the school, they were Joe Manna, Chuck Phillips, the late Carl Fox, and Jim Moore. The views and opinions expressed in this chapter are those of the authors and do not necessarily reflect those of the sponsoring agency.

References

1. Y. J. Tenney and L. C. Kurland, "The development of troubleshooting expertise in radar mechanics," in J. Psotka, L. D. Massey, and S. A. Mutter (eds.), *Intelligent tutoring systems: Lessons learned,* Hillsdale, NJ: Lawrence Erlbaum, 1988, 59–83.
2. L. C. Kurland and Y. J. Tenney, "Issues in developing an intelligent tutor for a real-world domain: Training in radar mechanics," in J. Psotka, L. D. Massey, and S. A. Mutter (eds.), *Intelligent tutoring systems: Lessons learned,* Hillsdale, NJ: Lawrence Earlbaum, 1988, 119–80.
3. A. M. Lesgold, S. P. Lajoie, D. Logan, and G. Eggan, "Applying cognitive task analysis and research methods to assessment," in N. Frederiksen, R. Glaser, A. M. Lesgold, and M. Shafto (eds.), *Diagnostic monitoring of skill and knowledge acquisition,* Hillsdale, NJ: Lawrence Earlbaum, in press.
4. B. Liskov and J. Guttag, *Abstraction and specification in program development,* Cambridge, MA: The MIT Press, 1986.
5. R. Davis and D. B. Lenat, *Knowledge-based systems in artificial intelligence,* New York: McGraw-Hill, 1982.
6. W. J. Clancey, "The epistemology of a rule-based expert system—A framework for explanation," *Artificial Intelligence,* 20:215–51, 1983.
7. M. Stefik, "Planning and meta-planning (MOLGEN: Part 2)," *Artificial Intelligence,* 16, 1981.
8. W. R. Swartout, *Producing explanations and justifications of expert consulting programs,* Cambridge, MA: MIT Laboratory for Computer Science, 1981.
9. W. A. Woods, "What's in a link: Foundations for semantic networks," in D. G. Bobrow and A. Collins (eds.), *Representation and understanding: Studies in cognitive science,* New York: Academic Press, 1975.
10. R. J. Brachman, "I lied about the trees' or, defaults and definitions in knowledge representation," *The AI Magazine,* 6:3, 1985.
11. D. S. Touretzky, J. F. Horty, and R. H. Thomason, "A clash of intuitions: The current state of nonmonotonic multiple inheritance systems," *Proceedings of the Tenth International Joint Conference on Artificial Intelligence,* 1987.
12. E. Hutchins, "The technology of team navigation," *UCSD Cognitive Science ICS Report 8804,* La Jolla, CA: University of California, 1988.
13. J. Lave, *Cognition in practice: Mind, mathematics and culture in everyday life,* Cambridge: Cambridge University Press, 1988.
14. D. Newman, P. Griffin, and M. Cole, *The construction zone: Working for cognitive change in school,* Cambridge: Cambridge University Press, 1989.
15. L. B. Resnick, "Learning in school and out," *Educational Researcher,* 16:9, 13–20, 1987.
16. C. Roberts-Gray and T. Gray, "Implementing innovations: A model to bridge the gap between diffusion and utilization," *Knowledge: Creation, diffusion, utilization,* 2:213–32, 1983.
17. B. C. Bruce, A. D. Rubin, and C. Barnhardt, *Electronic quills,* Hillsdale, NJ: Lawrence Erlbaum, in press.

Chapter 10

Intelligent Conduct of Fire Trainer: Intelligent Technology Applied to Simulator-Based Training

DENIS NEWMAN
MARIO GRIGNETTI
MARK GROSS
L. DAN MASSEY

BBN Systems and Technologies Corporation
Cambridge, MA

> **Abstract** *An intelligent tutoring system was developed to demonstrate how intelligent feedback can enhance conventional simulation-based training. The Intelligent Conduct of Fire Trainer (INCOFT) provides simulated exercises for soldiers training to operate the PATRIOT missile system, a modern automated air defense system. INCOFT presents scenarios in which students learn the PATRIOT aircraft identification procedures. After each scenario, INCOFT provides feedback on each identification action in a "fast forward" replay as well as in a summary data table. Students can also see expert demonstrations for any aircraft on which they had trouble. The feedback and demonstration capabilities of INCOFT made it very different from the conventional simulator. As a result of these differences, there were also differences in the kinds of scenarios provided. INCOFT scenarios were shorter and targeted particular concepts. While such scenarios could be presented on conventional systems, without the feedback and demonstration capabilities, students are unlikely to notice the concepts being illustrated by the scenario. INCOFT, thus, plays a different role from the conventional simulator in bridging between the theoretical lectures and the practical exercises.*

Intelligent tutoring systems (ITSs) are just beginning to enter real-world contexts as part of programs of instruction for complex tasks. We describe an ITS for training operators of a highly automated missile system as a case of study of how technology providing intelligent feedback can powerfully enhance conventional simulation-based training.

PATRIOT, a surface-to-air missile system, is deployed by the U.S. Army for air defense. Because of PATRIOT's automated decision logic, the operator's work requires understanding how *the system* decides which aircraft detected by its radar are friendly and which are hostile. The operator must be prepared to override the system in the case of local exceptions to the decision procedure, intelligence from other sources, or overriding commands from higher echelons. Our ITS, called INCOFT (Intelligent Conduct of Fire Trainer), is a simulation-based trainer for PATRIOT operators; it contains a console display simulation, a set of scenarios, and PATRIOT's decision logic provides a basis for feedback on the students' actions and for an articulate presentation of the decision process.

INCOFT grew out of earlier research and development projects at BBN on intelligent instructional systems. An earlier system, TRIO, also dealt with a *real-time* operational task, aircraft navigation [1]. Like INCOFT, TRIO was built on an AI workstation, made use of several modes of presentation—expert demonstration, replay, and critique—and of speech synthesis for feedback as well as for simulating communication during the running of a scenario. INCOFT takes on a more multifaceted task and has developed unique methods for presenting feedback. INCOFT also was intended for experimental use in a program of instruction in which a conventional simulator plays a major role. This study examines the specific ways that intelligent systems provide an advance over conventional programs of instruction.

We begin with a description of the PATRIOT system to provide background concerning the operator's task. Then the current program of instruction is described. Our description of INCOFT begins with findings concerning performance of graduates of the current program which motivated INCOFT's features. We then describe the intelligent feedback features of INCOFT and their potential impact on the program of instruction and role of the instructor. We conclude with observations on the practical applications of ITS technology that appear effective in this context.

The PATRIOT System

PATRIOT is an automated missile system in which radar information is processed and presented to the operators in highly abstract form. The system itself assigns identities to aircraft as friendly or hostile based on flight patterns and electronic signals. The great range of the PATRIOT missiles means that the system makes these decisions about aircraft which are well out of visual range.

A PATRIOT battalion consists of a command center and several batteries, each with a supply of missiles and its own engagement control station operated by two soldiers, generally an officer (Tactical Control Officer) and an enlisted soldier (Tactical Control Assistant). Each operator has his or her own console, but there is a division of labor such that the assistant (TCA) is in charge of firing the missiles and system maintenance, while the officer (TCO) is in charge of "friendly protection," that is, monitoring the development of the battle and assuring that friendly aircraft are not misidentified as hostile. The TCO must understand what is happening during the few minutes that a track takes to traverse the radar's area of coverage and be prepared to override the system in cases of local exceptions to the decision procedure, intelligence from other sources, or overriding commands from higher echelons.

PATRIOT uses a point system for classifying aircraft picked up on its radar as friendly or hostile. The airspace surrounding the battery and any assets it is defending is divided into volumes. Figure 1 shows a situation typical of scenarios used in our training. Aircraft that penetrate a defensive volume lose a certain number of points for each kind of volume they enter. Friendly aircraft presumably know the exact location of these volumes as well as of safe passage corridors which cut through them. Flying so that they are aligned with the corridor, friendly aircraft not only avoid losing points for defensive volume penetration, but actually accrue positive points as a friendly indicator. Depending on the specific tactical situation, there are also speed and altitude limits which cause points to be added or subtracted. Finally, there are codes (e.g., NATO Identification Friend Foe (IFF)) which friendly aircraft can transmit that supply additional evidence of friendly status. The sum total of the points any aircraft has accumulated allows the PATRIOT computer to assign an identity (Friend, Assumed Friend, Unknown, or Hostile) in accordance with certain predefined thresholds. Figure 2 shows the hypothetical weight set and thresholds that we used in our research.

Intelligent Feedback   241

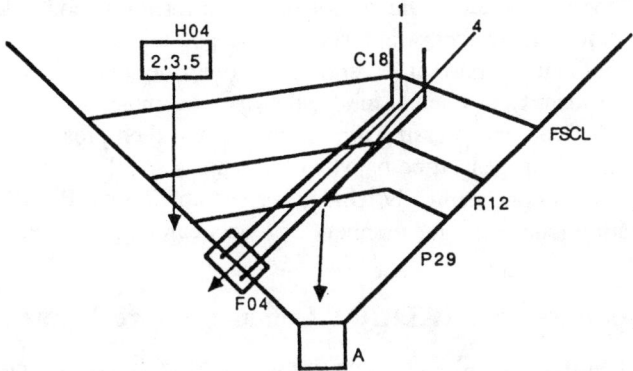

Figure 1 Schematic of PATRIOT situation display. The V-shaped area is the search sector for the battery. The line labeled FSCL is the "front line." Restricted (R12) and Prohibited (P29) volumes are shown. FO and HO indicate locations where friendly or hostile aircraft are likely to originate. C18 indicates a safe passage corridor. "A" is an asset, in this case the PATRIOT battery. The tracks for five aircraft are shown.

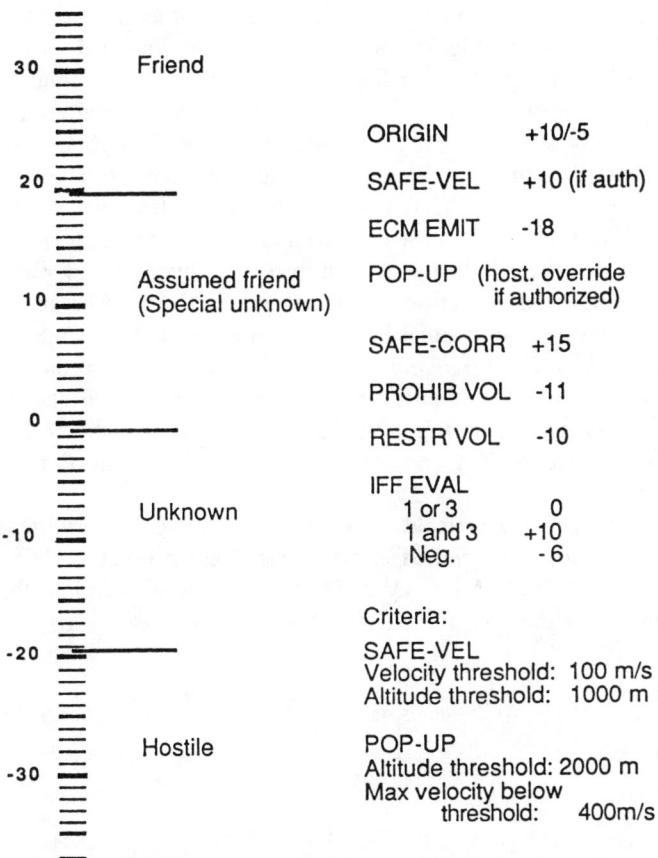

Figure 2 Hypothetical weight set and thresholds used in INCOFT research scenarios.

PATRIOT can operate in automatic or manual identification mode. In automatic mode, the tracks, indicated by icons representing the possible identity states, change their identification without operator intervention. To override an identity in automatic mode or to assign an identity in manual mode, the operator must "hook" the track by moving a cursor to the icon and pressing a hook button and then press a button for the specific identity intended. In peacetime or when there are a small number of actions to be taken, identification is done manually. Under combat conditions, PATRIOT is likely to be put in automatic mode so larger numbers of tracks can be processed.

Current Instruction with the PATRIOT Conduct of Fire Trainer (PCOFT)

Soldiers are trained in the operation of PATRIOT at the U.S. Army Air Defense Artillery School at Ft. Bliss, Texas, where there are two programs of instruction, one for the officers, and another for the enlisted soldiers. These programs cover a wide range of topics, including emplacement, initialization, maintenance, and higher echelon command functions. While the programs cover different job responsibility areas, they are quite similar with respect to the friendly protection function, because both operators are expected to be able to perform that function. We concentrated our own efforts on this portion of the program of instruction, because it was difficult as well as important.

The development of our intelligent training system was intended as a test of the incremental value of knowledge-based technology over conventional simulator technology. The existing program of instruction makes use of a conventional simulator called PCOFT, which is a relatively high fidelity replica of the PATRIOT operator console—the buttons and display are the same; the simulator software provides the same complex data bases and tabular displays. Thus, PCOFT serves to give students a detailed hands-on experience of PATRIOT through scenarios which illustrate air battle situations.

In order to teach a student how the system performs automatic identification, the basic information is presented in lectures, and then practical exercises are provided in which PCOFT runs a scenario in manual identification mode and the student makes the same identifications that PATRIOT would have made. The assumption is that if an operator can mimic the PATRIOT identification algorithm, then he or she must understand the algorithm. During the portion of practical exercise devoted to this aspect of operation, students sit at the PCOFT consoles and do manual identification on scenarios that last up to forty-five minutes.

For each track, the PATRIOT console display provides a track history in tabular form indicating what terms should be included in the sum giving the point total for the track. Figure 3 illustrates the "Track Amp. Data Tab" (Track Amplification Tabular Data Display). For example, if the aircraft is flying inside a safe passage corridor heading

```
             TGTNO      THRT  TLR   TLL   ENGST/M   ID/SZ/IDS   CONFLICT ID:
      -N     004              +00   +99             AF/  /OPR   RECOMMEND-ID:
                                                                      ORIGIN:
             GEOREF           ALT   SPEED  HDNG   RANGE ELEV    SAFE-VEL:
   00:01:44                  1.0 KM 280 M/S 222    056 KM       ECM EMIT:
             IFF CONDITION:                                     POP-UP:
                      MODE:    4     1      2        3          SAFE-CORR:   L
                      CODE:                                     PROHIB VOL:
                  RESPONSE:  NO RESP NO RESP NOT INIT  NO RESP  RESTR VOL:
                   QUALITY:                                     IFF EVAL:    N
```

Figure 3 The "Track Amp Data Tab" for track number 004 showing local (L) determination of safe corridor correlation and a negative (N) IFF evaluation.

with a certain number of degrees of the center line, an indicator will appear next to SAFE-CORR. In automatic identification mode, this will signal that a certain number of points have been added to the aircraft's score. In manual mode, the operator has to add the points mentally. Knowing the weight set, that is, how many points to add or subtract for the various terms, and comparing the resulting score with the given thresholds, an operator can calculate an identity. Some of the terms are conceptually complex. For example, "minimum safe velocity" (SAFE-VEL) may or may not be authorized, but if it is, an aircraft can *gain* points if it is flying below both altitude and speed limits (a flight characteristic not associated with an attack). It is not obvious to students that the indicator for SAFE-VEL means that points should be added (for meeting the criteria) or subtracted (for violating the criteria).

While the job of operator can be quite complex in other respects, the systems identification function is defined fairly narrowly in terms of this mental arithmetic. Cases where the system-assigned ID may be incorrect are not considered in school-based training. For example, friendly aircraft returning across the front line from an attack on enemy positions may be damaged, their transponder may be damaged, they may be flying off course, and so on. Such cases that may require judgment are specifically not considered on the grounds that these trainees must learn to follow a set of procedures which can be set down unambiguously. The commanders at higher levels have the responsibility for setting the criteria for firing the missiles and thus assume the risk themselves. This situation means that the identification task can be stated clearly. The task is not to acquire the expertise of a seasoned soldier. The task is to learn the algorithm used by the computer and be able to reproduce it.

The program of instruction is divided between lecture and practical exercise. In the lecture, students are presented with the list of duties to be performed by the operator, and the basic function of identification is explained (the officer course has two additional lectures on theory and tactics). The practical exercises on the PCOFT are intended to put the theoretical knowledge into practice. A group of eight students works simultaneously at the eight PCOFT consoles. They are supervised by two or three instructors who begin by walking them through procedures for bringing up the console in the proper mode, "hooking" a track, assigning an identification, consulting the many available tabular displays, and so on. Later, scenarios are used to practice the skills. Instructors continue to help students who are having trouble, answer questions, and present specific tasks such as checking something in the data base, setting a new authorization, etc. The PCOFT provides measures of overall performance at the end of each twenty-five- or forty-five-minute scenario concerning asset defense and threat attrition. At the end of the block of instruction on tactical operations, there are written and practical exams.

Design of an Intelligent Training System

From the beginning, INCOFT was intended for use within the ongoing program of instruction. Our intention was to replace part of the conventional simulator-based instruction. We developed approximately three hours of courseware consisting of eight scenarios intended for use as an alternative to PCOFT practice concerned with the duties of the Tactical Control Officer, i.e., identification and friendly protection [2]. Our choice of domain and our design of the system and its courseware was based on formative research with students and instructors. Instructors served as subject matter experts for both the officer's job and the instructional process at the school. Students were interviewed using early versions of INCOFT in order to get their feedback concerning what was difficult to learn and what kind of courseware was easy to learn from.

Where Students Had Trouble

Initial research examined student understanding of the identification process after working with a conventional simulator. Students were given short (three- to five-minute) scenarios on the INCOFT. They were interviewed beforehand on their understanding of identification principles and afterward, following up on the feedback provided by IN-COFT. Student understanding generally was poor. A majority of students did not recall the criteria the PATRIOT computer used in assigning positive or negative points for speed-based factors or corridor alignment. For example, getting points for a safe passage corridor requires that the aircraft be in the corridor *and* aligned with its centerline. Many students mistakenly believed that an aircraft simply crossing through a corridor would get points.

The identification criteria were covered in lectures and sometimes reviewed by instructors during practical exercises but were not forced into play by the exercises themselves. Calculating an identification does not require understanding the criteria PATRIOT uses—only checking to see whether the criteria are met or not. The student refers to the Track Amp. Data Tab for each track, checks the criteria, remembers the point values or looks them up on a "cribsheet," and does the mental arithmetic.

In addition to not understanding the criteria, many students did not even carry out the mental arithmetic. Tracks which had a positive indicator were declared friendly before sufficient points were accumulated, and tracks with a single negative indicator were declared hostile while the PATRIOT computer would have still considered them unknown according to the given point values and thresholds. Further, many students did not notice important identity changes in the middle of an aircraft's flight path. Their tendency was to find a single identity for an aircraft and leave it at that.

The student's tendencies were understandable given the kind of practice provided by the PCOFT. Most tracks in PCOFT scenarios do not contain both positive and negative indicators. They can be assigned an identification of either Friend or Hostile; seldom do tracks remain Assumed Friend or Unknown. Of course, some PCOFT scenarios do contain ambiguous tracks with positive and negative points. However, students rarely noticed such cases, and their errors were undetected. Rather, the student's task becomes to assign an identity of either Friend or Hostile to each track as it appears and assign it quickly. This task is in contrast to PATRIOT's identification process that assigns an identification only when sufficient evidence has accumulated and reevaluates the decision in the face of new evidence. In summary, the student's task is shaped by PCOFT scenarios, and the original intent of teaching the process by which PATRIOT assigns identification is obscured.

INCOFT's Intelligent Feedback

Like the PCOFT, INCOFT simulates the operator's console and presents scenarios on which the student can practice tactical operations. INCOFT consists of a stand-alone console containing a Symbolics AI workstation and a DECTALK speech synthesizer which serves as the voice of a "tutor." The Symbolics monitor is mounted in the console so that it can represent the circular PATRIOT CRT and use the edges of the screen (outside the circular representation of the CRT) for feedback and additional information.

INCOFT's simulation of PATRIOT was not as complete as the PCOFT's, because it served a narrower instructional purpose. The extent of the simulation of PATRIOT was driven by the needs of the scenarios which comprised the courseware set.

The most striking difference between INCOFT and PCOFT is the feedback provided by INCOFT by virtue of an "expert" embedded in the system. The expert in this case is largely a simulation of the PATRIOT program which assigned identities. The student's manual identification is compared to how and when PATRIOT would have done it. The articulate expert also provides demonstrations of how automatic identification mode would have handled the same situation.

Commentary on Student Actions. As in classical ITSs, INCOFT compares the student performance to an expert performance, in this case a simulation of the PATRIOT computer. The student is presented with a scenario of between two and thirteen minutes in which he or she must manually identify a number of tracks. A typical scenario is diagrammed in Figure 1. Some of the scenarios require that the student "bring up the console," i.e., set the switch/indicators to the correct position for the friendly protect function and to display the information appropriate to that function. Since the switch/indicators have to be set correctly for the remainder of the scenario to be valid, the tutor corrects any oversights before continuing with the exercise.

Once the scenario begins, the students are on their own until all the tracks have made their last significant movement. The scenario is then replayed, and each operator action is compared to the expert action and reviewed, relying heavily on synthetic speech and ostensive actions such as highlighting the track icon. This replay runs much faster than real time up to the point where an identification action was taken. At that point, the replay pauses, and a threshold indicator similar to that in Figure 2 is displayed on the screen next to the situation display along with the calculation that PATRIOT would have made. The synthesized voice of the tutor notes the student identification action and points out its own calculation as shown on the threshold indicator. When possible, the tutor points out sources of error that are logically consistent with the student's actions. This type of direct feedback, obtained by suspending the simulation right where the student's action took place, with the whole scenario context available, allows the visual cues and the synthetic speech to focus the student's attention to the precise spot, with nothing to distract the student or divert his or her visual attention.

Summary of ID Actions Table. After the replay, a table is presented summarizing all the tracks and indicating what the expert would have done compared with what the student did. Figure 4 illustrates the table entry for one track. The first column gives the track number and indicates (by a #) if the student's battery would have shot it down on the basis of his or her ID. The next column is the time at which the PATRIOT program (called EDWA, which stands for Evaluation Decision and Weapons Assignment) would have made an ID change in automatic mode. The "EDWA ID" is the ID that PATRIOT would have assigned. This is the ideal expert action. "Your ID" is the ID that the student assigned at some point during the time that PATRIOT would have had the track as a particular ID. "Your ID (secs)" is the lag time between the time that PATRIOT would have made the ID and the time when the student made the ID. The next set of columns indicate how urgent it is that the track be assigned an ID. These values are important indicators of the amount of time that the officer is giving the assistant to

Track	EDWA Time	ID	Your ID	Your ID (secs)	URGENCY TLR	TLL	ATC	# of Tracks
004#	01:10	U	same	----	+05	+99	1	5
	01:39	AF	same	10	+00	+99	1	5
	02:00	U	AF	----	+00	+99	1	5
	02:29	H	same	65	+00	+87	1	1
	Average		75%	37				

Figure 4 One entry on the Summary of ID Actions table showing that the student made three of the four identifications the same as PATRIOT decision logic ("EDWA") but took sixty-five seconds to make the critical hostile (H) identification.

engage the trace. For example, a low TLL (time to last launch) and a high "Your ID (secs)" indicates a potentially dangerous situation, because the identification could have been made too late to successfully defend the asset. The final column indicates the number of tracks that are displayed on the screen at the time of the ID. The student's time to identify a track is expected to be affected by the number of tracks that must be attended to. Summary "scores" for correct ID and time to identify are presented at the bottom of the table.

The feedback provided by the table organizes and integrates the feedback provided by the preceding replay. It is a very useful summary for both students and instructors. First, it groups actions by track so that all the actions that were taken or should have been taken for that track can be considered as a unit. Second, it lists all the actions that PATRIOT would have taken and matches up the student action and nonaction (e.g., interim identifications that the student neglected). Third, it gives the lag time between the point that PATRIOT would have changed a track's identification and when the student did. It is always expected that the student will lag behind the computer, but lags greater than thirty seconds can indicate that the student did not notice the relevant event (e.g., the track's crossing into a volume or leaving a safe passage corridor). Fourth, the table relates the identification actions to other aspects of the situation which affect the urgency of the identification action and ought to be considered in relation to the lag time. Finally, the table can be used as a dispatcher for obtaining expert demonstrations.

Expert Demonstrations. The summary table serves as a menu for selecting demonstrations of specific tracks in addition to providing detailed performance data. The student can choose an "expert demonstration" for any track on which he or she made an error or late identification or for which he or she did not understand the identification process. For example, interim identifications, e.g., where a track is reclassified twice before obtaining a final identification, are often not noticed by the student. For identifications that are done correctly, the lag time may have been much greater than the student thought due to factors that were not noticed. The demon replays the scenario in "fast forward," focusing on the track that the student requested. The selected track is highlighted, and the replay pauses each time the track reaches or is about to reach a spot where the point total will change. The tutor (synthesized speech) points to the cause (volume penetration, corridor dealignment, speed or altitude change, etc.) by highlighting it and explains the specific cause and how the total points are affected. The threshold indicator displays the increase or decrease in points. In cases involving specific criteria, the tutor explains the criteria. For example, the points for minimum safe velocity can be lost if the aircraft goes too high or if it speeds up. In this case, the specific violated criterion is ostensively displayed on the screen, and an appropriate comment is produced by the speech synthesizer.

This expert demonstration presents the identification task from the perspective of a single track rather than from the perspective of an expert operator dealing with a large number of tracks simultaneously. The decomposition of the problem into the separate problems was presented in diagrams in an initial lecture in the conventional program of instruction but was not used in the practical exercises with PCOFT. Including review of the lecture material in the practical exercises was an important way that INCOFT changed the nature of instruction beyond simply giving more detailed feedback.

Implications for Courseware and the Program of Instruction

There are critical differences between INCOFT and PCOFT which arise as a consequence of the intelligent feedback provided by INCOFT. The examples described in this section illustrate some of the changes that are possible in the program of instruction when this kind of technology is introduced. The set of scenarios constituting the courseware is more responsive to the conceptual issues in understanding PATRIOT operations and therefore can lead to changes in the sequence of instruction, the role of instructors, and ultimately the kinds of skills that can be trained.

Characteristics of the INCOFT Scenarios. Making use of the feedback and expert demo facilities of INCOFT led us to create scenarios which were very different from those used in PCOFT. We often targeted a single concept, such as corridor alignment, using a single track as the basic illustration but including several other more routine tracks as distractors. While PCOFT scenarios typically ran thirty-five or forty-five minutes, our scenarios were much shorter: seven minutes seemed to be a reasonable limit before giving a replay. While relatively short, some of the INCOFT scenarios nevertheless required intense concentration. With the speech synthesis of INCOFT we were able to increase the scenario complexity by including communication events in which commands from higher echelons request changes in authorizations and overrides of local identifications.

INCOFT's application of intelligent technology makes it possible to present scenarios which contain cases more subtle than those presented with PCOFT. For example, an INCOFT scenario presents an aircraft that does not give off the correct NATO IFF code but nevertheless aligns with the corridor for a minute. This particular aircraft (track 4 in Figure 1) starts out as an Unknown, changes momentarily to Assumed Friend when it aligns with the corridor but goes back to Unknown when it dealigns but is still within the corridor. While still within the volume of the corridor, it becomes hostile as it correlates with the Prohibited volume. When it finally leaves the corridor, it has been Hostile for more than a minute. Unless the student's actions can each be critiqued in relation to each track, there is little point in presenting complex cases such as this. The student is unlikely to notice all the phases of the PATRIOT identification process unless specific feedback is given. This is especially true if the student does not realize that a track can be declared Hostile while still within the safe passage corridor. The student may identify the track correctly as Hostile once it leaves the corridor, but this is after a critical minute has already elapsed.

The Impact on Instruction. INCOFT was designed to fit into the existing program of instruction but potentially impacts on the organization of instruction. For example, the summary of ID actions table presents the point of view of the Tactical Control Assistant in charge of firing the missiles by listing the urgency of the identification action in terms of seconds remaining before it is too late to engage the track. The assistant's job is not normally covered until later lectures. Working with an intelligent simulation begins to allow for more comprehensive presentation of the whole task with its interlocking relevancies.

The usual division between theoretical lecture and practical exercise also began to be broken down. INCOFT reviewed and reinforced the theoretical information that had been presented. In some cases, INCOFT appears to be a better means for getting across the information, e.g., the process of identification calculation. The use of INCOFT may lead to a different division of instruction between lecture and practical exercise.

The role of the instructor in the practical exercises will also change. INCOFT was not designed to replace the instructors; instead, it changes the kind of work they do. In our test of the system in actual instruction, we found that instructors helped students interpret the data table and suggested tracks that they should review. Rather than giving students their own feedback on errors, instructors discussed the feedback that the system was giving to the students, using it as a basis for instruction. INCOFT freed the instructors from having to attend to the correctness of the student's actions, allowing them to focus on more subtle aspects of performance.

New Perceptions of PATRIOT Tracks. A powerful pedagogical function of INCOFT's expert system was revealed in the students' responses to the single track expert demos helped to make clear the function of the PATRIOT computer and the perceptual skill of the experienced Tactical Control Officer. While assigning an identification to any single track is just a matter of straightforward adding and subtracting, PATRIOT is able to do that for hundreds of tracks simultaneously. A novice operator faced with, for example, fifteen tracks, must look at each in some sequence and make identifications. This snapshot approach makes it hard to take into consideration the past trajectory of each track, yet it is the track's intentions as inferred from patterns of motion which are the real issue. Interestingly, interviews with experienced air defense operators who were being reassigned from different systems indicated that it is the perception of these patterns for particular tracks that seems to mark expertise in air defense operation.

Conclusions for the Design and Use of ITS Technology

Classically, ITSs are built around the concepts of expert knowledge and a model of students' partial understanding of the expert knowledge which is used to direct instruction. The system we have discussed has neither of these features in the ordinary sense. Using an articulate simulation of PATRIOT's identification logic as the model expert was a practically useful simplification of ITS technology which usually attempts to embed a model of *human* expertise.

The use of the ITS in the context of instructor-student interaction provides extensive human resources (instructor and student) for monitoring progress and directing next steps. For example, the student or instructor can choose to see a single track demonstration. Our field test of INCOFT in actual instruction—supervised by instructors rather than researchers—has begun to demonstrate the reasonableness of putting this power at the disposal of real instructors and their students rather than attempting to build the presentation of the computer identification process into an automatic tutor. This simplification is proving effective and usable in instruction. Rather than puzzling over how to get an ITS to decide how much feedback to provide and what kind of scenario to present next, these decisions are handed over to the people involved.

Our work on INCOFT demonstrates some simplifications of the classic ITS model that make the concept practically useful in instruction. Educational applications which involve well-defined procedures can benefit from our findings about how to use AI technology to model relatively mechanical processes with no pretense to modeling actual human cognition. By putting various decompositions or representations of the processes in the hands of students and teachers, we might expect useful instructional interactions to ensue. In fact, the resulting instruction went beyond learning the PATRIOT decision logic. The opportunity to interact with conceptually complex cases and to see, in comparison, how the PATRIOT computer would have "perceived" and handled them, leads to a different way of approaching the task involving a greater sensitivity to critical changes in the track that can clarify the aircraft's intentions.

Acknowledgments

For their comments on earlier drafts, we are grateful to Wally Feurzeig, Laurel Allender, Joe Psotka, and Marshall Farr. INCOFT was sponsored by the Joint Services Manpower and Training Technology Development (JS/MTTD) Program and monitored by the Army Research Institute (ARI) under Contract MDA903-86-C-0382. The views, opinions, and findings contained in this report are those of the authors and should not be construed as an official Department of the Army position, policy, or decision, unless so designated by other official documentation. Address requests for reprints to the first author at BBN, 10 Moulton St., Cambridge, MA 02138.

References

1. F. Ritter and W. Feurzeig, "Teaching real-time tactical thinking," in J. Psotka, L. D. Massey, and S. A. Mutter (eds.), *Intelligent tutoring systems: Lessons learned*, Hillsdale, NJ: Lawrence Erlbaum Associates, 1988.
2. D. Newman and B. A. Lubetsky, *Instructor's guide to INCOFT courseware*, Report No. 6885, BBN Systems and Technologies Corporation, 1988.

Chapter 11

An Intelligent System for Training Space Shuttle Flight Controllers in Satellite Deployment Procedures

R. BOWEN LOFTIN

University of Houston-Downtown
One Main Street
Houston, TX 77002

LUI WANG
PAUL BAFFES

Artificial Intelligence Section, FM72
NASA/Johnson Space Center
Houston, TX 77058

GRACE HUA

Computer Sciences Corp.
16511 Space Center Blvd.
Houston, TX 77058

Abstract *An intelligent training system has been developed for use by NASA/Johnson Space Center flight controllers in learning to perform satellite deployments from the Space Shuttle. The system is modular and segregates domain-dependent knowledge from generic training knowledge. Its significant features include four cooperating expert systems communicating via a blackboard, support for multiple solution paths, error handling appropriate to the trainee's demonstrated level of skill, and automatic generation of unique training scenarios based on training objectives and a trainee's history of interaction with the system. The system architecture has been used for the development of additional intelligent training systems for use by astronauts and space shuttle ground-support personnel.*

Introduction

Flight controllers at NASA/Johnson Space Center (JSC) are responsible for the ground control of all space shuttle operations. Those operations which involve alterations in the characteristics of the space shuttle's orbit are under the direction of a flight dynamics officer (FDO) who sits at a console in the "front room" of the mission control center (MCC). Most FDOs have backgrounds in engineering, physics, and/or mathematics. They acquire the skills needed to perform their jobs through the study of flight rules, training manuals, and on-the-job training (OJT) in integrated simulations. Two to four years is normally required for a trainee FDO to be certified for many of the tasks for which he or she is responsible during space shuttle missions. OJT is highly labor-intensive and presupposes the availability of experienced personnel with both the time and ability to train novices. As the number of experienced FDOs has been reduced through retirement, transfer (especially of Air Force personnel), and promotion and as the preparation for and actual control of missions occupies most of the MCC's available schedule, OJT has become increasingly difficult to deliver to novice FDOs. As a supplement to the existing modes of training, the Artificial Intelligence Section (AIS) at NASA/JSC has developed a training system based on artificial intelligence technology. The system trains inexperienced flight controllers in the deployment of a payload-assist module (PAM) satellite from the space shuttle. This procedural task is complex and requires many of the skills used by the experienced FDO in performing many other on-orbit operations.

Background

Training is a major endeavor in all modern societies: new personnel must be trained to perform the task(s) which they were hired to perform, and continuing personnel must be trained to upgrade or update their ability to perform assigned tasks and to tackle new tasks. Within industry, government, and educational institutions a great number of training methodologies are employed, singly or in concert. These methods include training manuals, formal classes, procedural computer programs, simulations, and on-the-job training. The last method is particularly effective in complex tasks where a great deal of independence is granted to the task performer. Of course, this training method is also the most expensive and may be impractical when there are many trainees and few experienced personnel to conduct on-the-job training.

Since the 1970s a number of academic and industrial researchers have explored the application of artificial intelligence concepts to the task of teaching a variety of subjects [1–5] (e.g., computer programming in Lisp [6, 7] and Pascal [8], economics [9], geography [10], and geometry [11]). The earliest published reports that suggested the application of artificial intelligence concepts to teaching tasks appeared in the early 1970s [10, 12]. Hartley and Sleeman [12] actually proposed an architecture for an intelligent tutoring system. In the sixteen years since the Hartley and Sleeman paper, a consensus has been reached on the major components of an architecture for intelligent tutoring systems. However, there exists a great diversity in details of implemented systems [9, 13, 14].

Along with this extensive work on intelligent tutoring systems for academic settings has come the development of systems directed at training personnel to perform complex procedural tasks. The design of such training systems must account for a number of distinctions which exist between tutoring in academic subjects and training in an operational environment [15]. Among these training systems are Recovery Boiler Tutor [16], SOPHIE [17], STEAMER [18], and MACH-III [19]. These differ from the tutoring systems mentioned above in providing a simulation model with which the student or trainee interacts. Although these intelligent training systems each use the interactive simulation approach, they have very different internal architectures. The system described below employs a unique architecture which serves to isolate domain-dependent from domain-independent knowledge. This isolation is accomplished through the application of cooperating expert systems and a blackboard approach to intersystem communication.

System Overview

The Payload-assist module Deploys/Intelligent Computer-Aided Training (PD/ICAT) system is an autonomous intelligent training system which integrates expert system technology with training/teaching methodologies. The system was designed for novice FDOs who have completed a study of the PAM deploy procedures as described in available training documents. The training system is designed to aid these trainees in acquiring the experience necessary to carry out a PAM deploy in an integrated simulation. It is intended to permit extensive practice with both nominal deploy exercises and others containing typical problems. After successfully completing training exercises which contain the most difficult problems, together with realistic time constraints and distractions, the trainee should be able to complete successfully an integrated simulation of a PAM deploy without the aid of an experienced FDO. The philosophy of the PD/ICAT system is to emulate as closely as possible the behavior of an experienced FDO devoting his or her full time and attention to the training of a novice: proposing challenging training scenarios, monitoring and evaluating the actions of the trainee, providing meaningful comments in response to trainee errors, responding to trainee requests for information and hints (if appropriate), and remembering the strengths and weaknesses displayed by the trainee so that appropriate future exercises can be designed.

The PD/ICAT system (see Figure 1) consists of five components:

1. A user interface permits the trainee to access the same information available to him or her in the MCC and serves as a means for the trainee to take actions and receive feedback from the training session manager.
2. A domain expert (DeplEx) is capable of carrying out the satellite deployment process using the same information available to the trainee. DeplEx also contains a list of a "mal-rules" (explicitly identified errors that novice trainees commonly

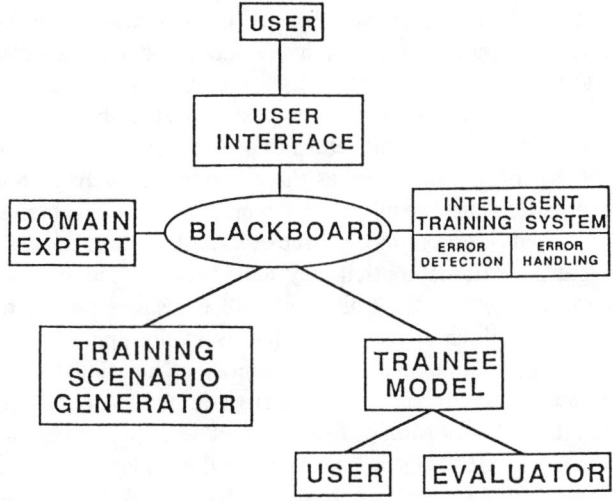

Figure 1 The PD/ICAT architecture.

make) so that the trainee can be provided with feedback specifically designed to help him or her overcome any anticipated conceptual or procedural problems.
3. A training session manager (TSM) includes an error detection component which compares the assertions made by DeplEx (of both correct and incorrect actions in a particular context) with those made by the trainee. In addition, the error-handling component of the TSM decides on the appropriate level of guidance based on the trainee's skill level.
4. A trainee model contains a history of the individual trainee's interactions with the system together with summary evaluative data. The model also provides a report generation feature that produces a formatted trace of each trainee session and generates a high-level description of the trainee's current skill level and progress for the trainee's supervisor.
5. A training scenario generator (TSG) designs increasingly complex training exercises based on the current skill level contained in the trainee model and on any weaknesses or deficiencies that the trainee has exhibited in previous interactions.

All of the expert system components of PD/ICAT (DeplEx, TSM, and TSG) communicate via a common blackboard (see Murray, this volume). In addition, the blackboard provides the means for the TSM to control which expert system is allowed to alter the facts stored in the blackboard. The blackboard also contains a representation of the current trainee action(s) and is a source of data for the updating of the trainee model. Figure 1 contains an illustration of this architecture.

Detailed Description of System Components

User Interface

The primary factor influencing the interface design was fidelity to the task environment. To avoid negative training, it is essential that the functionality and, to the extent possible, the appearance of the training environment duplicate the MCC console. Figure 2 contains a view of the typical display seen by a trainee on a Symbolics 3600 series Lisp machine. The upper right corner of the display contains menus that allow the FDO trainee to communicate with other (simulated) flight controllers, obtain needed displays of data, obtain information about the current or previous step in the deploy process, request help from the training system, and return to a previous step in the process. This menu may lead the trainee through as many as three levels, depending on the nature of the action taken. Some actions are completely accomplished through menu interaction, while others require the input of one or more parameters. All actions taken by the trainee through these menus and the parameters that they may require become assertions to the blackboard. All requests directed to the trainee and all messages sent to the trainee in response to his requests or actions appear in a window in the upper left corner of the screen. These two portions of the screen serve to represent functionally the voice loop interactions that characterize the current FDO task environment. Any displays requested by the trainee appear in the lower portion of the screen. The displays replicate those seen by an FDO at a console in the MCC. Data is supplied to these displays (from a dedicated ephemeris-generating program) so that what is seen by the trainee is reasonable and negative training does not occur. Finally, a pop-up window appears approximately in the

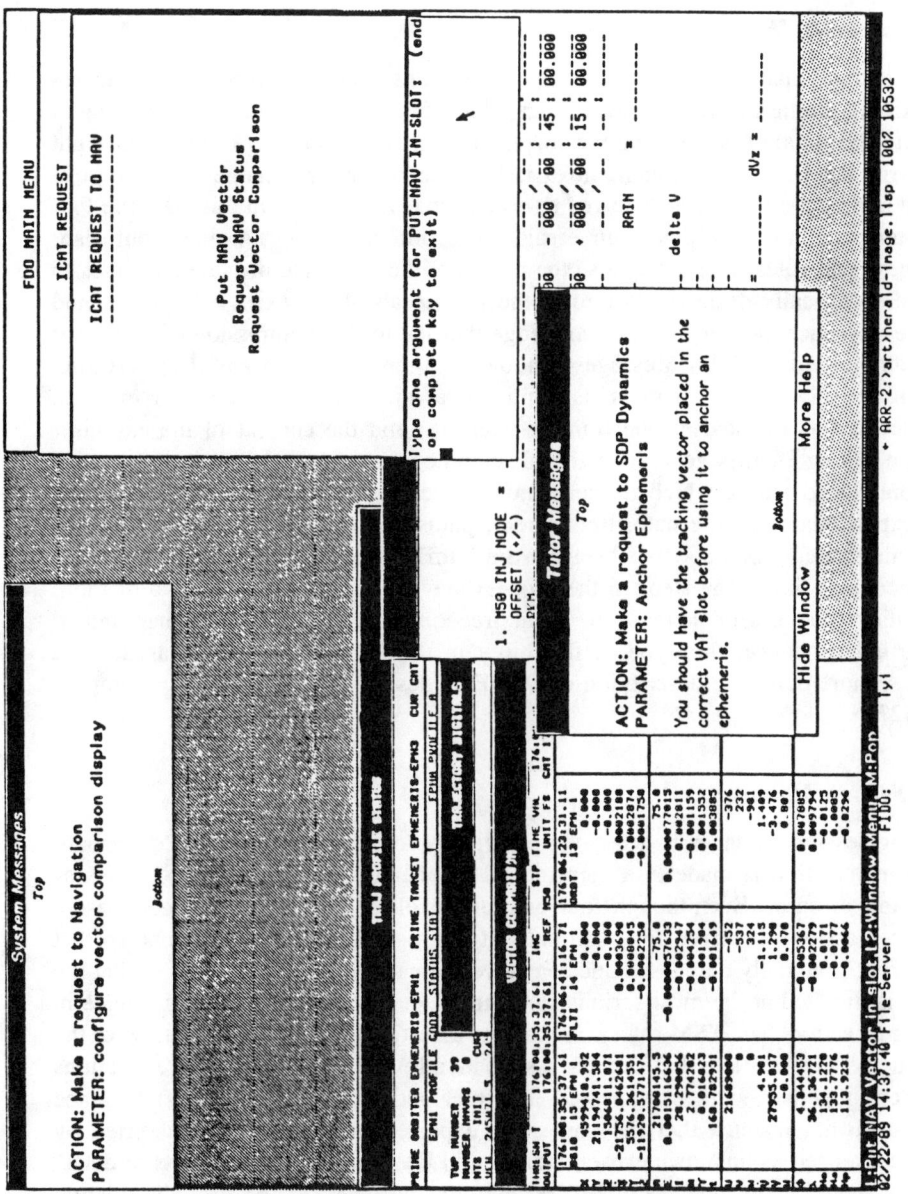

Figure 2 Typical screen display seen by a user of the PD/ICAT system. The menus shown in the upper right are used by the trainee to interact with the training system. The system message window in the upper left confirms actions taken and provides responses from simulated flight controllers. Displays in the lower left are identical to those used in the Mission Control Center. A worksheet (lower right) similar to that used by flight controllers at their console position may be filled in by the trainee. Error messages, explanations, and system status are reported in a pop-up window.

center of the screen to provide tutor messages, context information, and help. Experienced FDOs using PD/ICAT have expressed satisfaction with the user interface.

Deploy Expert (DeplEx)

DeplEx is a traditional expert system in that it contains production rules which access data describing the task environment. It is capable of executing the task and arriving at the correct solution(s) or performing the correct actions. In addition to knowing the right way to carry out the task, the domain expert also contains knowledge of the typical errors that are made by novices (referred to here as mal-rules [20]). In this way, the PD/ICAT system can not only detect an erroneous action made by a trainee, but also, through these mal-rules, it can diagnose the nature of the error and provide messages to the trainee specifically designed to inform the trainee about the exact error made and correct the misconception or lack of knowledge that led to the commission of that error (for an example, see the Tutor Messages window in Figure 2). Another of the interesting features of the PD/ICAT system is its continual awareness of the environment (the external constraints dictated by the training exercise) and the context of the exercise. Rather than having DeplEx generate a complete and correct set of actions to accomplish the task, only those actions which are germane to the current context are asserted. In this way the expert adapts to alternate, but correct, paths that the trainee might choose to follow. This strategy was adopted because the human experts that perform the PAM deploys recognize that many steps in the process may be accomplished by two or more equally valid sequences of actions. To grant freedom of choice to the trainee and to encourage independence, this type of flexibility in the PD/ICAT system was deemed essential. A more complete description of DeplEx's design may be found in references [21] and [22].

Training Session Manager

The training session manager is dedicated principally to error detection and error handling. Error detection is made in a hierarchical manner by defining trainee actions as action/parameter pairs. Even though four or more levels of such pairs may comprise a single trainee action, the error detection halts at the highest level. This is done so that only one error is actually diagnosed and remedied for each trainee action and so that the correction is directed at the most serious, or highest level, error made by the trainee in completing an action. The TSM rules compare the assertions of DeplEx with those of the trainee to detect errors. Note that DeplEx asserts all allowed (and known) correct actions at each step, together with all expected incorrect actions (from mal-rules). Trainee actions that do not match either the correct or expected incorrect actions asserted by DeplEx are matched as unknown errors, stored, and later examined by experts to determine if they should be included as mal-rules. Subsequently, the DeplEx provides the text that allows the TSM to write appropriate tutor messages to the trainee through the user interface. In addition, the TSM is sensitive to the skill level of the trainee as represented by the trainee model. As a result, the detail and tone of tutor messages are chosen to match the current trainee. For example, an error made by a first-time user of the training system may require a verbose explanation to make certain that the trainee will have all of the knowledge and concepts needed to proceed. On the other hand, an experienced trainee may have momentarily forgotten a particular procedure or may have lost his or her place. In this latter case, a terse tutor message would be adequate to allow the trainee to resume the exercise. The TSM also encodes all trainee actions, both correct and incorrect, and passes them to the trainee model.

Trainee Model

Successful intelligent tutors incorporate student models to aid in error diagnosis and to guide the student's progress through the tutor's curriculum (see, for example, a number of papers on student models in reference [1]). The trainee model in the PD/ICAT system stores assertions made by the TSM as a result of trainee actions. Thus, at its most fundamental level, the trainee model contains a complete record of the correct and incorrect actions taken by the trainee during each training session. At the conclusion of each training session, the model updates a training summary containing information about the trainee's progress, such as skill-level designator, number of sessions completed, number of errors made (by error type and session), and the time taken to complete the session. After completing a session, the trainee can obtain a session report containing a comprehensive list of correct and incorrect actions together with an evaluative commentary. A supervisor can access each trainee's model to obtain this same report or to obtain summary data, at a higher level, on the trainee's progress. Finally, the training scenario generator uses the trainee model to produce new training exercises.

Training Scenario Generator

The training scenario generator (TSG) relies on a data base of task problems associated with DeplEx to structure unique exercises each time a trainee interacts with the system. The initial exercises provided to a new trainee are based on variants of the purely nominal task with no time constraints, distractions, or problems. Once the trainee has demonstrated an acceptable level of competence with the nominal task, the generator draws upon the problem data base to insert selected problems into the training environment. In addition, time constraints are tightened as the trainee gains more experience, and distractions (such as calls from other flight controllers) are presented at inconvenient points during the task. The generator also examines the trainee model for particular types of errors committed by the trainee in previous and current sessions. The trainee is then given the opportunity to demonstrate that he or she will not make that error again. Ultimately, the trainee is presented with exercises that embody the most difficult problems with time constraints and distractions comparable to those encountered in the actual task environment. The TSG contains production rules that generate a frame-based representation of parameters required for each scenario [23].

Special Features of the System

Although a number of intelligent training systems have been developed (for example, SOPHIE and STEAMER), few, if any, can be said to have been deployed. The PD/ICAT system has been deployed and is in use for the training of novice flight controllers and for practice and refreshment by experienced personnel. Many features of the PD/ICAT system are innovative in their own right or are innovative in their application to intelligent training systems.

- PD/ICAT is composed, in part, of four expert systems that cooperate through and communicate by means of a common blackboard. This approach was used to

permit the segregation of domain-independent knowledge so that the system architecture could easily be adapted to different training tasks.
- Unlike most intelligent tutoring/training systems, PD/ICAT does not require the trainee to follow a single correct path to the solution of a problem. Rather, a trainee is permitted to select any correct path, as determined by the training context. The method used to accomplish this flexibility, without generating a combinatorial explosion of solution paths, is believed to be unique.
- Error detection occurs through the comparison of the trainee's actions with those of an expert. Frequently, an action is comprised of a number of sequential steps, any one of which may be incorrect. For example, the trainee may proceed through three levels of menu selections, culminating in keyboard input. Rather than identifying all steps in the sequence as incorrect, the error detection is made at the first step that fails to match that anticipated by the expert. This process avoids confusion about the nature a given error and permits much better assistance to be provided the trainee.
- Error handling may be accomplished through the matching of trainee actions with mal-rules containing errors commonly made by novices. In addition, the TSM error-handling component decides, based on the trainee model, what type of feedback to give the trainee. Explanations or hints may be detailed for novices and quite terse for more experienced personnel. In some cases, the TSM may decide not to call attention to the error if a reasonable probability exists that the trainee will catch his or her own mistake.
- The training scenario generator examines the trainee model and creates a unique scenario for each trainee whenever a new session begins. This scenario is built from a database containing a range of typical parameters describing the training context as well as problems of graded difficulty. Scenarios evolve to greater difficulty as the trainee demonstrates the acquisition of greater skills in solving the training problems.
- At the conclusion of each session, the trainee is provided with a formatted trace of the session highlighting the correct and incorrect actions, the time taken to complete the exercise, and the type of assistance provided by the system. In addition, the trainee's supervisor may view a global history of each trainee's interaction with the system and may even generate graphs of trainee performance measured against a number of variables.

Development and Delivery

From the inception of the PD/ICAT project, a number of factors were recognized as essential for its success at meeting its objectives and its acceptance by its intended audience.

1. The involvement of the ultimate users of the system throughout the development process was certainly the most important factor in PD/ICAT's successful deployment. This involvement allowed the development team to have at hand the experts in the targeted procedure and to quickly test competing approaches to the solution of specific development problems.
2. The intended audience for the training system was asked to provide commentary at each stage of the interface development. Thus, at the project's conclusion, the interface had already achieved acceptance by those it was intended to serve.

3. The coding, in a production-rule system, of those components of PD/ICAT that will require alteration as the procedures it was designed to teach are altered provides built-in maintainability. After fielding a number of production rule systems, the AIS has found that such systems may be easily altered and verified as the task they were designed to accomplish changes.
4. Sufficient training in production-rule coding and detailed documentation were provided to the users so that they were able to provide their own long-term support for PD/ICAT.
5. Although PD/ICAT was developed on a Symbolics Lisp machine using ART and Lisp, it has been ported to a unix-based workstation using C and CLIPs (a production-rule system, written in C, and developed by the AIS at JSC). Such workstations were already in use by PD/ICAT's intended users, and they were not required to purchase and learn to use a different hardware platform.
6. An important motivation for the managers from the intended user community was the ability of PD/ICAT to capture the expertise of personnel who were to be transferred to other areas. This was a key factor in the ready availability of the experts and in the management's support of their dedication of time to this project.

Conclusions

Training of astronauts and ground-based flight controllers and system engineers is a massive task. The best training, and the mechanism for certification that personnel have met training objectives, occurs through large-scale, integrated simulations. Unfortunately, these simulations require the support of hundreds of people while delivering training to only one person in each position. The ability of a given trainee to get significant exposure to a particular process is, therefore, quite limited. The PD/ICAT system, on the other hand, can provide a trainee with virtually unlimited exposure to training in a specific procedure and ensure that the integrated simulation environment can be used to maximum effect.

The PD/ICAT system has demonstrated the capability of intelligent training systems in NASA's operational environment. As a result, a number of similar systems are under development at JSC and other NASA operational centers (Marshall Space Flight Center and Kennedy Space Center). In addition to the impact of this technology on Space Shuttle training, NASA is supporting its application to future Space Station training. Since Space Station training may be a task at least an order of magnitude larger than current Space Shuttle training, the use of intelligent training systems may be the only way to meet Space Station training objectives with the available resources. To this end, the Space Station program is supporting the refinement of the PD/ICAT architecture into a general-purpose training architecture and the creation of a general-purpose development environment for the rapid creation and adaptation of intelligent training systems. This latter activity will have a profound impact on the nature of training, not only within NASA, but within other government agencies, industry, and the education establishment.

Acknowledgements

The authors wish to acknowledge the invaluable contributions of expertise from three FDOs: Capt. Wes Jones, USAF; Maj. Doug Rask, USAF (Ret.); and Kerry Soileau. Various students assisted with the knowledge engineering and coding of portions of the user interface and TSM: Tom Blinn, Joe Franz, Bebe Ly, Wayne Parrott, and Chou Pham. Finally, the encouragement and guidance of Chirold Epp (Head, ODS) and Bob Savely (Head, AIS) are gratefully acknowledged. Financial support for this endeavor has been provided by the Mission Planning and Analysis Division, NASA/Johnson Space Center, the NASA Office of Space Flight, and, for RBL, by NASA/American Society for Engineering Education Summer Faculty Fellowships and a NASA/National Research Council Senior Resident Research Associateship.

References

1. D. Sleeman and J. S. Brown (eds.), *Intelligent tutoring systems,* London: Academic Press, 1982.
2. M. Yazdani, "Intelligent tutoring systems survey," *Artificial Intelligence Review,* 1:1, 43, 1986.
3. E. Wenger, *Artificial intelligence and tutoring systems,* Los Altos, CA: Morgan Kaufmann Publishers, 1987.
4. G. Kearsley (ed.), *Artificial intelligence and instruction,* Reading, MA: Addison-Wesley, 1987.
5. J. Psotka, L. D. Massey, and S. A. Mutter (eds.), *Intelligent tutoring systems: Lessons learned,* Hillsdale, NJ: Lawrence Erlbaum Associates, 1988.
6. J. R. Anderson and B. J. Reiser, "The LISP Tutor," *Byte,* 10:4, 159, 1985.
7. J. R. Anderson, C. F. Boyle, and B. J. Reiser, "Intelligent tutoring systems," *Science,* 225:4698, 456, 1985.
8. W. L. Johnson and E. Soloway, "PROUST," *Byte,* 10:4, 179. 1985.
9. V. Shute and J. G. Bonar, "An intelligent tutoring system for scientific inquiry skills," *Proceedings of the Eighth Cognitive Science Society Conference,* Amherst, MA, 353, 1986.
10. J. R. Carbonell, "AI in CAI: An artificial intelligence approach to CAI," *IEEE Transactions on Man-Machine Systems,* 11:4, 190, 1970.
11. J. R. Anderson, C. F. Boyle, and G. Yost, "The geometry tutor," *Proceedings of the Ninth International Joint Conference on Artificial Intelligence,* Menlo Park, CA: American Association for Artificial Intelligence, 1, 1985.
12. J. R. Hartley and D. H. Sleeman, "Towards intelligent teaching systems," *International Journal of Man-Machine Studies,* 5:215, 1973.
13. M. C. Polson and J. J. Richardson (eds.), *Foundations of intelligent tutoring systems,* Hillsdale, NJ: Lawrence Erlbaum Associates, 1988.
14. H. Mandl and A. Lesgold (eds.), *Learning issues for intelligent tutoring systems,* New York: Springer-Verlag, 1988.
15. P. Harmon, "Intelligent job aids: How AI will change training in the next five years," in G. Kearsley (ed.), *Artificial intelligence and instruction: Applications and methods,* Reading, MA: Addison-Wesley Publishing Co., 1987.
16. B. P. Woolf, D. Blegen, J. H. Jansen, and A. Verloop, "Teaching a complex industrial process," *Proceedings of the National Conference on Artificial Intelligence,* Menlo Park, CA: American Association for Artificial Intelligence, 722, 1986.
17. J. S. Brown, R. R. Burton, and J. de Kleer, "Pedagogical, natural language and knowledge engineering techniques in SOPHIE I, II, and III," in D. Sleeman and J. S. Brown (eds.), *Intelligent tutoring systems,* London: Academic Press, 227, 1982.
18. H. D. Hollan, E. L. Hutchins, and L. Weitzman, "Steamer: An interactive inspectable simulation-based training system," *AI Magazine,* 5:2, 15, 1984.
19. L. D. Massey, J. de Bruin, and B. Roberts, "A training system for system maintenance," in J. Psotka, L D. Massey, and S. A. Mutter (eds.), *Intelligent tutoring systems: Lessons learned,* Hillsdale, NJ: Lawrence Erlbaum Associates, 1988.

20. D. H. Sleeman, "Inferring (mal) rules from pupils' protocols," *Proceedings of the European Conference on Artificial Intelligence,* Orsay, France, 160, 1982.
21. R. B. Loftin, L. Wang, P. Baffes, and M. Rua, "An intelligent computer-aided training system for payload-assist module deploys," *Proceedings of the First Annual Workshop on Space Operations, Automation & Robotics* (SOAR '87), Houston, TX: NASA/Johnson Space Center, 53, 1987.
22. R. B. Loftin, L. Wang, P. Baffes, and G. Hua, "An intelligent training system for space shuttle flight controllers," *Telematics and Informatics,* 5:3, 151, 1988.
23. R. B. Loftin, L. Wang, and P. Baffes, "Simulation scenario generation for intelligent training systems," *Proceedings of the Third Artificial Intelligence and Simulation Workshop,* Menlo Park, CA: American Association for Artificial Intelligence, 69, 1988.

IV
Implications for Future Research and Development

Chapter 12

An Emerging Technology Gears Up

ELLIOT SOLOWAY

Department of Electrical Engineering & Computer Science
University of Michigan
Ann Arbor, MI 48109-2110

The promise of computers in education has always been individualized instruction. A computer, with its limitless patience, and essentially limitless knowledge, should be able to tailor a learning environment for the needs and goals of each student. The goal is most laudable, since we know that there are as many learning styles as there are learners. The intelligent tutoring system (ITS) model ratchets up the level to which we can approach the tailorability goal. The computer-based instruction (CAI) systems of the 1960s and 1970s provided a first-level flexibility in how the student could learn a corpus of material. The ITS of the 1980s and 1990s permits the next order of magnitude of flexibility; with its model of the student, its expertise in the task domain and in instruction, and its task-facilitating interface, an ITS appears able to support a wider variety of learning experiences while producing a substantial improvement in learning effectiveness.

Now, when the ITS community moved into gear 10 years ago, the cost of providing such performance was considerable. We did not know what really went into making an effective ITS. The computer needed to run the system was in the range of $50–100 thousand, and the language used to program an ITS was only several steps above the machine's 0s and 1s. However, cost should no longer be a factor. ITS-based research has been productive, and there is a growing understanding of how to build them effectively. The computer capable of running sophisticated ITSs will shortly be in the $2–5 thousand range, and we are seeing expert system shells and hypertext-style programming that allow a much wider range of individuals to be developers. Thus, as I am about to argue, ITS is becoming a technology with the potential to significantly influence education. As the personal computer increases its presence in our everyday lives, we can see ITSs routinely embedded in software packages and environments.

The present volume presents an important snapshot of ITS efforts: chapters here report on third/fourth generation systems that are proving genuinely effective in the field. These efforts provide the field definite legitimacy. How many other educational technologies can measure improvements in "standard deviations"?! In what follows, I sketch how ITS is moving toward "technology" status, and briefly, what some of the next-generation questions are that have come to light in response to ITS technology.

As ITS Matures, Technology Matures

For a long time, computer-types have been saying, "Wait for next year's computer, then we will have lots of power." And next year came, and things changed, but not as dramatically as we expected. The 1990s is the decade in which we computer-types will no longer be crying wolf. Figure 1 shows quite clearly that we are finally on the elbow of the "MIPS exponential." Those who translate the line graph onto log paper, get a

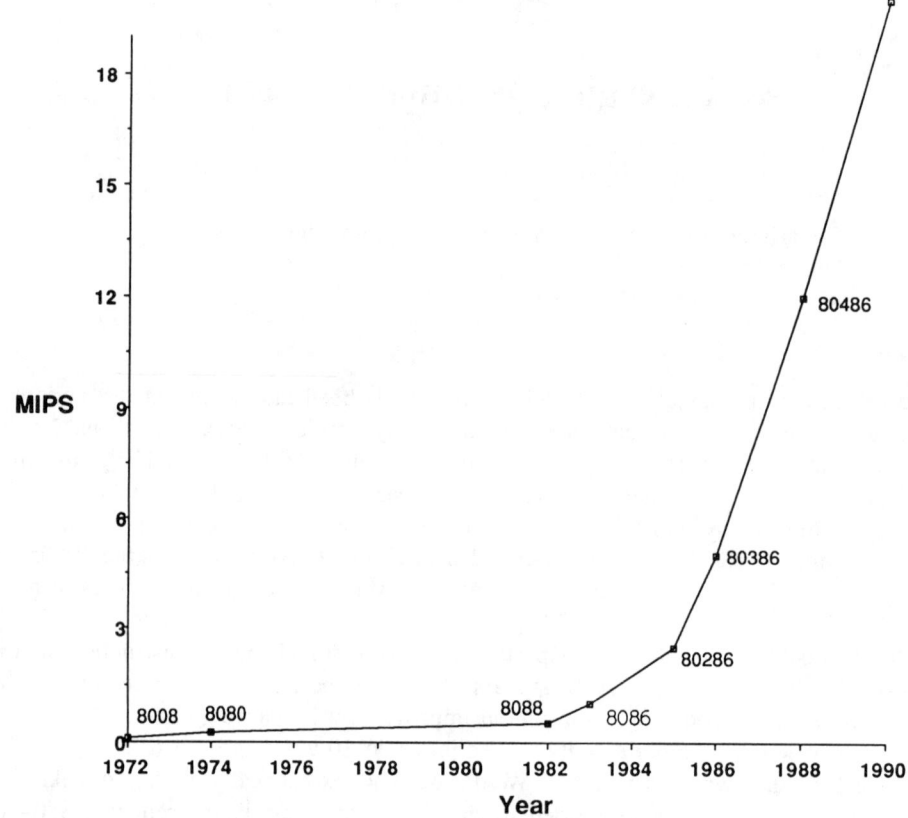

Figure 1. The accelerating power of common CPUs.

straight line, and then say "so what" are missing the point: having 200 MIPS on the desktop will be *qualitatively* different than having 2 MIPS on the desktop. Also, the price of that 200 MIP machine will be close to what we are currently paying for the 2 MIP machine.

We will need this level of horsepower to run the graphically oriented, computationally intensive ITSs of the 1990s. Those "fancy" interfaces that many desire are not computationally cheap; likewise, the decisions that must be made by the tutoring module, in concert with the student model and expert model, require considerable computation. It is interesting to speculate where CAI would have gone in the 1960s and 1970s if researchers then could have transferred their ideas to low-cost, high-powered personal computers, instead of being tied down to mainframe/time-sharing environments. Frankly, ITS workers should consider themselves lucky.

Beyond raw horsepower, there are three other factors that will help to transition ITS work into a technology.

We know what should go into an ITS (well, more or less). Cognitive task analysis, for example, is moving from art to standard practice; the work of Lesgold and colleagues [1] have set the pace here. Anderson and his colleagues [2] have demonstrated that a detailed student model can be used to drive a tutoring environment, and they have demonstrated that student models are readily buildable. In concert with the Educational Testing Service, my colleagues and I are developing a "bug cataloguing technique" and a "bug-matching technique" that should allow us to routinely build bug analysis systems. STEAMER-like systems [3] have demonstrated the clear win of good graphical displays. We also have a clear sense of what benefits will result from using an ITS. For example, the work in this volume documents the order of magnitude in improvement, at reasonable cost, that one can count on in using an ITS. Moreover, ITSs do not require an integration of all the traditional components (tutoring expert, domain expert, student model, interaction manager); rather, we are seeing how individual components can be exploited to provide an enhanced learning environment.

Software tools are reducing the "buy in" for ITS developers. In earlier days, you had to learn LISP if you wanted to build ITSs. Although techies do not flinch, those interested in education per se were essentially locked out; there was just too much overhead in learning to use LISP effectively. However, we are now seeing two types of software tools in the marketplace: (1) shells that allow the encoding of expert knowledge, bug catalogs, and so forth and (2) hypertext-style programming environments that enable the rapid development of interfaces. Thus, a degree in computer science is no longer necessary to become an ITS developer; the future will see a range of tools that will open up development to a much wider range of educationally oriented individuals.

People who know ITSs are finally moving into positions of responsibility within the educational community. Although tools will help reduce the level of expertise needed to produce an ITS, there is no magic in them. Thus, students who cut their teeth on the front line research, who are now heading out of academe and into "real jobs" will need to lead the ITS development teams.

In sum, ITS is becoming a technology: (1) development of ITSs is no longer the sole province of academics, but rather ITSs are being produced by educators and deployed in real world settings, (2) the computing infrastructure to support this transition is almost in place, and (3) ITS technology can produce an acceptable level of educational improvement. Thus, there is the clear potential for ITS technology to have a significant impact on education and training over the next 10 years.

What to Do Next

The aforementioned rosy predictions are a safe bet, because there is now a substantial base of solid results. However, though similar predictions were made 10 years ago, when the whole ITS enterprise really got off the ground, there was little support. It has been the vision and chutzpah of people like Marshall Farr and Henry Halff at the Office of Naval Research, and Harry O'Neil and Joe Psotka at the U.S. Army Research Institute, who cornered a hefty chunk of resources and seeded the researchers at a variety of institutions. Moreover, these same risk takers stayed active in the field, organizing yearly workshops, conferences, books, panel sessions at conferences, special issues of journals, white papers, and so on. We owe them an unrepayable debt; and in fact the next generation of investigators should be so lucky as to have people of the caliber and commitment of Farr and colleagues.

I myself stated out 10 years ago with support from the Farr, Halff, O'Neil, and Psotka regime. So, it is all their fault. For the record, though, I will admit I was definitely a true believer, and because I thought that learning was about learning "stuff," what better delivery vehicle did I have than the computer—with its unending patience and essentially limitless knowledge?

I now squirm at my naivete and hubris; seeing real kids in real classrooms trying to use the computers I built has been most . . . illuminating. "Stuff" is only one facet of what happens in learning; "essentially limitless knowledge" is not the real point, and the kids used my "boxes" in ways that *they* saw fit. After some reflection it became clear that the computers could contribute much more if only I understood better all the facets of the learning environment. The techie in me got an infusion of the real world; it was so much richer than the idealized, simplified world that I had created for myself.

Although the details will certainly vary, it seems that lots of people in the computer community are recognizing the need to broaden their research. Although ITS technology can have an impact, there is more to discover. Rather, the techies who started out in the ITS community are now (finally?) seeing the need to seriously work with "that other culture"—and vice versa. With this two-way communication will come a redistribution of the roles that many diverse participants in the learning environment can play, and a better sense of what, besides stuffing kids with stuff, learning is about. The 1990s can be a most exciting decade: the widespread dissatisfaction with the current educational system will continue to build and bring with it opportunities to try new things. Maybe this time all the components for truly restructuring education are coming together at the same time.

References

1. A. Lesgold, S. Lajoie, E. Logan, and G. Eggan, "Applying cognitive task analysis and research methods to assessment," in N. Frederiksen, R. Glaser, A. Lesgold, and M. Shafto (eds.), *Diagnostic monitoring of skill and knowledge acquisition.* Hillsdale, NJ: Lawrence Erlbaum, 1989.
2. J. R. Anderson, C. F. Boyle, and B. J. Reiser, "Intelligent tutoring systems," *Science,* 228:456–462, 1985.
3. J. D. Hollan, E. L. Hutchins, and L. Weitzman, "STEAMER: An interactive inspectable simulation-based training system," *The IA Magazine,* 2:15–27, 1984.

Index

Abnormal symptoms, 109
Abstracted subgoal lattice, 23
ACT*, 71, 78, 81, 179, 180, 182, 185–187, 192, 194, 198, 199
Action Part (of a Rule) (*see* Consequent)
Adaptive instruction, 26, 31
Address space (*see* Memory address space)
Addressing instructional needs, 208
Advice, 222
Advisor, 213, 225, 232
AI (*see* Artificial intelligence)
Air Force, 147, 153
Algorithms, 159, 164–168
Analogy, 77–85
Anderson, J. R., 179, 180, 182, 186, 188, 200, 202, 203
Answer judging, 148
Antecedent, 134, 141, 151
Apple Computer, Inc., 153
Apprentice, 154
AR15 (*see* Rifle)
Architecture(s), 127, 137
ARI (*see* Army Research Institute)
Arithmetic, 153–176
Army Research Institute (ARI), 1, 153, 267
Army (*see* U.S. Army)
ART, 259
Artificial intelligence (AI), 128, 151, 153, 252
Assessment, 25–26
Authoring, 127, 129, 130, 139, 148
Authorware Professional, 149

Backward chaining, 133, 134
Blackboard, 251, 253, 254, 257
Blackboard architecture, 3, 10, 11, 87, 96
BLACKBOARD INSTRUCTIONAL PLANNER, 96
Bladefold, 107
Brown, A. L., 17
Brown, J. S., 16

Bugs, 67, 267
Byte-code, 132

C programming language, 139, 259
C*T (*see* Computer* Thought Corporation)
CAI, 88, 265
Caramazza, 15
Carnegie Mellon University (CMU), 180, 182, 187, 188, 190, 191, 194, 198, 202
Case method, 135
Case study, 179
Categorizing problems, 225
CBT (*see* Computer-based training)
CHI, 179, 180, 187, 198, 200, 202
Chunking, 67
CLIPS, 259
CMU (*see* Carnegie-Mellon University)
Coaching, 16, 24–27, 33
COBOL, 179–183, 185–188, 190, 192–202
Cognitive apprenticeship, 16, 20
Cognitive interface, 187, 193–195, 198, 200
Cognitive research, 206, 224
Cognitive task analysis, 21, 31, 206, 266
Coinland, 154, 159
Collaboration, 225, 236
Collaboratories, 4
Collins, 17, 31
Competence model, 20, 26–27
Complex cognitive skill, 182, 183
Complex-device, 216
Component interactions, 218
Computer programming, 2, 179
Computer-based coached practice environment, 15, 18
Conceptual understanding, 26
Computer* Thought Corporation (C*T), 132
Computer-assisted instruction (CAI), 153
  (*See also* Computer-based training)
Computer-based instruction (CBI), 135, 149, 154
  (*See also* Computer-based training)

269

Computer-based training (CBT), 127–133, 136
Condition part (of a rule) (*see* Antecedent)
Conduit program, 214, 217
Conflict resolution, 151
Consequent, 151
Conservation (of quantity), 154
Constraint (semantic), 153–154, 169
Consulting, automatic, 38
Context, 135, 148
Context sensitive menus, 197
    (*See also* Menus)
Cooperating Expert Systems, 251, 253
Cooperative environment, 18
Corporate training, 130, 152, 179, 198, 199
    (*See also* Industrial training)
Courseware, 127–138, 147, 148, 152
Courseware Design Phase, 135
Critique, 222, 225
Critiquer, 213, 232
Cronbach, 30
Current context, 47
Curriculum, 4, 179, 182, 185, 186, 192, 193, 198, 200, 202
Curriculum support materials, 226

DEC (*see* Digital Equipment Corporation)
Decision tree, 223
Declarative knowledge, 183
Deep simulation, 105, 107, 112
Default reasoning, 224
Defense Language Institute, 147
Dependency graph, 58
Deploy Expert (DeplEx), 253, 254, 256, 257
Depot(s), 128, 133, 135, 144
Derivative simulations, 114
Design Principles, 136, 139
Development and Implementation Phase, 135, 136
Device instance, 216
Device behavior, 216
Device type, 215
Diagnosis, 23, 130, 151, 152
Diagrams, animated, 208
Diagrams, physical, 208
Diagrams, functional, 208
Diagrams, physical-functional, 208
Didactic, 19
Digital Equipment Corporation (DEC), 129
Direct manipulation, 10
Discourse management network, 10, 87, 90

Discovery, 9
Discovery learning, 149
Documentation use, 208
Domain, 3, 147–151, 154, 164
Domain expert, 2, 253, 256
Domain expertise, 23
Dynamic instructional planning, 88
Dynamic student model, 15

EEMT, 117
Effective problem space (EPS), 9, 23–25, 28
EIDS (*see* Electronic Information Delivery System)
Electronic Information Delivery System (EIDS), 131, 147, 150, 152
EMYCIN, 133
Environment, 153, 159
Error detection, 254, 256, 258
Error handling, 251, 254, 256, 258
ES (*see* Expert system)
Evaluation, 6, 11, 28–30, 175–176
Exercises, 226
Expertise, representing, 46
Expert system (ES), 2, 11, 127–141, 151, 251, 253, 254, 256, 257
Expert troubleshooting system, 205, 217–218
Explanation, 208, 209, 219
Explanation, canned, 210
Explanation, generated, 210, 222
Explanation-based learning, 80
Exploration, 153, 159

Facts (*see* Working memory [elements])
Fading, 16, 26
Farr, Marshall, 267
Fault, 139, 141, 143, 148, 152
Fault diagnosis, 106, 107, 118, 124
Fault Isolation Procedures (FIP), 206
Federal training, 130, 152
Feedback, 180, 185–187, 190, 191, 196, 201, 215
Field trial observations, 233
Formative research, 224
Forward chaining, 133, 134
Frame, 257
Frame(s), knowledge, 132, 151
Full motion video (*see* Interactive video)
Functional block, 207
Functional hierarchy, 218, 221, 222, 223

Index

Gaming environments, 152
General hint(s), 138, 151
Generalization, 78
Generic objects, 108, 110
Genetic graph, 57, 154
GENIE, 51–53
Geometry Tutor, 182, 203
Gitomer, 21
Gott, 19, 23, 27
Grace Tutor, 179–181, 183, 185–187, 190, 192, 193, 195, 196, 198–200
Grain-size, 190, 192, 197, 201, 202
GRAPES, 67, 75, 76
Graphics-over-video (*see* Video-graphics overlay)
Guidance, 163–168
GUIDON, 93

Halff, Henry, 267
HAWK MACH-III, 6, 224
HAWK Radar Tutor, 129
Heuristic (*see* Guidance)
Heuristics, 23, 25
Hierarchical knowledge, 8, 10, 207
Hint(s), 134, 138, 142–144, 147, 151, 152, 166–168
Hughes Simulation Systems, Inc., 132
Human–computer interaction, 179, 180
   (*See also* Computer–human interaction)
Hybrid Instructional System(s), 127, 149, 152
HyperCard, 152, 153
Hypermedia, 6, 8, 9, 210
   (*See also* Multimedia)
Hypertext, 267
Hypotheses, 210, 222

IBM PC, 131
ICAI (*see* Intelligent computer-aided instruction)
IDE-INTERPRETER, 97
Ideal models, 7, 67
Ideal student model, 185–187, 189–199
IMTS, 105
Industrial training, 192
   (*See also* Corporate training)
Inference engine, 206, 214, 217
Information overload, 207
Inheritance, 215, 224
Instructional design, 4
Instructional technology, 127, 128, 149

Instructional (systems) design, 128, 130, 153
Instructor questionnaires, 228
Integrated instructional system(s), 152
Integrating MACH-III in classroom, 224
Intel 80×86, 131, 150
Intelligent computer-aided instruction (*see* Intelligent tutoring systems)
Intelligent instructional systems (*see* Intelligent tutoring systems)
Intelligent learning environments (*see* Intelligent tutoring systems)
Intelligent micro-worlds (*see* Intelligent tutoring systems)
Intelligent training system, 251, 252, 253, 257, 258, 259
   (*See also* ITS)
Intelligent tutoring systems, 2, 5, 127–134, 136, 141, 147, 148, 252, 257, 258, 265
   (*See also* ITS)
Interactive video, 127–132, 136, 143–148, 152
Interface, 155–157, 161, 179, 180, 185, 187, 188, 190, 193–195, 200, 201
ISA, 215
ISD (*see* Instructional [systems] design)
ITS, 1, 2, 179–182, 187, 193, 195, 196, 198–200, 265
   (*See also* Intelligent tutoring systems)
ITS architecture, 266
ITS technology, 265, 267, 268
IVD (*see* Interactive video)

Knowledge acquisition, 133
Knowledge acquisition phase, 135
Knowledge base, 217
Knowledge compilation, 67, 78
Knowledge engineer(ing), 2, 133–135, 148, 149, 153
Knowledge representation, 8, 11
Knowledge source, 96
Knowledge tracing, 1, 3, 9, 10, 11, 13, 17, 18, 21, 179, 180, 185–188, 191, 194

Lajoie, 28
Learned state, 26
Learning by discovery, 17
Learning by (guided) doing, 138
Learning by imitating, 138
Learning environment(s), 132, 149

Left-hand side (*see* Antecedent)
Lesgold, 15, 21, 28, 29
LISP programming language, 129, 132, 133, 252, 254, 259
LISP Tutor, 182, 188, 189, 203
Local expertise, 39
Loops, 207, 215

M16, M16A1 (*see* Rifle)
MACH-III, 6, 224, 252
Macintosh (Mac), 150, 152, 153
Maintenance training, 127–131, 137, 150–154
Mal-Rules, 253, 256, 258
Malfunctions, 210
Marksmanship, 130, 133, 150
Matrox Electronic Systems LTD, 131
McDermott, 15
Medical diagnosis, 2
Memory address space, 131, 147, 150, 152
MENO-TUTOR, 91
Mental models, 4, 19, 25
Menu(s), 132, 136–143, 151, 183, 185, 187, 194–197, 201
Mestre, 27
Met Life, 199–202
Meta-Rule(s), 133
Microsoft, 131
MIPS exponential, 266
Mixed-initiative dialogue, 1, 179
Model based simulation, 205, 214, 217
Model-directed reasoning, 151
Model tracing, 68, 150
Modeling, 16, 26
Morris, 19
Mouse (pointing device), 131, 136–142, 149, 152
MS-DOS, 131, 137, 147, 150, 152
MultiFinder, 152
Multimedia (hypermedia), 129, 130, 136, 138, 148, 152

NASA, 251, 259, 260
Newell & Card, 180
Nichols, 15, 29
Nodes, 219, 222
Non-resident amnesia, 138
Numberland, 156, 161–163
NYNEX, 179–182, 192, 198, 202

O'Neil, Harry, 267
Object oriented libraries, 110, 111
Object-oriented programming, 24
Office of Naval Research (ONR), 1, 267
On-the-job training, 11, 252
ONLINE video-graphics overlay board, 131
Operational readiness, 128
OPS5, OPS5+, 132, 133, 138, 139, 150, 152
OS/2, 150
Overlay, 179
    (*See also* Knowledge tracing)
Overlay model, 150
Overlays (memory), 150

Palincsar, 33
Patriot missile, 239–249
Payload-assist module Deploys/Intelligent Computer-Aided Training System (PD/ICAT), 253, 254, 256, 257, 258, 259
PCOFT, 242, 244
Pedagogical, 24, 26, 28, 30
Pedagogy, 157
Performance model, 20, 26–27
Personal computer (PC), 127, 131, 132, 138, 147, 152
Physical interface, 187, 190, 193–196, 198, 200
Piaget, 135
Place-value, 154, 156
Placeholder symbol, 183, 185, 194, 196, 197
PLATO, 130, 132
Post-mortem, 28, 31
Practice tutor, 9, 183
Prerequisite, 171–172
Presenting symptom(s), 134, 135, 139, 141, 148, 151
Primitive actions, 219
Principles, 174–175
Problem categories, 225, 233
Problem difficulty levels, 225
Problem selection, 171–173
Problem-solving, 154
Problem space, 10, 208, 213
Problem taxonomy, 169–171
Procedural, 26
Procedural abstraction, 218
Procedural environments, 39
Producer/critique, 16, 32, 34

Production system theory of transfer, 69–73, 79
Production rules, 219, 222, 256, 257, 259
  (*See also* Rule)
Production system, 182, 187–190, 194
  (*See also* Rule-base system)
Profile, 107
Programming, 67–69
Programming language, 179, 182
Programming strategies, 76
Propagation of effects, 108, 109
Proximal development, 26
Psotka, 267
Pull-down menu(s) (*see* Menu)
PUPS, 75, 76, 78, 79
PUPS-TUTOR, 182, 187–191

Qualitative simulation, 6
Question intent, 48

R1 (*see* XCON)
Radar maintenance, 6
Radar system, 5, 7
RAM (*see* Random access memory)
Random access memory (RAM), 131, 137–139, 150
Reasoning, 128, 145, 148, 151
Reasoning with exceptions, 222
Recapitulation, 28
Reciprocal teaching, 16
Record-keeping, 28
Recovery Boiler Tutor, 252
Rediffusion Simulation, Inc. (RSI), 132
Reflective, 27–28, 31
Remedial sequence (*see* Remediation)
Remediation (remedial sequence), 127, 148, 151
Representation, 156, 160–162
Rifle(ry) (AR15, M16, M16A1), 127, 130, 131–140, 154
Right-hand side (of a rule) (*see* Consequent)
RSI (*see* Rediffusion-Simulation, Inc.)
Rule(s), 132–134, 138, 139, 141, 147, 148, 150, 151, 154, 256
Rule-based system, 127, 132, 153
Rule systems, 8, 10, 173–174

Salomon, 27
SAT (*see* Systems approach to training; ISD)
Scaffolding, 207

Scenario(s), 135–152
Schoenfeld, 16, 28, 31, 33
SCHOLAR, 129
Search space, 158
Self-explanation, 79
Self-reflection, 9
Self-regulation, 28
Semantic, 9
  (*See also* Constraint [semantic])
Sherlock, 15
Simulation, 10, 19–21, 144, 145, 149, 152, 252, 253, 259
Simulation engine, 214
Simulation system models, 213, 215, 217
Situated evaluation, 7
Situated learning, 16, 31
Situated symbol systems, 4
Skills gap, 127, 128
SME (*see* Subject matter expert)
Snow, 30
Social facilitation, 31–34
Sony, 131
SOPHIE, 129, 153, 252, 257
Space Shuttle, 251, 252, 259
Space Station, 259
Specific hint(s), 138, 143, 147, 151
SPORTS, 144–146
Spreading activation, 76
STEAMER, 107, 252, 257
Still-frame(s), 129, 131, 136, 153
Still(s) (*see* Still-frame)
Strategies, 23–24, 28, 207
Strengthening, 79
Structured editor, 180, 185, 191, 195, 197–199, 201
Student exam performance, 234
Student model, 2, 7, 257, 266
Student modeling, 26, 132–134, 138, 147–152
Student questionnaires, 227
Student response, 135, 148
Subactions, 219
Subgoal, 169–171
Subject matter expert (SME), 128, 133, 134, 138, 152, 153
Subtraction, 153–176
Surface simulation, 105, 116
Syllabus, 168–175
SYMBOLICS, 129
Symptoms, 210, 222
System and device models, 207
Systems approach to training (SAT), 130, 133, 135, 148, 149, 152

T-rule, 93, 99
Table lookup, as a basis for simulation, 105, 116
TALUS, 121
Task analysis, 3–4, 133, 135
Task analysis phase, 135
Tasks, functional, 210
Taxonomy, 24, 215
Technical documentation, 134
Technical personnel, 127
Technological webs, 4
TENCORE, TENCORE+, 132–141, 150, 152
Testing, 119, 121
Theorem proving, 59
TICCIT, 130
Tool evaluation stage, 132
Touch panel (pointing device), 136
Trainee model, 254, 256, 257, 258
(*See also* Student model)
Training scenario, 251, 253
Training scenario generator, 254, 257, 258
Training session manager, 253, 254, 256, 257, 258, 260
Transfer, 29, 179–182, 185–187, 191, 194, 199, 202, 203
Transfer of procedural knowledge, 69–73
Transfer of skills, 161–163
TRIO, 239
Troubleshooting, 6–9, 11, 127, 130–140, 143, 145, 149–151
Troubleshooting expert system, 213, 214, 217
Troubleshooting mode, demo, 213, 218
Troubleshooting mode, magic mode, 213, 218
Troubleshooting mode, real-life mode, 213, 218
Troubleshooting tree, 210, 212, 217, 218
TUTOR (programming language), 132
Tutorial strategies, 128, 134, 138, 147
Tutoring strategies, 38, 40–46
Tutoring strategies, rule for choosing, 49–51, 59

U.S. Army, 127–131, 153, 154
Underlying cause (*see* Fault)
User interface, 129, 134, 137, 139, 142, 148, 224, 253, 254, 256, 260
User model, 46, 47

Validation and evaluation phase, 135, 152
Video(disc) (*see* Interactive video)
Videodisk (limitations), 124
Video-graphics overlay, 129, 131, 136, 137, 152
Vygotsky, 26, 27

Working memory element (WMELT), 187–190, 194
Working memory (elements), 133, 134, 141
Workshops, 226

X, 75–79
XCON, 129, 153

Yoking, 114

About the Editors and Contributors

Editors

Marshall J. Farr is President of Farr-Sight Co., specializing in consulting in instructional and cognitive sciences. He is also a senior research associate of the National Research Council. His Ph.D. was awarded in 1967 in experimental psychology from the Graduate Faculty of the New School for Social Research in New York City. From 1971 to 1985, Dr. Farr managed the basic research program of the U.S. Office of Naval Research in psychometrics, education, and training (particularly computer-based instruction), and cognitive psychology. He recently (1988) coedited, with Robert Glaser and Michelene Chi, *The Nature of Expertise,* published by Lawrence Erlbaum Associates.

Joseph Psotka is chief of the Smart Technology Team at the U.S. Army Research Institute. He received a Ph.D. in cognitive psychology from Yale University in 1975. He is preparing a book and hypertext on artificial intelligence, hypertext, and instruction.

Contributors

Michael E. Atwood, of NYNEX Science and Technology Center, is Supervisor of the Intelligent Interfaces Group, where intelligent tutoring systems are being investigated as a means of improving language training programs. Other research activities include the development of integrated learning environments to support using, as well as learning, programming languages, the development and use of analytic models to design interactive systems, and multimedia communication systems. Dr. Atwood received his Ph.D. in psychology from the University of Colorado in 1976.

Paul Baffes, formerly with NASA/Lyndon B. Johnson Space Center, is currently a knowledge engineer with GHG Corporation in Houston. Mr. Baffes was awarded the A.B. degree in computer science from Harvard University.

Robert Alan Granville is a scientist at Bolt Beranek and Newman, Inc., in Cambridge, Massachusetts, with whom he has been since 1987, concentrating on natural language generation, intelligent tutoring systems, and advanced simulation. His latest publication, "The Role of Underlying Structure in Text Generation," appeared in the *Proceedings of the Fifth International Workshop on Natural Language Generation,* 1990.

Wayne D. Gray received his Ph.D. from the University of California at Berkeley. He is currently an associate professor with the Graduate School of Education in Fordham University at Lincoln Center. The development of intelligent tutoring systems is the most recent manifestation of his interests in the psychology of programming. Other research interests include the development of analytic models of human-system interaction, end-user programming, the cognitive validity of consistency as a guideline in interface design, and the cognitive characteristics of a good example. Recent papers include a validation of GOMS (with Bonnie John and Michael Atwood) and chapters on "Change Episodes in Coding" (with John R. Anderson) and on the "Transfer of Cognitive Skills" (with Judith Orasanu).

Mario Grignetti received his M.S. degree in electrical engineering from Massachusetts Institute of Technology in 1962. He is a senior scientist with the Advanced Simulation Group of Bolt Beranek and Newman, Inc.

Mark Gross is a simulation software engineer in the Advanced Simulation Group of Bolt Beranek and Newman, Inc. He has an M.S. in computer science from Washington University.

Henry Hamburger received his Ph.D. in computer and communication sciences from the University of Michigan in 1971. He is currently associate professor of computer science at George Mason University and has served three years at the National Science Foundation, most recently as Director of the Program in Knowledge Models and Cognitive Systems. His recent research in tutoring systems merges with his career-long interest in human language in recent papers including "Foreign Language Tutoring and Learning Environment," a paper at the NATO Workshop on Intelligent Tutoring Systems for Second Language Learning in Washington, DC, 1990.

Grace Hua is a computer scientist with Computer Sciences Corporation. She has an M.S. in Oceanography from Texas A&M University.

Gail E. Kaiser, associate professor of computer science at Columbia University, received her Ph.D. in 1985. She was honored in 1988 by being selected as a National Science Foundation Presidential Young Investigator. She has published over 50 papers in a wide range of areas, including software-development environments, programming support for distributed systems, and application of artificial intelligence technology to software engineering.

Laura C. Kurland, currently a senior scientist at Bolt Beranek and Newman, Inc., has been an educational and training software designer, a special education teacher, a technical writer, and a formative and cognitive researcher. In addition to the intelligent maintenance trainer described here, other AI-based technology she has also developed include an operator trainer and the interactive scenario generator portion of a long-haul distributed simulation network for real-time tactical training. She has also developed a variety of commercial software for children 3–13 at Children's Television Workshop (Sesame Street). She is currently developing simulation-based science software for middle and high school students. Her latest publication, "Tutor, Tool and Training Environment: Training Troubleshooting Using AI," appears in the *Proceedings of the 4th International Conference on AI and Education,* May 24–26, 1989, Amsterdam, Netherlands, edited by D. Bierman, J. Breuker, and J. Sandberg, Burke, VA: IOS Press.

Susanne P. Lajoie received her Ph.D. from Stanford University in 1986. In 1987 she received the American Psychological Association Outstanding Dissertation Award in Educational Psychology and an Honorable Mention for Outstanding Dissertation from the American Educational Research Association. She completed a postdoctoral fellowship at the Learning Research and Development Center at the University of Pittsburgh and was later appointed as a research scientist. She is now an Assistant Professor at the University of Wisconsin-Madison in the Cognitive Science and Human Learning Program. In her research she uses a cognitive approach to skill identification and applies her research to the design of computer-coached practice environments for domains such as spatial visualization, electronics troubleshooting, biology problem solving, and medical diagnosis.

Alan Lesgold is professor of psychology and intelligent systems at the University of Pittsburgh, Codirector of the Intelligent Systems Program, and Associate director of the Learning Research and Development Center. He received his Ph.D. in experimental psychology from Stanford University in 1971. He is the coeditor, with Robert Glaser, of *Foundations for a Psychology of Education,* published by Lawrence Erlbaum Associates in 1989.

Akhtar Lodgher completed his B.E. in production engineering from Anna University, his M.S. in computer science from South Dakota School of Mines and Technology, and his Ph.D. in information technology from George Mason University. Currently he is a faculty member at Marshall University in the Department of Computer and Information Sciences. His research interest is primarily in the field of artificial intelligence with special interest in the fields of intelligent tutoring systems, cognitive science, and knowledge-based systems.

R. Bowen Loftin is professor of physics at the University of Houston-Downtown and holds a Ph.D. from Rice University. The work reported in this volume's article received the Innovative Application of Artificial Intelligence Award from the American Association for Artificial Intelligence in 1989.

Scott R. Lucado is a courseware developer recognized for his expertise in using the TENCORE, PLATO, and Regency authoring systems. Lucado has been developing computer-based training materials for the past decade, including applications in medical training, aerospace, and defense.

Dawn M. MacLaughlin received a B.S. in computer science and engineering in 1986 from the Massachusetts Institute of Technology. From 1986–1990 she worked at Bolt Beranek and Newman, Inc., in Cambridge, Massachusetts in the Speech and Natural Language Processing department on projects in a variety of areas, including natural language understanding, natural language generation, and intelligent tutoring systems. Currently, she is a Ph.D. student in the Applied Linguistics Program at Boston University, where she is interested in studying first and second language acquisition. Her latest publication is "Towards Understanding Text with a Very Large Vocabulary," with Damaris Ayuso, R. Bobrow, D. MacLaughlin, M. Meteer, L. Ramshaw, R. Schwartz, and R. Weischedel, in *Proceedings of the Speech and Natural Language Workshop,* Morgan Kaufmann Publishers, 1990.

L. Dan Massey, Jr. holds an M.S. degree in computer science from Harvard University. He is currently employed by Bolt Beranek and Newman, Inc., where he is manager of the Systems and Software Engineering Department. His most recent publication is "Intelligent Tutoring Systems: Lessons Learned," coedited with J. Psotka and S. A. Mutter, Lawrence Erlbaum Associates, 1988.

Kathleen R. McKeown is an associate professor of computer science at Columbia University in New York City. Her Ph.D. in computer science was awarded by the University of Pennsylvania in 1982. Her work there focused on natural language generation. In 1985 she received a National Science Foundation Presidential Young

Investigator's Award for her contributions in that area. She has recently worked on techniques for producing multimedia explanations, represented in a recent book chapter of which she is senior author: "Language Generation in COMET," coauthored with M. Elhadad, Y. Fukumoto, J. Lim, C. Lombardi, J. Robin, and F. Smadja, in *Current Research in Language Generation,* edited by R. Dale, C. Mellish, and M. Zock, Academic Press, 1990.

Mark L. Miller is manager of the Business Learning Research Department within the Advanced Technology Group at Apple Computer, Inc., where his group has been investigating advanced learning environments and authoring tools for corporate and federal training applications. Miller has been involved in research on intelligent tutoring systems since the early 1970s, when he worked at Bolt Beranek and Newman, Inc., on SCHOLAR, WEST, and related systems. His doctoral dissertation, completed at MIT in 1979, involved an intelligent tutoring system for elementary computer programming using the LOGO language.

Allen Munro is associate director of Behavioral Technology Laboratories at the University of Southern California. He received his Ph.D. in cognitive psychology from the University of California at San Diego. His current research and development interests are in the area of model-based interactive graphical simulation composition for technical training. Recent publications include "Model-Building Tools for Simulation-Based Training" in the journal *Interactive Learning Environments.*

William R. Murray received his Ph.D. in computer science in 1986 from the University of Texas at Austin. His dissertation investigated the automated debugging of student programs using heuristic algorithm-recognition techniques and formal program verification techniques. He has been employed as a member of the technical staff since that time in FMC's Artificial Intelligence Department. His recent work applying dynamic planning to control intelligent tutoring systems appeared in the *Proceedings of the 4th International Conference on Artificial Intelligence and Education* in 1989 and won the best paper award for the conference; a more recent description of this work appears in the AAAI-90 proceedings.

Denis Newman, who has a Ph.D. in psychology from the City University of New York, is a senior scientist at Bolt Beranek and Newman, Inc. He is author (with Griffin and Cole) of *The Construction Zone: Working for Cognitive Change in School,* Cambridge University Press, 1989.

Peter Pirolli is assistant professor in the School of Education at the University of California, Berkeley, and an affiliate of the Xerox Palo Alto Research Center, and the Institute for Research on Learning. He received his Ph.D. in cognitive psychology at Carnegie-Mellon University.

Elliot Soloway is an associate professor in the Department of Electrical Engineering and Computer Science at the University of Michigan. He received his Ph.D. from the University of Massachusetts, Amherst, in 1978. Dr. Soloway is the executive editor of a new journal on education and technology: *Interactive Learning Environments,* published by Ablex Press.

Douglas M. Towne is the director of Behavioral Technology Laboratories at the University of Southern California. He received his Ph.D. degree in engineering from the University of Southern California in 1969. Dr. Towne has been conducting research in the areas of simulation-oriented instruction, diagnostic reasoning, and design of systems for ease of operability and maintainability for 25 years.

Lui Wang is a computer engineer with the NASA/Lyndon B. Johnson Space Center in Houston. He holds the B.S.M.E. degree from Texas A&M University.

Ursula Wolz is assistant professor of computer sciences at Trenton State College, Trenton, New Jersey. She received an M.S. in computer science from Columbia University in 1985. The research presented here is part of her doctoral work at Columbia University.